Water-Energy Interactions in Water Reuse

Water-Energy Interactions in Water Reuse

Valentina Lazarova, Kwang-Ho Choo and Peter Cornel

Publishing
London • New York

Published by **IWA Publishing**
Alliance House
12 Caxton Street
London SW1H 0QS, UK
Telephone: +44 (0)20 7654 5500
Fax: +44 (0)20 7654 5555
Email: publications@iwap.co.uk
Web: www.iwapublishing.com

First published 2012

British Library Cataloguing in Publication Data
A CIP catalogue record for this book is available from the British Library

Library of Congress Cataloging-in-Publication Data
A catalog record for this book is available from the Library of Congress

ISBN: 9781843395416 (Paperback)
ISBN: 9781780400662 (eBook)

Contents

Chapter 7

Chapter 8

List of contributors

Editors

Dr. Valentina LAZAROVA
Suez Environnement - CIRSEE
38 rue du president Wilson
78230 Le Pecq,
France
E-mail : valentina.lazarova@suez-env.com

Prof. Kwang-Ho Choo, Ph.D.
Department of Environmental Engineering
Center for Blue Gold Cluster
Kyungpook National University
1370 Sankyeok-Dong, Buk-Gu
Daegu, 702-701,
Korea
E-mail: chookh@knu.ac.kr

Prof. Dr.-Ing. Peter Cornel
Institut IWAR
Technische Universität Darmstadt
Petersenstr. 13,
64287 Darmstadt,
Germany
E-mail: p.cornel@iwar.tu-darmstadt.de

Major Contributors

Dr. Steve Conrad
Simon Fraser University

School of Resource and Environmental Management
Burnaby, BC
Canada
E-mail: steve_conrad@sfu.ca

Dr. Olivier Lefebvre
Division of Civil and Environmental Engineering
National University of Singapore
Block E2, #02-16, 1 Engineering Drive 2,
Singapore 117576
E-mail: ceelop@nus.edu.sg

Dr. Yanjin LIU
Environmental Engineer
Innovation and Environmental Stewardship,
American Water
1025 Laurel Oak RoadVoorhees,
NJ 08043,
USA
E-mail: Yanjin.Liu@amwater.com

Prof. Dr. H.C. Helmut Kroiss
Institute for water quality management
Vienna University of Technology
Karlsplatz 13
A 1040 Vienna
Austria
E-mail: hkroiss@iwag.tuwien.ac.at

Contributors

Chapter 1
Glen T. Daigger, Ph.D., P.E., BCEE, NAE
President
International Water Association (IWA)
Senior Vice President and Chief Technology Officer
CH2M HILL
9191 South Jamaica Street
Englewood, CO 80112,
USA
E-mail: gdaigger@ch2m.com

Chapter 2
Dott. Ing. Alessandro Meda,
Dipl.-Wi.-Ing. Dorothee Lensch,
Dr.-Ing. Christian Schaum,
Prof. Dr.-Ing. Peter Cornel
Technische Universität Darmstadt

Petersenstr. 13
64287 Darmstadt,
Germany
E-mail: p.cornel@iwar.tu-darmstadt.de

Chapter 3
Vladimir Novotny, Ph.D, P.E., BCEE
Professor Emeritus
Marquette University (Milwaukee, WI)
and Northeastern University (Boston, MA)
Partner AquaNova LLC
780 Centre Street, Newton, MA,
USA
E-mail: vnovotny@aquanovaLLC.com

Chapter 4
Michael Wilson, P.E.
CH2M HILL
18 Tremont Street | Suite 700 |
Boston MA 02108
USA
E-mail: mwilson@ch2m.com

Chapter 5
Prof. Bruce Logan, Ph.D.
Director, Hydrogen Energy (H2E) Center &
Engineering Energy & Environmental Institute (E3I)
231Q Sackett Bldg
Penn State University
University Park, PA 16802,
USA
E-mail: blogan@psu.edu

Chapter 6
Dr. Valentina Lazarova,
Dr. Carlos Peregrina,
Dr. Pascal Dauthuille
Suez Environnement - CIRSEE
38 rue du president Wilson
78230 Le Pecq,
France
E-mail: valentina.lazarova@suez-env.com

Chapter 7
Christoph Brepols, Dipl.-Ing.
Erftverband, Am Erftverband 6,
D 50126 Bergheim,
Germany
E-mail: christoph.brepols@erftverband.de

Chapter 8
Dr.-Ing. Laurence Palmowski
Dipl.-Ing. Katrin Veltmann
Prof. Johannes Pinnekamp
Institute of Environmental Engineering
RWTH Aachen University
Mies-van-der-Rohe Str. 1,
52056 Aachen,
Germany
E-mail: palmowski@isa.rwth-aachen.de

Chapter 9
Sudhan Paranjape, P.E.,
Rod Reardon, P.E., BCEE
Carollo Engineers, P.C.
1089 W. Morse Blvd., Suite A
Winter Park, FL 32789,
USA
E-mail: SParanjape@carollo.com

Chapter 10
Dr. ir. Marlies J. Kampschreur,
Prof. Dr. M.C.M. van Loosdrecht
Innovatietechnoloog
Waterschap Aa en Maas
Postbus 5049
5201 GA 's-Hertogenbosch
The Netherlands
E-mail: mkampschreur@aaenmaas.nl

Chapter 11
Dr.-Ing. Susanne Bieker,
Dipl.-Wirtsch.-Ing. Carola Zeig
Prof. Dr.-Ing. Peter Cornel
Technische Universität Darmstadt
Petersenstr. 13,
64287 Darmstadt,
Germany
E-mail: s.bieker@iwar.tu-darmstadt.de

Chapter 12
Mehul V. Patel, P.E.
GWRS Program Manager
Orange County Water District
18700 Ward Street, Fountain Valley,
CA 92708,
USA
E-mail: mpatel@ocwd.com

Chapter 13
Laura Delgado,
Yvan Poussade,
Pere Aguiló,
Veolia Water Systems Ibérica S.L.
Pol. Ind. Santa Ana
c/ Electrodo, 52 28529 Rivas Vaciamadrid
Madrid, Spain
E-mail: Laura.delgado@veoliawater.com

Chapter 14
Dott. Ing. Alessandro Meda,
Dr.-Ing. Jochen Henkel,
Dr.-Ing. Yue Chang,
Prof. Dr.-Ing. Peter Cornel
Technische Universität Darmstadt
Petersenstr. 13,
64287 Darmstadt,
Germany
E-mail: p.cornel@iwar.tu-darmstadt.de

Chapter 15
Dr. Roumiana Hranova, Ph.D
University of Botswana
Private bag 0061,
Gaborone,
Botswana
E-mail: rhranova@gmail.com

Chapter 16
Nikolay Voutchkov, P.E., BCEE
Water Globe Consulting, LLC
200 Broad Street
Suite 2450
Stamford, CT 06901,
USA
E-mail: nvoutchkov@water-g.com

Chapter 17
Dr.-Ing. Süleyman Yüce,
Dr. Christian Kazner,
Dr. Rita Hochstrat,
Prof. Thomas Wintgens,
Prof. Thomas Melin
RWTH Aachen University
Aachener Verfahrenstechnik, Chemical Process Engineering
Turmstr. 46, D-52056 Aachen
Germany
E-mail: sueleyman.yuece@avt.rwth-aachen.de

Chapter 18
Dr. Graeme K Pearce, Ph.D.
Membrane Consultancy Associates
Little Eastfield
Eastfield Lane
Whitchurch
Reading RG8 7EJ,
UK
E-mail: graemekpearce@btinternet.com

Chapter 19
Dr. Sarah Ward
Dr. Fayyaz Ali Memon,
Prof. David Butler
Centre for Water Systems
College of Engineering, Mathematics and Physical Sciences (CEMPS)
University of Exeter
Harrison Building, North Park Road
Exeter, EX4 4QF
UK
E-mail: sw278@exeter.ac.uk

Chapter 20

Katharina M. Tarnacki,
Prof. Thomas Melin
Montse Meneses
Aachener Verfahrenstechnik - RWTH Aachen Universtity
Chemical Process Engineering
Turmstr. 46,
52056 Aachen
Germany
E-mail: Katharina.Tarnacki@avt.rwth-aachen.de

Chapter 21

Winnie P. Gerbens-Leenes,
Arjen Y. Hoekstra,
Theo H. van der Meer
Faculty of Engineering and Technology
Water Engineering and Management
University of Twente
Drienerlolaan 5
P.O. Box 217
7500 AE Enschede
The Netherlands
E-mail: P.W.Gerbens-Leenes@ctw.utwente.nl

Chapter 22

Prof. Dr.-Ing. Karl-Heinz Rosenwinkel,
Linda Hinken,
Selin Ay
Institut für Siedlungswasserwirtschaft und Abfalltechnik
Leibniz Universität Hannover
Welfengarten 1
30167 Hannover,
Germany
E-mail: rosenwinkel@isah.uni-hannover.de

Foreword

Helmut Kroiss

This book is devoted to reveal the relationship between energy and water with emphasis on water reuse for engineers and decision makers. Both, energy and water represent basic resources for all living systems irrespective of their size and complexity. As human population has become a dominant user of both, the understanding of this relationship plays an increasing role for political and managerial decisions on a local, regional as well as on a global scale, which stresses the relevance of this book.

The following data on energy and water shall elucidate only the order of magnitude and relevance of natural and anthropogenic water and energy flows, they are related to one inhabitant (capita) and 1 year.

We need every day about 2 to 4 litres of water for drinking as part of our nutrition, which is about 1 m^3 per year. We use about 30 to 100 m^3/year for hygiene and pollution transport and several hundred m^3/year for industry and cooling water at least in the developed world. All the used water (wastewater) is not lost in a region and can be reused.

Food production consumes 1000 to 2500 m^3 of water per capita and year. The water needed for agriculture cannot be reused in the region as it is evaporated. In moderate climatic regions, this does not have a great relevance as natural vegetation would have similar effects and natural rainfall covers most of the water demand. In semi-arid and arid regions, natural vegetation often needs much less water than intensive agriculture withdrawing water for irrigation from fresh water resources. With increasing water scarcity or population density, water reuse becomes increasingly relevant in an increasingly number of regions on the globe.

Total solar power supply to the earth is in the range of 10 million W per capita, most of this power is driving the global natural fresh water cycle which is the natural water reuse system, corresponding to a global mean of more than 10,000 m^3 fresh water per capita and year. The problem is that solar irradiation and fresh water availability are extremely variable over time and region. In addition, population development in many regions is not any more in agreement with the regional water availability. This is mainly a consequence of the scientific, technological, economic and fossil energy related development during the last 200 years.

Water management is always primarily a regional or local problem. Where the natural regional water availability does not allow a reliable nutrition for the inhabitants by regional agriculture, reliable food supply can only be achieved by food (virtual water) transport on a global scale. Water reuse can help to decrease regional water scarcity problems at lower costs and lower environmental impact than long distance water and/or food transfer or energy intensive fresh water production from seawater. Also a change to a more healthy nutrition with a low consumption of animal protein can markedly contribute to reduce regional water stress if it is complemented by a lower cattle production in the regional agriculture.

Our nutrition provides us with a continuous power of $\geq$100 W, of which our brain needs 10–15 W. The actual mean global primary power consumption will soon reach 3,000 W/capita which corresponds to about 30 slaves continuously working for us and never happened before in history.

Energy requirements for water supply are extremely different depending on the specific local situation. In Vienna/Austria, drinking water transport from the mountains is used for energy production, while sea water desalination needs $\geq$4 kWh/m^3 and treated wastewater "desalination" $\geq$2 kWh/m^3. With around 50 to 100 m^3 per capita and year, these values can be converted to a continuous power for water supply of 10 to 50 W/capita. Highly efficient biological nutrient removal wastewater treatment plants need a power input of 0 to 20 W/capita. Several large full scale energy self-sufficient plants are already in operation. Pumping energy requirements for water systems and energy requirements for wastewater polishing before recycling strongly vary depending on the specific local situation but are in the same order of magnitude.

From these data the following conclusions can be drawn:

Reliable water and food supply get more and more dependent on continuous reliable power supply. Minimisation of energy requirements and the use of renewable energy sources to run the water systems, including drinking water supply, sanitation, water recycling, irrigation and water protection, is mainly a sustainability and reliability question. The predicted population growth going along with increasing urbanisation will aggravate the water problems in an increasing number of regions and agglomerations worldwide for at least the next 40 decades.

In most urban but also in rural areas, the anthropogenic changes influencing on water systems have been much faster in the past and will remain faster than climate change. This applies also to many agricultural areas and enhances competition for water between agriculture and water supply. Adaptation of the water systems to all the changes represents a regional problem and has much higher priority than measures for climate change abatement in the water industry by reducing carbon dioxide, methane and nitrous oxide emissions. Also the economic, social and ecologic optimisation of existing and/or new water supply and sanitation systems are also of higher priority than minimisation of energy requirements for these systems. But wasting energy should never be allowed irrespective of the water system.

Reliable and safe drinking and wastewater systems are recognised as the main contributors to increased life expectancy; this hygienic aspect therefore has to be considered first irrespective of the questions regarding the pros and cons of centralised versus decentralised, traditional or novel systems. Beyond the fulfilment of the hygienic requirements, decision making will be strongly influenced by economic and social aspects.

As the natural fresh water circuit alone cannot fulfil any more the water quantity and quality requirements for humans and the water environment in many regions, water reuse will become more and more relevant in the coming decades. As on a global scale up to ~3/4 of the water systems in the developing world will have to be newly built, and a great part of existing water infrastructure in the developed world approaches the end of their lifetime, the next decades offer a unique opportunity for new developments of which water reuse systems – adapted to the specific local situation – will have increasing relevance.

Thanks to the authors, this book will help to enhance the technological and managerial development in the field of water use and reuse systems worldwide based on the actual state of scientific background and experience of full scale operation.

Introduction

Peter Cornel, Kwang-Ho Choo and Valentina Lazarova

The sun is the source of all energy on Earth and water is the basis of all life on our plant. The relationship between these vital media only becomes clear on closer inspection: solar radiation energy fuels the natural water cycle. Gigantic quantities of sea water are desalinated via evaporation then transported by the clouds, and return to earth as fresh water precipitation. Without the input of solar radiation energy, the entire water cycle would quickly come to a standstill.

Solar energy is the driving force for photosynthesis too: the photo-biological reaction – converts CO_2 and water into organic molecules that form the components of biomass and living cells. Photosynthesis represents a highly endothermic energy transformation process that converts solar radiation energy into chemical energy. For example, 2,876 kJ/mol are absorbed and respectively stored by the photosynthetic production of sugar:

$$6CO_2 + 6H_2O \longrightarrow C_6H_{12}O_6 + 6O_2$$
$$\Delta_R H = +2{,}876.4 \text{ kJ/mol}$$

With terrestrial plants, the necessary water is taken up via the roots, transported inside the plant and thus made available for these reactions. Besides the stoichiometrically necessary amount of water, the water's function is temperature control via evapotranspiration. Furthermore, water is the transport medium for dissolved nutrients inside the plant. Therefore, besides solar energy, water is indispensable for the growth of plants.

These considerations show that all fossil fuels that are based on plant growth (gas, petroleum, coal) "store" significant quantities of water and solar energy. The same holds true for all plant-based renewable energy sources, as substantial quantities of water are necessary for their generation. Depending on the boundary conditions, several thousand litres of water might be consumed to produce 1 L biodiesel with an energy content of approximately 9 kWh (33 MJ) (see Chapter 21).

When water experts talk about the nexus of water and energy, they usually do not include these global natural processes, without which life on Earth would not be possible. It is because of this common omission that, within the last decade, the terms "water footprint" and "virtual water" have aroused more and more interest. However, in general, technical processes used for energy conversion require water, while energy is needed for water utilization. But even at this level, the nexus of water and energy is inextricable and more strongly linked with each other than might appear at first sight.

Directly and indirectly, water is essential for the conversion of energy. Water is used for the generation of electricity by hydropower, wave power or tidal power. Huge quantities of water are needed for the exploration and extraction of fossil fuels, as well as for power plants that use steam for driving turbines and cooling water for heat dissipation. Furthermore, the cultivation of energy crops for biogas and biofuel production requires water, too.

Vice versa, there is manifold use of energy in the water cycle: energy is required for water and wastewater distribution, usage and treatment, with households requiring the largest share. Finally, water and wastewater contain energy in several forms, which can be used in the generation of electricity and/or heat.

These links also indicate that water and energy depend on and influence each other. Therefore, water saving and reuse contribute to the protection of water resources and preservation of energy supplies. However, on the other hand, additional treatment steps for water reuse may increase energy consumption. In order to make technical decisions, to balance the pros and cons of alternative solutions, and to benchmark new resource-efficient supply and disposal concepts, the knowledge of quantitative relationships and boundary conditions is decisive:

- What kind of water quality is needed in each respective case of application? How much energy has to be spent to reach the required quality standard? In this context, *Fit for Purpose* is the key solution.
- Water reuse versus desalination and versus long-distance transport: how much energy per m^3 water is required to supply water with the *Fit for Purpose* standard at the tap or point of use?
- Intra-urban water reuse versus one-time water use: there will be no universal response – depending on the boundary conditions, for example arid regions with expensive water treatment and transport, megacities or water-rich rural areas.
- Centralised versus decentralised versus semi-centralised.

While in their infancy at present, sustainable wastewater treatment and reuse have come of age and are likely to gain a greater importance in the planning of integrated water and wastewater management systems in the future.

Similar to water reuse, the unlocking and the enhanced reuse of the energy contained in wastewater, is a key tool to solving the water-energy nexus which urbanized centres face worldwide.

The following chapters in this edited volume, compiled in five sections with different settings of priority, will give initial answers to the questions raised – from very different points of view and objectives – from different countries with different stages of technical development – from both water-rich and arid regions.

The overriding objective of *Water-Energy Interactions in Water Reuse* is to generate awareness of the questions raised and to gather ideas for future-oriented solutions, as well as to initiate interdisciplinary discussions.

The editors hope you will gain many insights from the issues raised within this book and look forward to your feedback on the materials presented.

Part 1

The Water-Energy Nexus

By The Editors

FOREWORD

The traditional linear approach to water management often lacks a necessary recognition and understanding of the interactions between water and energy. With the increase of water scarcity and energy cost, the strong links between water and energy are becoming increasingly obvious. Energy drives every element of the water cycle, while water is necessary for energy production. The effects of climate change and global warming reinforce the need for a holistic approach to the management of energy and water – along with nutrients – into an integrated system that fits to the growing urban development.

Part 1 of this book "The Nexus of Energy and Water" (e.g. Chapters 1 to 5), presents several integrated approaches for the management of water and energy in urbanized societies, enabling the saving of water and energy and the move towards more sustainable development with reduced carbon and environmental footprints for the cities of the future.

The energy required to treat water to acceptable levels is in the range of 0.03–7 kWh/m^3, depending on water sources (freshwater, seawater, wastewater) and specific regional parameters, such as climate conditions, water availability, water usage and population density. In each situation, there are possibilities for energy savings in both water production and distribution. Additionally, energy can be recovered from wastewater streams in the form of calorific heat and biogas. Achieving optimum energy recovery is a function of carbon content and sewage systems, for example centralized or decentralized urban infrastructures.

The most promising technology for the recovery of the chemically bound energy in wastewater is still anaerobic digestion, either treating wastewater, particularly in upflow anaerobic sludge blanket reactors, or recovering energy from sludge digestion. Emerging energy efficient and energy producing technologies, such as microbial fuel cell, microbial electrolysis cell, and microbial desalination cells, are expected to be feasible in near future.

Sustainability in water resource planning requires the consideration of the embodied energy in the water cycle, and in particular in water recycling. Water reuse is enabling the optimization of energy intensity within the water cycle, especially within decentralized and semi-autonomous urban systems; and with a treatment level which adheres to the Fit for Purpose principle. Recycled water needs to be delivered at a cost justifiable by its purpose, therefore energy-intensive processes should be limited when exceptionally high water quality is required.

Implementing such high performance integrated urban water management systems is a challenging task which will require new policies, proactive planning, innovative technologies, institutional and financial incentives, as well as performance standards and management.

Chapter 1

Designing and implementing urban water and resource management systems which recover water, energy, and nutrients

Glen T. Daigger

1.1 INTRODUCTION

As currently configured, urban water management systems result in the consumption of significant quantities of water and energy and inefficient management of nutrients (Daigger, 2007, 2008, 2009, 2010). This arises because of the "linear" approach that has historically been and is generally used today. Relatively pristine surface and/or groundwater is removed from the environment, often from remote locations, and transported to the urban area where it is used once, collected and treated (to a quality less than that removed from the environment), and subsequently transported and discharged back into the environment at a location remote from the urban area. This linear system must be converted into a recovery and recycle system which makes optimum use of the water, energy, and nutrients in the urban water system to meet evolving needs. These evolving urban water and resource management systems can significantly reduce the quantity of water which must be removed from the environment, be energy neutral (or even energy positive), and recycle significant quantities of nutrients from the system (Daigger, 2010, 2009). The result will be improved protection of public health, environmental enhancement through reduced water abstraction from the environment and reduced environmental discharges, reduced energy consumption, an enhanced urban environment, a more robust system, and an easier system to upgrade and expand. Fortunately, technologies and approaches exist to accomplish these goals today, and on-going advances will further enhance these capabilities (Daigger 2003, 2008, 2009, 2010).

This chapter will first describe evolving urban water and resource management systems requirements, contrasting historical and future approaches. Components of these evolving water, energy, and nutrient management systems are then presented, and these components are subsequently combined into integrated systems. Finally, requirements to implement these integrated systems are presented.

1.2 EVOLVING URBAN WATER AND RESOURCE MANAGEMENT SYSTEM REQUIREMENTS

1.2.1 The necessity for change

Several factors combine to create the necessity for change by the urban water sector. The most significant is population growth. As illustrated in Figure 1.1, population growth began to accelerate in the middle of the

20th century and is continuing in the 21st century. The projection to the middle of the 21st century is the "mid-range" one which will lead to a stable population of about 10 billion in the second half of the 21st century. Achieving this requires decreasing birth rates in the developing countries, which in turn depends on an increased standard of living in these countries (which can increase resource consumption – more about that later). Thus, the mid-range population scenario may or may not occur. Consequently, upper and lower bounds for the estimates are also provided. Population growth increases water demand, not only for urban residents but also for agriculture, energy, and industry, unless it is compensated for by changes such as those discussed subsequently in this chapter. Since the global water supply does not increase with population, this results in increased competition for water, especially when it is recognized that water is also needed to preserve the environment.

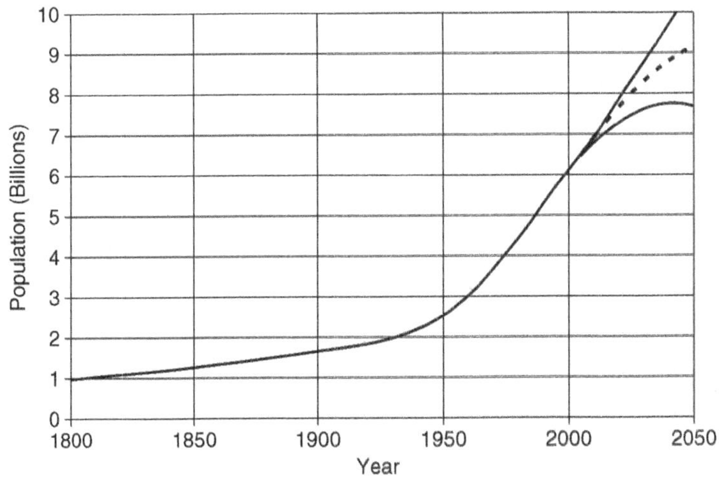

Figure 1.1 Historical (solid line) and projected (dotted line) growth in global population, along with upper and lower bounds for the projections (*Source*: United Nations, 2011)

Population growth, by itself, results in increased resource consumption. However, per capita (per person) consumption has also increased significantly in the recent past, and further increases in the developing world are necessary if the global population is to stabilize as illustrated in Figure 1.1. It is estimated that in about the year 2000, the global population began consuming the equivalent of the entire renewable resources of planet earth (Wallace, 2005). Moreover, if present consumption trends continue consistent with the population trend presented in Figure 1.1 (significant increase in living standards in the developing world), human renewable resource consumption will be the equivalent of three planet earths by the year 2050. Obviously this is not sustainable. Scientists have been assessing the impact of increased consumption on global environmental resources and find, as illustrated in Figure 1.2, that we have crossed planetary boundaries in the areas of biodiversity, nutrients (nitrogen and phosphorus), and climate change (Rockström *et al.* 2009). Urban water and resource management affects each of these as water withdrawal from the environment adversely affects biodiversity, nutrient dispersion increases nutrient discharges, and the urban water cycle is a source of greenhouse gas (GHG) emissions. Chemical pollution has not yet been quantified by these researchers, but urban water and resource management is also part of this cycle and contributes to chemical pollution.

Figure 1.3 illustrates the resulting effects of increased environmental discharges by presenting coastal hypoxic "hot spots" (Daiz & Rosenberg, 2008). These "hot spots" are zones of oxygen depletion (hypoxia), generally resulting from the discharge of nutrients (especially nitrogen) into the aquatic environment, resulting in the growth of excessive quantities of algae whose respiration depletes dissolved oxygen. To date nearly 400 such zones have been identified, nearly all clustered around areas of high human population. Most importantly, such zones have been identified essentially every place scientists have looked, suggesting that many more exist and will be found.

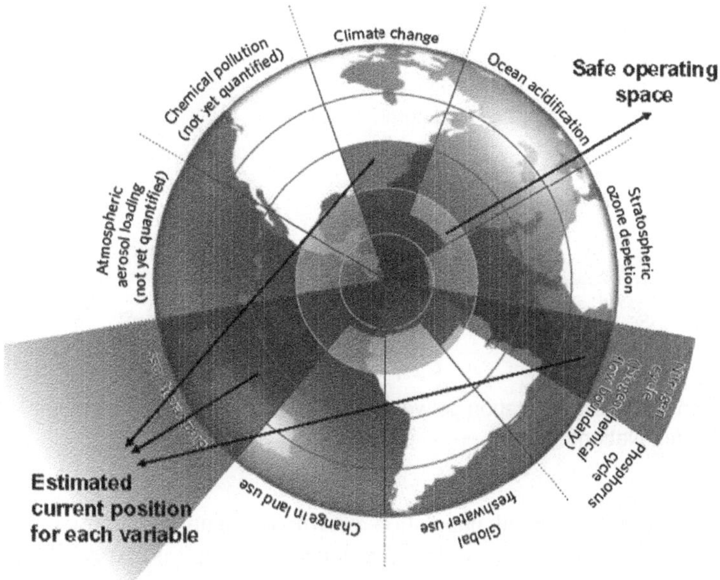

Figure 1.2 Comparison of the use of planetary resources with the renewable resources of planet Earth: the inner green shading represents the proposed safe operating space for nine planetary systems, while the red wedges represent estimates of the current position for each variable (*Source*: Rockström *et al.* 2009)

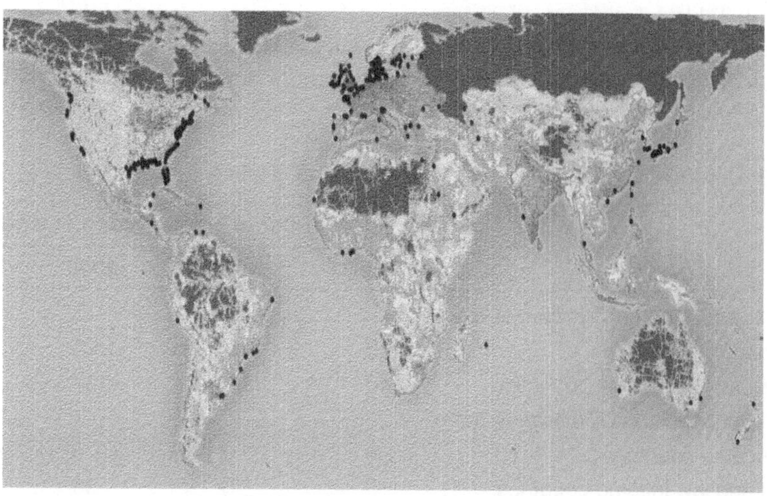

Figure 1.3 Occurrence of coastal eutrophication-associated dead zones (*Source*: Diaz & Rosenberg, 2008)

Urbanization of the human population increased in the 20th century and is expected to increase dramatically in the 21st century, as illustrated in Figure 1.4. More than half of the global population is estimated to already live in urban areas (developed countries are nearly 80% urban), and this is expected to increase to nearly 70% on a global basis by the middle of the 21st century.

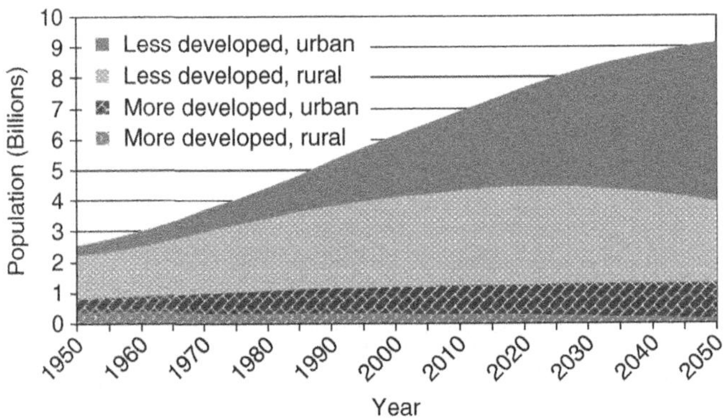

Figure 1.4 Distribution of human population between rural and urban areas in developed and less developed countries (*Source*: United Nations, 2011)

As also noted in Figure 1.4, global population growth is occurring almost exclusively in the less developed countries as the population of the developed countries is expected to remain relatively constant at essentially 1 billion, while that in less developed countries is expected to increase to nearly 8 billion. The rural population in both develop and less developed countries is expected to decline slightly during the first half of the 21st century, while the urban population in less developed countries is expected to increase significantly. Essentially, all of the growth in the global population between now and the middle of the 21st century will occur in urban areas in less developed countries.

As illustrated in Table 1.1, the size of urban areas is also increasing significantly, especially in the less developed countries. This dramatic shift in the human urban population (compared to historic trends) means that urban water management is becoming the dominant form of domestic (and industrial) water use.

Table 1.1 City development through the 20th and early 21st century.

Item	Number of cities in each category			
	1950	**1975**	**2000**	**2015**
Cities with population of 1 million or more				
High income countries	43	73	90	105
Middle and low income countries	40	122	297	449
Cities with population of 5 million or more				
High income countries	5	9	9	10
Middle and low income countries	3	12	30	47

(*Source*: National Research Council, 2003).

Urban water management also contributes to climate change, but only to a modest extent compared to sectors such as transportation and buildings. Thus, reduced emissions of GHGs from the urban water cycle will mitigate anthropogenic climate change to only a limited extent. Climate change is expected to result in more variable climate patterns (more droughts and more floods), thereby requiring more robust urban water management systems to accommodate these more variable conditions. Thus, the impacts of climate change on water management will be significant, and significant adaptation will be required.

The net effect of these factors is that, by 2025, more than half of the human population and approaching 2 out of every three people will live in areas experiencing water stress, compared to historically much less than 10% of the human population (Daigger, 2007; UNEP, 2009). Water stress will become "the norm", rather than the exception. Resource constraints will also restrict options for mitigating these effects, thereby further constraining utilities in their ability to provide reliable and robust service, not only during normal conditions but especially during the more frequent extreme weather conditions. At the same time, even under current conditions many utilities are unsustainable, resulting in failures to properly maintain aging infrastructure and to adapt quickly enough to changing realities (Monsma *et al.* 2009). Clearly change is needed in the practices used and approaches to urban water and resource management.

1.2.2 Contrasting historical and future approaches

Fortunately a consolidated picture of the evolving approach to urban water and resource management needed to address the issues listed above is evolving, as illustrated in Table 1.2 which compares the historical to the evolving approach (Daigger, 2011). Approaches to access local water supplies, such as rainwater harvesting and water reuse, are available and are being used more widely, resulting in the use of local water supplies rather than remote sources. Changing economics are also changing the basis for optimizing urban water and resource management systems, from optimizing the cost of infrastructure (larger pipes and centralized treatment facilities to capture the benefits of economies of scale) to minimizing operating costs by optimizing water use and minimizing energy consumption. The historical practice of implementing separate, single purpose systems for drinking water, storm water, and recycled water is being replaced by multipurpose systems which integrate these three functions. System configurations are transitioning from centralized to hybrid systems incorporating both centralized and decentralized components to optimize water use and minimize energy consumption rather than infrastructure costs. While implementation of these changes is evolutionary, the contrast between the historical and evolving approach is dramatic. This evolution is enabled by a dramatic increase in the technologies and tools available, as discussed in the next section.

Table 1.2 Historical and evolving approaches to urban water and resource management (abstracted from Daigger, 2011).

Item	Approach	
	Historical	**Evolving**
Water supply	Remote sources	Local sources
Optimized costs	Infrastructure	Water use, energy consumption
Functions	Single purpose systems for drinking water, stormwater, and recycled water	Multipurpose systems which integrate functions
Configuration	Centralized systems	Hybrid systems (centralized and decentralized components)

1.3 COMPONENTS OF EVOLVING URBAN WATER AND RESOURCE MANAGEMENT SYSTEMS

The evolving approach to urban water and resource management summarized in Table 1.2 is enabled by a series of water, energy, and nutrient management approaches with supporting technologies, as summarized in this section. They represent the elements of a growing toolkit that is enabling the evolving urban water and resource management systems to be described in the following section.

1.3.1 Water management

Table 1.3 summarizes approaches to urban water management which are enabling improved urban water use efficiency and reducing the quantity of water which must be removed from the environment to meet urban water needs (Daigger, 2008, 2009, 2010). Stormwater management and rainwater harvesting include a range of technologies which intercept and retain rainwater and either infiltrate it into the groundwater or store it in constructed facilities. Many of these use plant-based components to provide filtration and surfaces for chemical and microbial reactions to improve water quality prior to storage and/or direct use of the collected rainwater. Examples include green roofs, bio-retention systems, which direct collected stormwater into infiltration basins, rain gardens, and permeable pavement, as well as storages, which are designed to directly receive and store rainwater for subsequent use (Novotny & Brown, 2007; Daigger, 2011). These systems are generally designed to process the high frequency, low intensity storms while more conventional stormwater systems are still needed to manage the infrequent, high-volume storms. Since a significant volume of rainfall is represented by these high-frequency, low volume storms, and they represent the most polluted stormwater, as they are most frequently removing the material deposited on surfaces, they can provide significant stormwater pollution control as well as capture of a significant fraction of the rainfall from an urban area. As illustrated in Figure 1.5, urban stormwater can provide a significant urban water supply, even in relatively dense urban areas.

Table 1.3 Water management approaches enabling improved urban water use efficiency and reduced demand.

Approach	Contribution to water and resource efficiency	Examples
Stormwater management and rainwater harvesting	Captures water to serve as a local water supply and for environmental enhancement	Permeable pavements, green roofs, rain gardens, rainwater tanks
Water conservation	Reduces urban water demand	Water saving appliances, xeriscape, water pricing
Water reuse	Contributes to local water supply	Sewer "mining", greywater recycling, indirect potable reuse

Water conservation is a traditional approach which is receiving increasing attention and is enabled by a growing range of technologies. Water efficient fixtures such as low-flush toilets, water saving shower heads, and faucets; water efficient appliances (washing machines, dishwashers); and drip irrigation and xeriscape approaches to landscaping are allowing domestic water-using functions to be accomplished with less water. Commercial and industrial water users are also increasingly adopting low water use approaches. These physical changes are coupled with pricing and other behaviour-modification strategies which incentivize users to increased efficiency. The result is that domestic, commercial, and industrial water use per customer is decreasing in many locations. Obviously, as water is used more efficiently more is available for other uses and less must be taken from the environment.

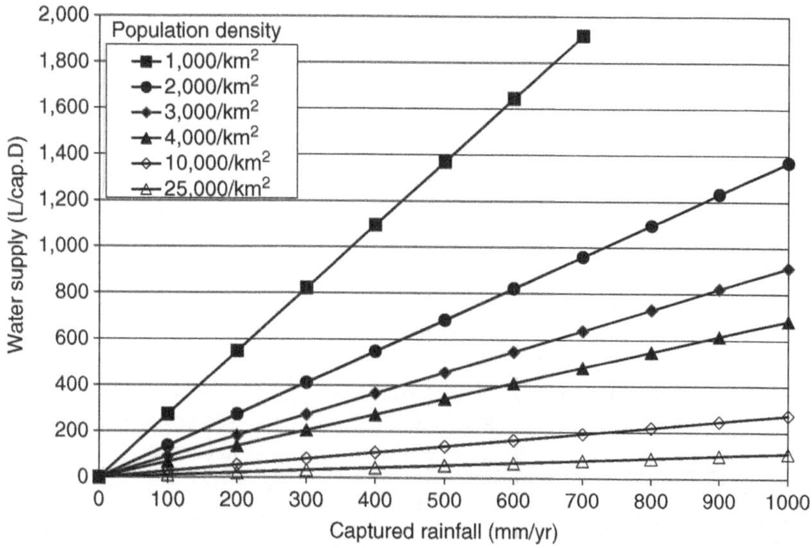

Figure 1.5 Water supply provided by rainfall capture as a function of urban population density

Water reuse is a well accepted and implemented approach which continues to expand as further increases in urban water are required (Jimenez & Asano, 2008). Recycled urban water can be reclaimed for non-potable and/or potable uses and can be distributed for a number of uses, including domestic, industrial, irrigation, and agriculture. Non-potable recycled water is generally distributed in a separate distribution systems (often referred to as dual distribution, as discussed below). Potable recycled water can be reintroduced into the water supply (either surface or groundwater) in a practice referred to as indirect potable reuse, or directly into the distribution system (referred to as direct potable reuse). The reuse of purified urban water directly reduces the quantify of water removed from the environment.

1.3.2 Energy management

Table 1.4 summarizes energy management approaches which are reducing energy use for urban water and resource management. Energy conservation is an obvious approach which continues to be enabled by advances in technology in general and specifically by water management technology.

Heat recovery is an evolving approach which takes advantage of the high heat content of water to provide heat for a variety of domestic and industrial uses. As illustrated in Figure 1.6, wastewater provides a significant source of heat energy, especially when it is recognized that total energy use for the urban water and wastewater system is on the order of 15 to 20 W/capita (Daigger, 2009). One may think of this as recovering the heat which is added to wastewater through the production and use of hot water (showers, clothes and dish washing). Heat exchangers are used to remove heat from wastewater, and heat pumps can be used if a higher temperature is needed. The recovered heat can be used, for example, in district heating systems. Reclaimed water can also be used for cooling.

Table 1.4 Energy management approaches.

Approach	Contribution to energy efficiency	Examples
Energy conservation	Directly reduces energy requirements by reducing energy needed for specific tasks	Energy-efficient motors and mechanical devices, fine bubble diffusers, optimization algorithms
Heat recovery	Heat recovered directly from water stream reduces residential and commercial energy use	In-sewer heat exchangers, heat pumps, district heating systems
Anaerobic treatment	Stabilizes biodegradable organic matter by conversion to biogas, produces energy source	Anaerobic sludge digestion, direct anaerobic treatment of wastewater, pre-treatment to increase digestability
Thermal treatment	Combustion of organic matter produces heat to be used directly or for energy production	Thermal oxidation, gassification
Microbial fuel cells	Direct conversion of organic matter in wastewater to electricity	Various technologies being scaled-up from lab- to pilot- to full-scale

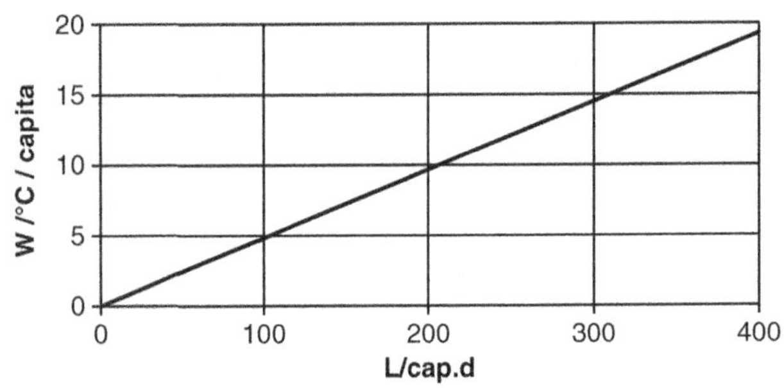

Figure 1.6 Heat recovery potential for wastewater (used water): reported values are for energy available – recovery efficiency must be considered to determine actual recovery (*Source*: calculations documented in Daigger, 2009)

Energy can also be produced from the organic matter contained in wastewater. The energy available ranges from about 15 to 30 W/capita (Watts per person), depending on whether kitchen waste (garbage grinders) and the energy available from nitrogen compounds are included (Daigger, 2009). Energy can be extracted by biological means (anaerobic treatment of either the wastewater directly or sludge produced form wastewater treatment, or by microbial fuel cells), or by thermal treatment of particulate organics removed from or produced by treatment of wastewater.

1.3.4 Nutrient management

Table 1.5 summarizes nutrient management approaches which are enabling improved nutrient management and recovery. Biosolids reuse represents a historical approach which continues. Wastewater treatment results in the production of biosolids which contain nitrogen, phosphorus, and other nutrients which can replace artificial fertilizers when applied to the land. Biosolids products of various types are produced, including digested biosolids which are applied to agricultural land either in liquid form or after dewatering, thermally dried and pelletized biosolids which are included in various fertilizer products, and compost products. These approaches allow reuse of nutrients contained in wastewater residuals.

Table 1.5 Nutrient management approaches.

Approach	Contribution to nutrient management and recovery	Examples
Biosolids reuse	Nutrient content of biosolids replaces fertilizer use	Land application of biosolids, sludge pelletization, compost production
Chemical and biological nutrient removal	Transfers phosphorus from liquid stream into solids stream, transfers nitrogen to atmosphere as nitrogen gas	Metal Salt precipitation of phosphorus, biological nitrogen and phosphorus removal
Nutrient recovery	Removal of nutrients from liquid stream and recovery in usable form	Ammonia stripping and adsorption, struvite precipitation

The removal of nitrogen and phosphorus from the liquid stream is a standard, well accepted and widely implemented practice (WEF, 2009). Phosphorus is transferred into the solid phase by either chemical precipitation or enhanced biological uptake, while reduced nitrogen (ammonia and organic nitrogen) compounds in the raw wastewater are oxidized to nitrate before being reduced to nitrogen gas and released into the atmosphere. Phosphorus can be recovered from the biosolids by a number of technologies (Sartorius *et al.* 2011). If the biosolids are thermally oxidized (for energy production), phosphorus can be extracted from the ash using technologies similar to those used to extract phosphorus from phosphate ore. Alternately, since the phosphate concentration is elevated if it is released from the biosolids, it can be precipitated as struvite ($MgNH_4PO_4$), which is an ideal component of slow release fertilizer, or calcium phosphate, which is similar in composition to phosphate ore. Fertilizer grade ammonium sulphate can also be produced from the high ammonia-nitrogen concentration sidestreams from sludge digestion processes by stripping and adsorption. This stream can also be treated biologically by nitritation (conversion to nitrite) and anammox (autotrophic denitrification). While not resulting in nutrient recovery, this approach significantly reduces energy requirements compared to conventional nitrification/denitrification and eliminates the carbon requirement for heterotrophic denitrification.

1.4 INTEGRATED SYSTEMS

The water, energy, and nutrient management approaches described in the previous section are organized into integrated urban water and resource management systems which provide superior performance relative to historical "linear" systems.

1.4.1 System approaches

Table 1.6 summarizes three elements of these integrated systems which offer significant advantages and are increasingly being adopted (Daigger & Crawford, 2007; Daigger, 2008, 2009, 2010). Examples of integrated systems are subsequently provided, along with procedures for developing and evaluating them.

Table 1.6 Evolving elements of integrated urban water and resource management systems.

Approach	Description	Benefits
Dual distribution	Potable and non-potable water separately distributed to user	Only potable water needs to be treated to potable standards; lower quality water sources usable for non-potable water
Source separation	Greywater, blackwater, and yellowater separately collected and processed	Individual streams can be separately processed for water, energy and nutrient recovery
Distributed water treatment	Potable water produced from non-potable water source using advanced treatment technology on a localized basis	Increases water supplies because allows high quality non-potable water for most uses

Dual distribution systems separately provide potable and non-potable water, which allows different water sources to be used and different levels of treatment. For example, a limited source of pristine water can be supplied for potable purposes while a poorer quality source water (for example, recycled water or rainwater) is used with an appropriate level of treatment to provide non-potable water, thereby allowing water demands to be met while avoiding expensive and energy-intensive treatment of the poorer quality water to potable standards. Energy savings can also be realized through reduced energy requirements for water distribution and reduced water consumption, especially heated water consumption (water efficient showers, washing machines, dishwashers, etc.). An alternate approach is distributed water treatment where non-potable water is distributed throughout the urban area and only the relatively small fraction required for potable purposes is further treated to potable standards using a series of distributed potable water treatment facilities employing, for example, advanced treatment technologies such as membranes and advanced oxidation. Significant energy savings can result because these energy-requiring technologies need only be applied to treat the relatively small potable water flow.

Source separation of domestic wastewater into greywater, blackwater, and yellowater offers significant opportunities to reduce the cost to recover water, energy and nutrients. Greywater consists of water used for bathing, laundry, and similar purposes and represents the largest contribution to the volume of wastewater. As indicated in Table 1.7, greywater is relatively uncontaminated (contains relatively low proportions of the five-day biochemical oxygen demand BOD_5, total nitrogen TN, total phosphorus TP, and potassium K) and, consequently, can be treated with relatively low energy input. It also contains much of the heat added to water as it is used, thereby offering increased opportunities for heat recovery. Blackwater includes feces and may also include water from the kitchen and includes the majority of the biodegradable organic matter which can be treated using anaerobic processes to produce biogas. The low volume of water used for toilet flushing and to convey kitchen waste means that this stream is highly concentrated, further facilitating its treatment. Yellowater is urine, which is a very small volume (about 1 to 2 $L/cap \cdot d$, liters per person per day, compared to between 135 and 400 $L/cap \cdot d$ for the total wastewater volume) but contains the vast majority of the nutrients (nitrogen, phosphorus, and potassium). Separate collection of yellowater facilitates capture and reuse of the nutrients in the wastewater stream.

Table 1.7 Organic matter and nutrient distribution in typical european wastewater.

Source	BOD_5 g/cap · d*	Total nitrogen g-N/cap · d	Total phosphorus g-P/cap · d	Potassium g-K/cap · d
Toilet waste				
Feces	20	1.1	0.6	1.1
Urine	5	11.0	1.4	2.5
Kitchen	30	0.8	0.3	0.4
Bath/Laundry	5	1.1	0.3	0.4
Total	60	14.0	2.6	4.4

*cap · d: capita per day or person per day
(*Source*: Henze & Ledin, 2001).

1.4.2 An example system

While the individual components described in the previous sections offer important opportunities to recover water, energy, and nutrients, the greatest opportunity exists when they are coupled together into integrated systems to supply water and manage storm and wastewater. This occurs for a number of reasons. Separately providing potable and non-potable water allows a variety of water sources to be used, including harvested rainwater and reclaimed water. Non-potable water can provide the "backbone" water supply, meeting all needs except direct potable consumption. One important benefit of this approach is that these water supplies are distributed throughout the urban area, thereby minimizing the pumping requirements to distribute it. Significant energy savings are possible as water distribution represents a significant energy demand in most conventional urban water and resource management systems (Daigger, 2009). Further savings result if source separation of the wastewater components is practiced. The high volume, relatively uncontaminated greywater can be easily treated to a high quality, certainly sufficient for non-potable uses, and combined with captured rainwater to provide the non-potable water supply. Several options are available to provide the relatively small volume of potable water required. Distributed water treatment approaches can be used to further upgrade the quality of the non-potable water supply using advanced water treatment technologies (membranes, advanced oxidation), or it can be supplied from a relatively pristine source if it is available. Heat can be recovered from the relatively warm reclaimed water stream and used for a variety of purposes, such as district heating. Organic matter contained in the blackwater can be used for energy production (such as by conversion to biogas using anaerobic processes), and the fact that it is captured in a relatively small flow makes such conversions more cost-effective. Nutrients can be recovered from the yellowater.

Figure 1.7 provides a schematic illustration of a system incorporating these features. In this example a local aquifer is considered to be available to provide seasonal storage of non-potable water, and also to partially contribute to its distribution throughout the urban area. Supply to the aquifer is provided by recharge of captured rainfall and reclaimed water. Water supply can also be provided directly by captured rainwater and reclaimed water when this is cost-effective. Potable water is provided by a separate, pristine water source. Because of the relatively low volume of potable water required, a water supply which would have been insufficient if it was the entire water supply would now provide a robust and reliable supply of potable water. Blackwater and yellowater are collected separately for energy production and nutrient recovery. Heat is recovered from both the reclaimed water and the non-potable water stored in the aquifer. Salt accumulation is certainly a possibility in a system with a high degree of reclamation and reuse, resulting in the need to remove salt to maintain appropriate levels in the reclaimed water. A variety of technologies, such as reverse osmosis, can be used to remove excess salt. This is considered to be disposed of in a saline aquifer in the example presented in Figure 1.7, but other options are available and more will develop in the future. Note that management of water, organic matter, and nutrients occurs at different locations in such a system, and at different scales. Water management is relatively distributed throughout the urban area, while the management or organic matter and nutrients may be more centralized.

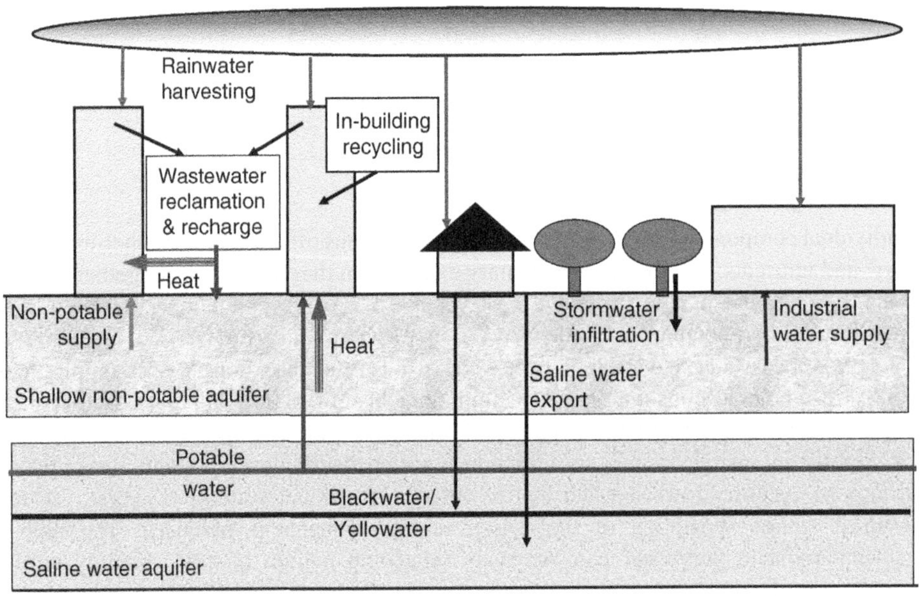

Figure 1.7 Example integrated urban water and resource management system

A system such as the one illustrated in Figure 1.7 will dramatically reduce the quantity of water which must be removed from the environment while maintaining a high level of service. Such a system would not only be energy neutral, but would likely result in net energy production. Significant nutrient recovery could also be accomplished. Public health protection could be increased with such a system compared to current approaches. Non-potable water would be of high quality – fully oxidized with low turbidity and with essentially no pathogens – and quite appropriate for most uses. Potable water could be of higher quality than is currently provided as the highest quality water can be preserved for potable purposes, extensive treatment can be provided cost-effectively due to the relatively low volume, and deterioration in the distribution system can be minimized by distributing treatment closer to the user. The distributed nature of such a system makes it more robust since difficulties with an individual component affect fewer customers and distributed components can be linked to provide system redundancy. System upgrade and expansion is also made easier due to its modular nature. All of this can be accomplished using technology and approaches which are widely available today. Just think of what can be accomplished using technologies and approaches which will be available in the future!

1.4.3 Developing and analyzing integrated urban water and resource management systems

While the example presented in Figure 1.7 is illustrative, the specific set of components to be incorporated into a particular urban area will depend on site-specific conditions. Thus, procedures are needed to conceptualize, evaluate, and select the most appropriate approach for a particular location. Relevant procedures are available (Daigger & Crawford, 2005; Guest *et al.* 2009, 2010), and some key principals are summarized here.

Candidate integrated systems can be formulated using the guiding principles and constraints listed in Table 1.8. Each system should protect and use locally available water resources, maintaining a water balance over a wide range of hydrologic conditions, and it will ideally mimic local hydrologic patterns before development occurred. Multiple barrier approaches are used to protect public health. The consumption of all resources is considered, along with the potential for resource recovery. Each system must maintain salt and nutrient balances over the long-term, and residuals must be effectively managed.

Table 1.8 Guiding principles and constraints used to formulate integrated systems (adapted from Daigger, 2011).

Guiding principles	Constraints
1. Protect and subsequently use locally available water resources whenever possible	1. Maintain water balance for typical and extreme (wet and dry) conditions
2. Mimic local hydrologic patterns before development; generally means infiltration of high-frequency, low-intensity storms and allowing high-intensity storms to create run-off	2. Maintain long-term salt balance
3. Public health is protected by incorporating multiple barriers, especially considering pathogens and trace constituents	3. Maintain nutrient balance
4. Consider total resource consumption and potential for resource recovery when formulating systems	4. Manage residuals, both over short and long time scales

Systems which comply with these principles and constraints can then be compared to determine which is the most sustainable. As illustrated in Figure 1.8, this can be accomplished using decision analysis concepts and evaluating options using a decision hierarchy based on the triple bottom line concept of balancing economic, environmental, and social factors. Table 1.9 illustrates criteria which could be considered to compare alternatives. Utility scales are then established for each criterion, and each alternative is evaluated relative to the utility scale. This is a technical evaluation and is based on rational criteria, some of which are quantitative and some semi-quantitative. However, the relative importance of each criterion is a matter of values and is quantified by the weighting applied to each criterion. Weights should be established by the residents of the urban area as it is their values which are relevant in determining how water and resource management services are to be provided to them. The relative value of each alternative is computed by multiplying the value for each criterion times its relative weight and summing the values for each alternative, as illustrated in Table 1.10. The numerical scores are then used to compare each alternative and determine which is the most sustainable. Importantly, it is not just the numerical score but also the discussion that such an analysis creates that leads to the selection of the most desirable option for a particular community.

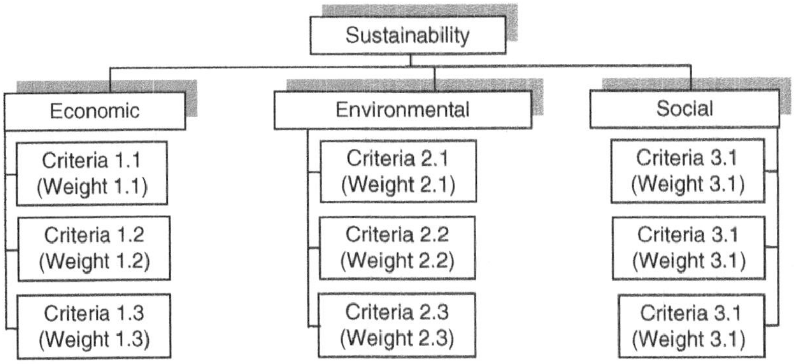

Figure 1.8 Sustainability can be quantified using a decision hierarchy based on the "Triple Bottom Line" concept

Table 1.9 Example "Triple Bottom Line" criteria.

Economic	Environmental	Social
• Affordability	• Water quality	• Public health
• Economic development	• Air quality	• Public impacts
• Financial management requirements	• Land quality	• Community development
	• Greenhouse gases	

Table 1.10 Example "Triple Bottom Line" analysis.

		Option 1		Option 2	
Criteria	**Weight**	**Score**	**Value**	**Score**	**Value**
Economic					
$C^{1}_{1.1}$	$W^{2}_{1.1}$	$S^{3}_{1,1.1}$	$W_{1.1} \times S_{1,1.1}$	$S_{2,1.1}$	$W_{1.1} \times S_{2,1.1}$
$C_{1.2}$	$W_{1.2}$	$S_{1,1.2}$	$W_{1.2} \times S_{1,1.2}$	$S_{2,1.2}$	$W_{1.2} \times S_{2,1.2}$
$C_{1.3}$	$W_{1.2}$	$S_{1,1.3}$	$W_{1.3} \times S_{1,1.3}$	$S_{2,1.3}$	$W_{1.3} \times S_{2,1.3}$
Environmental					
$C_{2.1}$	$W_{2.1}$	$S_{1,2.1}$	$W_{2.1} \times S_{1,2.1}$	$S_{2,2.1}$	$W_{2.1} \times S_{2,2.1}$
$C_{2.2}$	$W_{2.2}$	$S_{1,2.2}$	$W_{2.2} \times S_{1,2.2}$	$S_{2,2.2}$	$W_{2.2} \times S_{2,2.2}$
$C_{2.3}$	$W_{2.3}$	$S_{1,2.3}$	$W_{2.3} \times S_{1,2.3}$	$S_{2,2.3}$	$W_{2.3} \times S_{2,2.3}$
Social					
$C_{3.1}$	$W_{3.1}$	$S_{1,3.1}$	$W_{3.1} \times S_{1,3.1}$	$S_{2,3.1}$	$W_{3.1} \times S_{2,3.1}$
$C_{3.2}$	$W_{3.2}$	$S_{1,3.2}$	$W_{3.2} \times S_{1,3.2}$	$S_{2,3.2}$	$W_{3.2} \times S_{2,3.2}$
$C_{3.3}$	$W_{3.3}$	$S_{1,3.3}$	$W_{3.3} \times S_{1,3.3}$	$S_{2,3.3}$	$W_{3.3} \times S_{2,3.3}$
		T^{4}_{1}		T_2	

[1] $C_{i \cdot j}$ refers to criterion i · j
[2] $W_{i \cdot j}$ refers to weight for criterion i · j
[3] $S_{x,i \cdot j}$ refers to score for option x Relative to criterion i · j
[4] T_x refers to the total score for option x

1.5 IMPLEMENTING INTEGRATED URBAN WATER AND RESOURCE MANAGEMENT SYSTEMS

Implementing these higher performing integrated urban water and resource management systems will require significant changes in the manner these systems are implemented and managed, as illustrated in Table 1.11. Urban drinking water, stormwater and recycled water systems have historically been implemented and managed as separate, single-purpose systems. As is now obvious, future systems will be integrated, and they must be implemented and managed in an integrated system. This will require integration of the subject infrastructure, but more importantly the functional integration of system planning and management. Education of water professionals must also become more integrated so that they are prepared to implement and manage these more integrated systems. While detailed knowledge of individual system components is still needed, each professional must possess an overall understanding of how the system components operate together to achieve successful results.

Table 1.11 Implementing integrated urban water and resource management systems (abstracted from Daigger, 2011).

Item	Approach	
	Historical	**Evolving**
Institutions	Single purpose	Integrated
Management		
Conveyance	Distributed	Distributed
Treatment	Centralized	Distributed
Financing	Volume based	Service based
Planning	Independent, followed city planning	Integrated with city planning

System management will be quite different as treatment becomes much more widely distributed throughout the urban area. Management of water conveyance has always been distributed, but historically treatment has been centralized, thereby allowing centralized management of this function. As treatment becomes more distributed, different approaches to operate and manage the system must be developed. Ownership of treatment components may also pass from utilities to private hands, leaving utilities with decisions concerning the role they will play. Will the utility set performance standards for these distributed treatment components and simply monitor to determine they are performing as required? Will utilities operate and maintain these facilities, for a fee? It is likely that real-time control of the entire system, using smart grid concepts, will be required to ensure effective performance. In such instances, the utility must incorporate a significant number of distributed treatment components in their system. This raises both strategic and operational issues which will dramatically alter the operating mode for urban utilities.

The model for financing urban utilities must change from the current volume-based system where utilities are actually incentivized to produce and sell more water to a serviced based model where higher performance and greater efficiency are incentivized. Other utilities, such as electricity and gas, are faced with the same challenges, so significant experimentation and learning is expected in this area.

Perhaps the biggest change is the need for the water profession to engage proactively in planning and development of urban areas. City planning has traditionally proceeded independently, and water infrastructure is simply added after the fact. It is increasingly recognized, however, that significant opportunities exist to not only dramatically improve urban water and resource management but to also enhance urban life by integrating urban water and resource management with city planning. Green infrastructure components can significantly enhance the urban environment by reducing urban heat island effects and creating a more aesthetic and liveable environment. These benefits are best implemented as cities are being planned and constructed, rather than after the fact. The role and profile of water professionals in urban life will be transformed dramatically as the evolving urban water and resource management paradigm is implemented.

REFERENCES

Daigger G. T. (2003). Tools for success. *Water Environ. Tech.*, **15**(12), 38–45.

Daigger G. T. (2007). Wastewater management in the 21st century. *J. Environ. Eng.*, **133**(7), 671–680.

Daigger G. T. (2008). New approaches and technologies for wastewater management. *The Bridge*, **38**(3), 38–45.

Daigger G. T. (2009). Evolving urban water and residuals management paradigms: water reclamation and reuse, decentralization, resource recovery. *Water Env. Res.*, **81**(8), 809–823.

Daigger G. (2010). Integrating water and resource management for improved sustainability. In: Water Infrastructure for Sustainable Communities, X. Hao, V. Novotny and V. Nelson (eds), IWA Press, London.

Daigger G. T. (2011). Sustainable urban water and resource management. *The Bridge*, **41**(2), 13–18.

Daigger G. T. and Crawford G. V. (2005). Wastewater treatment plant of the future – Decision analysis approach for increased sustainability. *Proc. 2nd IWA Leading-Edge Conf. on Water and Wastewater Treatment Technology, Water and Environment Management Series*, Prague, Czech Republic, June 1–4, IWA Publishing, London, pp. 361–369.

Daigger G. T. and Crawford G. V. (2007). Enhanced water system security and sustainability by incorporating centralized and decentralized water reclamation and reuse into urban water management systems. *J. Environ. Eng. Manage.*, **17**(1), 1–10.

Diaz R. J. and Rosenberg R. (2008). Spreading dead zones and consequences for marine ecosystems. *Science*, **321**, 926–929.

Guest J. S., Skerlos S. J., Barnard J. L., Beck M. B., Daigger G. T., Hilger H., Jackson S. J., Karvazy K., Kelly L., Macpherson L., Mihelcic J. R., Pramanik A., Raskin L., van Loosdrecht M. C. M., Yeh D. and Love N. G. (2009). A new planning and design paradigm to achieve sustainable resource recovery from wastewater. *Environ. Sci. & Tech.*, **43**(16), 6126–6130.

Guest J. S., Skerlos S. J., Daigger G. T., Corbett J. R. E. and Love N. G. (2010). The use of qualitative system dynamics to identify sustainability characteristics of decentralized wastewater management alternatives. *Water Sci. Tech.*, **61** (6), 1637–1644.

Henze M. and Ledin A. (2001). Types, characteristics and quantities of classic, combined domestic wastewater. In: Decentralised Sanitation and Reuse: Concepts, Systems and Implementation, P. G. Lens, G. Zeeman and G. Lettinga (eds), IWA Publishing, London.

Jimenez B. and Asano T. (2008). Water Reuse: An International Survey of Current Practice. Issues and Needs. IWA Publishing, London.

Monsma D., Nelson R. and Bolger R. (2009). Sustainable Water Systems: Step One – Redefining the Nation's Infrastructure Challenge. A Report of the Aspen Institute's Dialogue on Sustainable Water Infrastructure in the U.S., Aspen Institute, Washington, DC.

National Research Council (2003). Cities Transformed: Demographic Change and Its Implications in the Developing World. National Academy Press, Washington, DC.

Novotny V. and Brown P. (2007). Cities of the Future: Towards Integrated Sustainable Water and Landscape Management. IWA Publishing, London.

Rockström J., Steffen W., Noone K., Persson A., Chapin F. S. III, Lambin E. F., Lenton T. M., Scheffer M., Folke C., Schellnhuber H. J., Nykvist B., de Witt C. A., Hughes T., Van der Leeuw S., Roldhe H., Sörline S., Snyder P. K., Costanza R., Svedin U., Falkenmark M., Karlberg L., Corell R. W., Fabry V. J., Hansen J., Walker

B., Liverman D., Richardson K., Crutzen P. and Foley J. A. (2009). A safe operating space for humanity. *Nature*, **461**(24), 472–475.

Sartorius C., van Horn J. and Tettenborn F. (2011). Phosphorus recovery from wastewater – state-of-the-art and future potential. *Proceedings of Nutrient Recovery and Management 2011*, January 9–12, 2011, Miami, FL, CD-ROM, Water Environment Federation, Alexandria, VA.

United Nations (2009). World Population Prospects: The 2009 Revision, Population Division of the Department of Economic and Social Affairs of the United Nations Secretariat. United Nations, New York. http://www.un.org/esa/population/unpop.htm (accessed April 2011).

United Nations Environment Programme (2008). Vital Water Graphics: An Overview of the State of the World's Fresh and marine Waters. 2nd edn, United Nations, NY, http://www.unep.org/dewa/vitalwater/article186.html (accessed April 2011).

Wallace B. (2005). Becoming Part of the Solution: The Engineer's Guide on Sustainable Development. American Council of Engineering Companies, Washington, DC.

Water Environment Federation (2009). Design of Municipal Wastewater Treatment Plants. Manual of Practice No. 8, 5th edn, Water Environment Federation, Alexandria, VA.

Weber W. J. Jr. (2004). Optimal use of advanced technologies for water and wastewater treatment in urban environments. *Water Sci. Tech.: Water Supply*, **4**(1), 7–12.

Chapter 2

Energy and water: relations and recovery potential

Alessandro Meda, Dorothee Lensch, Christian Schaum and Peter Cornel

2.1 ENERGY AND WATER – AN INTRINSICALLY BOUND COUPLE

There are far more interactions between water and energy than it might appear at first glance, water and energy is an intrinsically tied couple: Energy is required for water and wastewater distribution, usage and treatment, water and wastewater contains energy and water is directly and indirectly necessary for energy conversion. Each of these linkages is depicted in detail below.

2.1.1 Energy for water

Used energy for water supply and wastewater disposal can be divided into water supply, conveyance, treatment and distribution on the one hand, and wastewater collection, treatment and discharge on the other. Due to a complete representation of the energy consumption within the water cycle the energy of the private-water-end-consumer (mostly for hot water generation) has to be taken into account. Energy for industrial end water use cannot be quantified, therefore, it is neglected here.

Table 2.1 lists energy consumption ranges from California and Germany deducted from several literature sources. The values demonstrate the wide range, even while comparing industrialized countries. Specific values are chosen regarding per cubic metre water, respectively per person (cap) and year (yr). Due to different consumer behaviour in Germany and California, specific water consumption values (122 versus 274 L/cap · d) were computed according to the Federal Statistical Office, Germany (2009) and Asano *et al.* (2007). German values regarding water supply are determined by benchmarking studies of German water works (Roedl & Partner, 2006, 2007; ZVBWV, 2009; Plath *et al.* 2010). Californian values are results from the Integrated Energy Policy Report of the California Energy Commission (CEC, 2005). The determination of energy by water-end-consumers is based on the yearly energy demand for hot-water use and other water related energy consumption such as drying, washing, and so on (CEC, 2005; BDEW, 2008). Since municipal wastewater plants are constructed according to population equivalents (PE) this is the corresponding value for waste water. For German values, several benchmarking studies have been reviewed by Haberkern *et al.* (2008) and Keicher *et al.* (2008) with results displayed below. These values are average values of several plant size categories, therefore, the total range of values might be enlarged. For California this has been also done within the California Emission Report (CEC, 2005).

Table 2.1 Ranges of energy consumption within the water cycle in California and Germany (own calculations).

Water-end-consumers		California, USA		Germany	
		kWh/m³	kWh/cap · yr	kWh/m³	kWh/cap · yr
Water	Supply, conveyance	0.00–1.06	0–106	0.12–1.13	5–50
	Treatment	0.03–4.23	3–423		
	Distribution	0.18–0.32	18–32	0.03–0.58	1–26
End-consumer	Private households	~11	~2000	~21[1]	~950[1]
Wastewater	Collection, treatment	0.29–1.22	29–122	0.39–0.83	32–75

[1]Total energy for water heating in private households in Germany divided by municipal water consumption and population in 2007 (BDEW, 2008; Federal Statistical Office, Germany, 2009)

For the further purification of treated wastewater to be used as service water, the Californian energy report assumed that the energy consumed are already included in the wastewater treatment as advanced or tertiary treatment and the distribution of recycled wastewater requires 0.1–0.3 kWh/m^3. This is a fraction of the energy intensities for water supply, treatment, and distribution, which is 0.15–1.71 kWh/m^3 for German conditions and 0.21–5.61 kWh/m^3 for California. It is pointed up the fact that water reuse not only preserves water resources, but may be reasonable in terms of energy consumption and the emission of greenhouse gases.

Though, the range of the values is rather large and will vary according to the boundary conditions. Considering industrial processes, this fact is getting more significant. Depending on production process, energy consumption for water supply, use and treatment might be considerably increased. However, to display the general relations within the municipal water sector, Figure 2.1 shows typical consumption values. Noticeable is the high proportion of energy consumption by the end consumers.

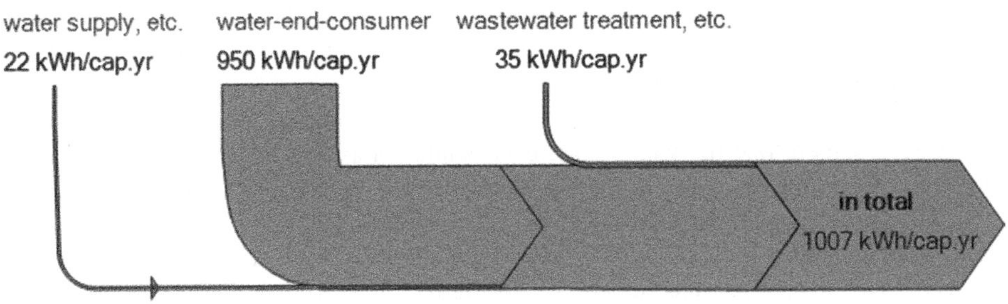

Figure 2.1 Typical values of specific energy consumption in the German municipal water sector (average values of benchmarking studies and own calculations)

2.1.2 Energy from water

Water, and especially wastewater, contains energy in different forms:

(1) Kinetic energy
(2) Potential energy

(3) Thermal energy
(4) Chemically-bound energy

Following estimation will show the order of magnitudes of energy content within water and wastewater.

2.1.2.1 Kinetic energy

The amount of kinetic energy depends mainly on flow rates, which are relatively low within water and wastewater streams. Depending on topography (large slope), these values might be higher at several points within the drinking water network and the sewer system. However, the use of pumps to equalize these slopes at other points equals this amount.

2.1.2.2 Potential energy

Neglecting frictional losses and so on, the potential energy is directly proportional to the height. But even in a height of 50 m, the potential energy content of water and wastewater ($E_{pot} = mgh$) is only 16 Wh/cap · d or about 6 kWh/cap · yr and hence almost to neglect compared to thermal and chemically bound energy.

2.1.2.3 Thermal energy

The thermal energy stored in wastewater mainly results from hot water generation. Therefore, greatest potential of heat recovery is within greywater, that is wastewater from showers, baths, washing machines, and possibly the kitchen. According to $E_{therm} = c_p \Delta T m$, maximum recoverable thermal energy is calculated via the specific heat capacity of water, the temperature gradient and the water quantity. Considering, for example, a greywater volume of 40 L/cap · d as typical for German households and an available ΔT of 30 K, the thermal energy amounts to 509 kWh/cap · yr. This is by far higher than the potential energy estimated above. In other words, the amount of energy which is released to the environment when one litre of water cools by one Kelvin, corresponds to the calculated potential energy of one litre of water retained at an altitude of 426 m.

Noticeable is also the gap to the calculated specific energy consumption by private households in Table 2.1. This can be explained due to losses during the generation process (efficiency of heating system), the general divergence of end energy and primary energy and by deviations of the assumed specific consumer behaviour, that is the population might consume more hot water or water with higher temperature levels.

2.1.2.4 Chemically bound energy

The chemically bound energy can be determined by the carbon content, estimated by the COD (chemical oxygen demand). The maximum potential amount of methane per kg COD can be calculated stoichiometrically:

$$CH_4 + 2O_2 = CO_2 + 2H_2O$$

that is the COD per mol methane is 64 g O_2.

With a molar volume of 22.41 L under standard conditions (0°C, 1 atmosphere), the CH_4 equivalent of COD (converted under anaerobic conditions) is 0.35 L methane/g COD respectively 0.35 m^3 methane/kg COD. Considering the heating value of methane of 35.89 MJ/m^3 (9.97 kWh/m^3), resulting chemically bound energy content is 3.49 kWh/kg COD.

Based on a daily COD load per capita of 110 to 120 g, the maximum theoretical energy content is on average 400 Wh/cap · d or 146 kWh/cap · yr, in case the entire COD could be transferred to methane and could be utilized.

For comparison, energy consumption of wastewater plants in Germany is about 30 to 80 kWh/cap · yr (see Table 2.1).

The comparison in Figure 2.2 shows that the major part of energy contained in wastewater is stored as thermal energy. The percentage of chemically bound energy is slightly less, whereby the potential energy is much smaller than thermal or chemically bound energy by several orders of magnitudes and kinetic energy content is negligible small.

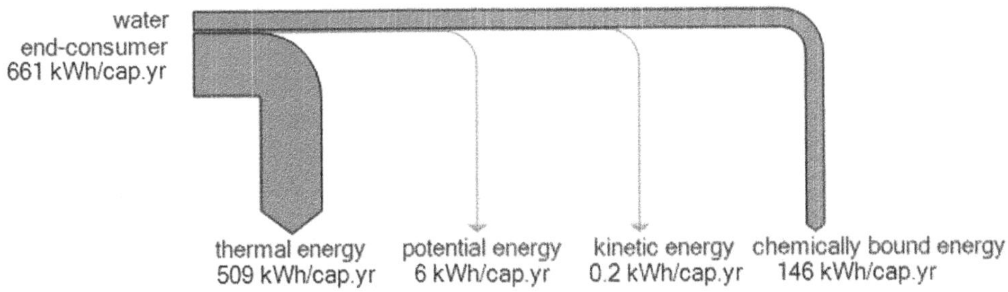

Figure 2.2 Theoretical energy potential of wastewater (assumptions: water consumption 122 L/cap · d, flow rate 5 m/s, altitude 50 m; greywater 40 L/cap · d, $\Delta T \sim 15$ K, 115 g COD/cap · d)

Due to losses during the transportation within the sewer, system thermal energy has to be reclaimed as close to the source as possible, whereas chemically bound energy can be transported in the wastewater via the sewer system almost without losses. The recovery of potential energy from wastewater seems promising only in high-rise building or appropriate topographies. Due to low flow rates, the amount of kinetic energy within wastewater can be neglected, and therefore, it is not considered any more.

These values do not allow any conclusion on the reclaimable and technically usable percentage, the necessary afford for its use or even the relation of electricity to unused waste heat.

2.1.3 Water for energy

Water is directly and indirectly necessary for energy conversion. Direct use of water for electricity generation by using hydropower or wave or tidal power is easily comprehensible and can be estimated by the energy content of water and wastewater, as mentioned previously. Additionally, evaporation losses have to be considered here, too.

But also huge quantities of water might be needed for the exploration, extraction and beneficiation of fossil fuels depending on the specific method. According to Gleick (1994), up to 20 m^3 per TJ_{th} are required for coal mining and due to enhanced oil recovery (e.g. water-flooding or steam injection), water requirements can increase up to several hundred m^3 per TJ_{th}.

Water is required also for renewable energy production. Here especially the energy crop cultivation has to be mentioned, whereas the so called bioelectricity (biogas for electricity and heat) is evaluated better than the fractional use for biofuel production (Gerbens-Leenes *et al.* 2009). Cornel & Meda (2008) estimate the specific water demand for biogas generation via maize silage in Germany at approximately 300 L/kWh. Gerbens-Leenes *et al.* (2009) quotes a range of 180 L/kWh (sugar beet, maize, and sugar crane for biogas) to 2,160 L/kWh (jatropha for biodiesel).

Last but not least, power plants are using steam for driving turbines and also cooling water for heat dissipation. This latter, when compared to the water consumption for bioenergy, is only a small amount (around 1 to 3 L/kWh according to Gleick, 1994). Nevertheless, the significance of this water requirement has been distinctly pointed out in the European "hundred-year-summer" in 2003, when numerous power plants have been forced to reduce output or even shut down due to an insufficient available amount of cooling water.

An important factor for water consumption in energy production is the quality and the availability of used water (wastewater). Some processes pollute water more than others and in different ways. For example, cooling towers mostly change the temperature of the water, whereas the water used for coal extraction or plant operation (ash handling or flue-gas desulfurization process make-up water) might be contaminated with suspended solids, acids, heavy metals, and so on. Regarding the availability, especially evaporation should be mentioned.

Furthermore, the water for crop cultivation is not directly available any more in contrast to water from once-trough cooling systems.

2.2 ENERGY AND WATER RECOVERY

The above depicted interactions have got two main influencing factors, the used technology (and its efficiency) and the amount of water. The latter one depends naturally on the specific water demand. Based on the knowledge of the demand, possibilities of water and energy recovery can be detected. Once again, considering the recovery, water and energy are tied together: water recovery might imply direct or indirect energy recovery. Indirect energy recovery means here energy savings due to water savings and decreasing energy consumption for treatment. In direct energy recovery, the utilization of the energy from water is considered.

2.2.1 Water demand

Figure 2.3 shows the annual water demand in Europe. It can be observed that by far the largest part is used for food production.

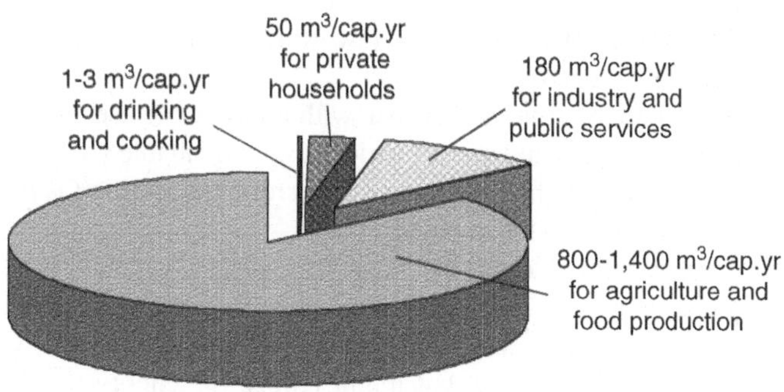

Figure 2.3 Annual water demand in Europe (Cornel & Meda, 2008)

One kilogram of bread needs approximately 2 kg of wheat as dry weight of total plant (Allan, 1997; Zehnder, 1997). In order to produce this quantity of plant material, the plants take up at least 1 m^3 of water which they mainly release to the atmosphere as transpiration losses. The rule of thumb "1 m^3 of water for 1 kg of bread" holds true only in case of optimum conditions. In reality, the consumed water quantity is much higher. American farmers, for example, consume about 4 m^3 of water per kilogram bread equivalent, and in the tropics, approximately 5 m^3 of water are used per 1 kg of rice, instead of the required 2 m^3 (Zehnder, 2001).

In order to generate foodstuff, which is essential for providing a sufficient nutrition with 2,500 kcal per day, currently 500–1,000 $m^3/cap \cdot yr$ with vegetarian diet and 1,200–1,500 $m^3/cap \cdot yr$ with a diet are needed, from which 500 kcal out of the 2500 kcal are supplied by meat (Zehnder, 2001, 2003). The higher demand for the nutrition of non-vegetarians derives from the approximately 10 times higher water demand per produced energy unit (kcal) for meat, compared with vegetarian food. Even with optimal irrigation, that is minimizing water losses, one has to calculate 250 $m^3/cap \cdot yr$ for vegetarians and 680 $m^3/cap \cdot yr$ for non-vegetarians.

2.2.2 Recovery by water reuse

Against this background, it seems that the recovery of water from private households can only make a small contribution towards the reduction of water shortage. However, looking at the drinking water management of large cities, one gets a completely different impression. Beijing, London, Los Angeles, Mexico City, Singapore, Tehran, Tokyo, cities of different development status and different locality, all have in common that the local drinking water demand by far exceeds the locally available resources. Drinking water supply can only be assured with large efforts, and occasionally, with severe impacts on the environment. Excessive exploitation of existing water resources, lowering of the groundwater level, energy-intensive and costly transport of water over many hundreds of miles and energy-intensive desalination of seawater are only few of the consequences of the current water supply of the megacities' population.

And hereby, most of the water is only used for the transport of pollutants. The demand of high quality, potable water could be significantly reduced by reusing reclaimed water.

One possibility would be intra-urban use of treated greywater for toilet flushing and other uses where potable water quality is not required. Another possibility is the direct or indirect reuse of recycled water as potable water, as practiced for example in Singapore, Windhoek (Namibia), Australia, and California (USA). Highly treated recycled water is introduced either directly into the potable water supply up- or downstream of the water treatment plant or into a raw water supply, such as potable water storage reservoirs or groundwater aquifers (Asano *et al.* 2007).

Accordingly, the quality requirements are higher than with non-potable use, and the required energy for treatment is higher as well. On the other hand, mass restrictions, unavoidable with greywater use, fall away, and a dual supply system, coupled with the potential risk of cross-connection (as exists in buildings with parallel use of potable and recycled water) is avoided.

"Water reuse is particularly attractive in the situation where available water supply is already overcommitted and cannot meet expanding water demands in a growing community. Increasingly, society no longer has the luxury of using water only once." (Asano *et al.* 2007) One might add: water reuse saves not only valuable water resources, but might reduce the energy demand of the water cycle and contributes to reduce the emission of greenhouse gases since the energy required for treating wastewater for non-potable water purposes and its distribution (see below) can be significantly lower than the consumption of conveyance, treatment and distribution of water (see Table 2.1).

However, in places where water is sufficiently available and can be supplied with low energy input, water recycling might be relevant only in case a significant cost reduction can be achieved. In such cases, water supply does not reduce water availability.

2.2.3 Possibilities of energy recovery from (waste)water

Direct energy recovery from water and wastewater can decrease the costs significantly by using or selling the generated energy. Therefore, this might be another possibility when water-reuse alone is not relevant.

2.2.3.1 Potential energy

Corresponding to hydropower plants, potential energy from wastewater can be used via turbines (e.g. Kaplan, Pelton, Francis turbines), archimedian screw turbines or water wheels. At present, water wheels are used within wastewater treatment plant (WWTP) effluents, whereas the overall potential is estimated to be small due to low altitude differences and seasonal fluctuations (DWA, 2009). Within high-rise buildings, it might be promising to treat wastewater from upper floors to be used as service (recycled) water in the lower floors and thereby minimize costs for pumping freshwater. Thereby, one can assume that the energy needed for pumping, in consideration of losses in the pipes and the efficiency of pumps and motors, is more than twice as much as the potential energy contained in the pumped water. Though, results for pilot experiments of such energy recovery are not yet published.

2.2.3.2 Thermal energy

Domestic wastewater, in particular greywater, presents a permanently available, year-round heat source with comparably high temperatures. The heat contained in greywater can be utilized already in-house, for example to preheat hot water via heat exchangers. Within the sewer system, warm greywater is mixed with relative cold rain water, infiltration water and remaining wastewater, causing distinctly lower temperature levels. Therefore, additional heat pumps are necessary.

To assess the reasonability of this method of heat recovery, it is important to consider the ratio between the useable heat capacity and the added electric power. The higher the temperature value of the water, that is the lower the temperature difference, the better is the heat recovery.

For example, two cases are compared to each other: on the one hand, mixed wastewater with a temperature of 12°C is used to generate hot water of 65°C ($\Delta T = 53K$). In this case about 3 times more heat is made available than electric energy has to be provided for the heat pump. However, taken into account the power plant efficiency in power generation and transmission losses in the power grid, almost the same amount of heat energy is released as in case of direct local use of primary energy for hot water heating, for example with a gas boiler instead of a heat pump. This means, there is no advantage in comparison to the use of primary energy.

In order to achieve optimum energy recovery, the process has to take place as close as possible to the origin of the hot water. Therefore, efficient heat recovery forces decentralized use of the generated heat, thus minimizing energy losses from the generated hot water on its way to the consumer, as explained below.

2.2.3.3 Chemically bound energy

Degradable organic constituents within wastewater are transported almost without losses and unused to the treatment plant, where they can be converted to usable energy by utilization of accruing sewage sludge. In principle two ways – alone or in combination – have been pursued (Schaum *et al.* 2010):

(1) Biogas production via anaerobic stabilization and usage of it by combined heat and power units (generation of electricity and heat).

(2) Steam generation via sewage sludge combustion and usage of it by turbines (generation of electricity and use of residual heat).

Though within the WWTP, the organic constituents are distributed throughout the process steps and are not completely available in the sewage sludge.

This is illustrated on the basis of a rough COD-balance for a typical treatment plant (with preliminary sedimentation, nitrification/denitrification and anaerobic sewage sludge stabilization with biogas production and utilization), which indicates the whereabouts of COD in terms of energy within the plant. Assuming a specific COD-load of 42 kg/cap · yr and conventional preliminary sedimentation rate about 10 kg COD/cap · yr, which remains in the primary sludge, only almost 9 kg COD/cap · yr gets to the excess sludge. Finally, almost half (45%) of the COD-load goes to the raw sludge, and around 55% is consumed by respiration processes or remains within the effluent. Assuming a degradation rate of about 55% of the organic fraction during digestion, roughly 11 kg COD/cap · yr are transformed to methane and 21% of the original COD-load remains within the digested sludge.

Expressing this in energy values (Figure 2.4), about 68 kWh/cap · yr of the total 146 kWh/cap · yr of the influent are transferred to the digestion unit and 38 kWh/cap.yr conveyed into methane. Considering a conversion efficiency rate of 32%, resulting power generation is about 12 kWh/cap · yr.

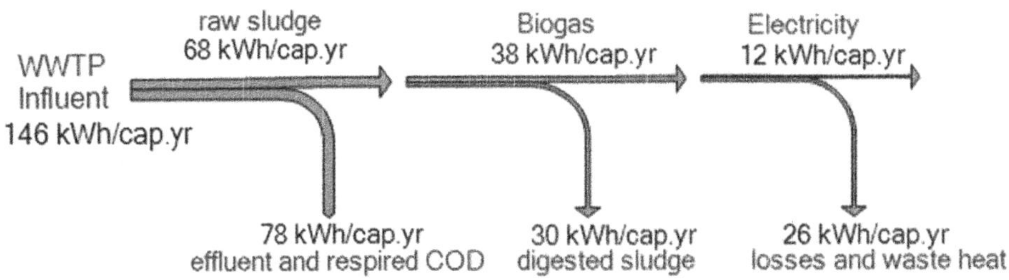

Figure 2.4 Simplified COD-balance within the wastewater treatment plant in specific energy values (Cornel *et al.* 2011)

A detailed description of this COD-cycle within the WWTP can be found in Cornel *et al.* (2011). In this publication, additional variables influencing the fate of COD such as the primary sedimentation and the sludge retention time are considered.

2.3 POTENTIAL OF WATER REUSE

Wastewater can be a valuable resource that contains the resource water in concentrations of more than 99.5%. After adequate treatment, that is adapted to its subsequent application, water can be a valuable product to be reused. Thus, water reuse is an essential component of integrated water resource management, not only in arid and in water deficient areas, but increasingly also in most of the densely populated urban areas, where water demand and supply diverge widely, at least regionally.

Water reuse opens up new water resources and reduces the demand for potable water. Potential regional or local lacks of water can be resolved. In addition, water reuse reduces the discharge of (treated) wastewater into water bodies. In regions where water supply may be energy-intensive and costly due to extensive transportation and/or pumping, water reuse can be a competitive alternative with lower energy consumption and lower costs than using freshwater. Moreover, finally yet importantly, valuable freshwater resources, such as high-quality groundwater, can be preserved via the alternative use of recycled water.

2.3.1 Reuse options

The fields of application for water reuse are manifold. According to Asano *et al.* (2007), the following categories of water reuse applications for recycled water originating from treated municipal wastewater can be established:

- Agricultural irrigation (crop irrigation; commercial nurseries)
- Non-potable intra-urban uses (toilet flushing, landscape irrigation like in parks, golf courses, greenbelts, residentials, cemeteries, freeway medians, school yards, fire protection, air conditioning)
- Industrial recycling and reuse (cooling water, boiler feed, process water)
- Recreational/environmental uses (lakes and ponds, stream flow augmentation, fisheries, snowmaking)
- Groundwater recharge (groundwater replenishment, salt water intrusion control, subsidence control)
- Potable reuse (blending in water supply reservoirs, blending in groundwater, direct pipe to pipe water supply)

Corresponding to the manifold applications, the requirements on the water quality, as well as on the treatment technology vary strongly. For detailed reviews, the publications of Asano *et al.* (2007) and Jimenez & Asano (2008) should be pointed out. In this chapter, the focus is set on intra-urban water-reuse and reuse in agriculture, since these are options with high potential.

The reuse of treated wastewater for agricultural irrigation presents by far the largest potential. Worldwide, more than 70% of the used freshwater are for agricultural irrigation (United Nations, 2003). Today, the reuse of wastewater for agricultural irrigation is practiced in almost all arid regions. Particularly in threshold and developing countries of Latin America, Asia and Africa, raw wastewater or insufficiently treated wastewater is used directly in crop irrigation. Often, this is done deliberately, in order to use the nutrients N and P as well as the organic load for forming humus, but often also without being aware of the health risks for farmers, farm workers and consumers. This is true also for the indirect and unplanned use of wastewater, where untreated wastewater is discharged into rivers and downstream river water is used for irrigation.

An additional challenge is water supply and disposal in large cities. According to UN-HABITAT (2006), until 2050 approximately 75% of the world's population will live in cities, 20% in urban areas and cities of 1 to 5 million inhabitants. The majority of about 80% will be living in threshold and developing countries, over half the world's urban population in Asia. "According to the conclusions of various water reuse surveys (Lazarova *et al.* 2001; Mantovani *et al.* 2001), the best water reuse projects, in terms of economic viability and public acceptance, are those that substitute recycled water in lieu of potable water for use in irrigation, environmental restoration, cleaning, toilet flushing, and industrial uses" (USEPA, 2004).

Intra-urban reuse of water for utilizations which do not require drinking water quality offer a high potential to save valuable water resources and reduce wastewater discharge. The freshwater consumption can be reduced by more than 30% to 40% when recycled water is used for toiled flushing, gardening, irrigation, and so on (Bieker *et al.* 2009).

But intra-urban water reuse fosters the transition from conventional centralized to nodal, semicentralized supply and treatment systems, with short distances from the firsthand user to the treatment units and back to the secondhand reusers. This will minimize the energy for transport and treatment and offers the chance to recover heat from wastewater, especially from greywater. But again, to avoid any health risks, proper planning and professional operation of water reclamation plants have to be guaranteed.

Intra-urban reuse and agricultural reuse are not mutually exclusive. On the contrary, they can be combined, as intra-urban water reuse is a so called non-consumptive use. Non-consumptive uses are all kinds of uses by which water is not physically lost but undergoes just a change in its quality, for example in power plant cooling, where most of the water is returned to the water body with almost no quality deterioration and can be used again. The same is true for most of the municipal and industrial waters that do not disappear by use but are returned to the rivers as treated wastewater and might be used again downstream. Agricultural irrigation, on the other hand, is a consumptive use, because the major part of the water is evaporated by the plants and is no longer available for other uses. Therefore, consumptive uses put higher pressure on water resources than non-consumptive uses and should be considered as the last step of a water reuse chain.

2.3.2 Water quality requirements

One precondition for unobjectionable and successful water reuse in the long term is the consistent compliance with defined quality standards. Here, various aspects have to be taken into account:

- Hygiene/protection of human health
- Soil protection/plant protection
- Groundwater protection
- Application of efficient irrigation techniques
- Public acceptance
- Storage during non-growing seasons

In order to protect human health, limiting values for pathogens (bacteria, viruses, helminth eggs) and toxic substances have to be defined and met. When defining these limiting values, the type of contact between humans and wastewater plays a decisive role. References can be taken from the WHO guidelines (WHO, 2006).

For soil protection, first of all it is important to restrict the concentration of salts and heavy metals. Well known, for example, is the potential of soil salinization due to irrigation with sodium-containing wastewater. This reduces the water conductivity of the soil, and therefore, leads to harvest depression. Decisive factors for defining the required water quality are type of soil (texture, grain size, permeability, chemistry, specific salt content), climate (aridity, precipitation quantity and distribution, humidity, wind), type of plant (nutrient balance in the soil) and irrigation technique (surface, sprinkler or subsurface irrigation, as well as irrigation quantity and frequency). Beside the total salt content, the concentration of several toxic ions to plants, such as boron, chloride, and sodium, must be considered. These ions are taken up via the root system, or via the leaves in case of sprinkler irrigation. Boron has a narrow tolerance range and can be toxic already in concentrations slightly above the essential concentration for plant growth. Boron reaches municipal wastewater in the form of perborate, that is as bleach in washing agents and disinfectants, whereas NaCl exists as regeneration salt for water-softening ion exchangers in dishwashers.

Although professional application of irrigation tries to avoid infiltration into groundwater, in practice it is hardly possible to prevent irrigation water from reaching the aquifer at least to a certain degree. However, in case the aquifer is required as seasonal storage for irrigation water, its protection should be of special interest. Irrigation techniques also require specific wastewater quality. Substances with corrosive impacts and insoluble substances (suspended matters) should be eliminated during wastewater treatment, in order to prevent clogging of pipes and damage of equipment.

In order to get public acceptance, also aesthetic aspects have to be considered. Wastewater for irrigation should therefore be as odorless as possible and colorless.

In case wastewater for irrigation has to be stored during non-growing seasons, specific quality standards for storage have to be met, for example standards for nutrient contents in order to prevent excessive algae growth in case of aboveground storage or to minimize nutrient input into the groundwater in case of subsurface storage.

2.3.3 Treatment options

Treatment options depend on numerous boundary conditions, which are also linked to each other. The purpose of reuse (irrigation water vs. toilet flushing vs. cooling water) involves specific quality requirements for use, storage and transport, which in turn can vary depending on legal requirements of the specific country. The climatic condition of the country can also influence the storage requirements due to the temperature depending microbial recontamination. Finally, the economic options are not to be neglected.

In accordance to conventional wastewater treatment, the treatment options can classified as physical and chemical methods, biological methods, and disinfection.

Compared with the use of raw wastewater for agricultural irrigation, the application of physical-chemical treatment techniques presents a large improvement, as these methods allow the significant reduction of solids and partial elimination of the organic load and pathogens. In addition, investment costs for these techniques are much lower compared to biological wastewater treatment plants, thus facilitating the implementation in economically weak countries as well. Hence, physical-chemical techniques for wastewater treatment (e.g. high rate clarification with flocculation/coagulation) represent a transitional solution for improving the water quality, and can therefore be the first step to be realized rapidly in those countries where raw wastewater is currently used for irrigation. However, the disinfection ability of only physically-chemically treated wastewater has to be viewed critically. High residual concentrations of organic substances and turbidity-causing substances put an efficient and ecologically sound disinfection into question. Depending on the disinfection method (for example chlorination), there is potential for the production of undesirable by-products.

The discharge quality of biological wastewater treatment plants normally fulfils the quality standards regarding the solids content, organic ingredients and nutrients. Deficits might be microbiological requirements, inorganic ingredients such as salts (sodium in particular) and boron as well as micropollutants.

In order to protect the health of persons who come in direct contact with irrigation water, such as farmers and irrigation technicians, but also uninvolved persons who remain in the vicinity of sprinkler irrigation units and consumers, who might get in contact with pathogens indirectly via the consumption of field crops, disinfection of irrigation water is "good practice" nowadays. One may do without, provided that health risks can be excluded by applying other measures, for example irrigation techniques, off-times for irrigation before the harvest, cleaning measures before consumption, and so on. The DWA topics "Assessment of Process Steps for the Treatment of Wastewater for Reuse" gives an overview on how different process steps affect the reduction of pathogens (DWA, 2008). The data are presented in log scale and they are "additive", that is by combining different process steps, the total reduction degree can be estimated. In addition, the WHO guideline (WHO, 2006) gives information on non-technical measures that reduce the risk of infection via irrigation water even further.

In case the salt concentration of the treated water is too high for direct use, desalination steps with reverse osmosis or ion exchanger can be installed downstream. This is necessary in particular in those cases when almost only municipal wastewater is used for irrigation, as practiced for example in Israel. Salts partly derive from drinking water. In households, the increase in salinity results from urine and faeces on the one hand (sodium 3.4 g/cap · d, potassium 3.9 g/cap · d, chloride 3.5 g/cap · d) (Schmidt *et al.* 2005), as well as from adding softening agents in washing machines and dishwashers. Moreover, there might be an additional increase due to evaporation in open reservoirs. With desalination, one has to take into account that approximately 20–30% of the water volume are lost as concentrate, and that the concentrate or regenerate needs further treatment or disposal. As well, those salts are eliminated, which have to be added again later as plant nutrients or trace elements.

2.3.4 Energy requirements

The energy demand varies widely for the different treatment options. For mechanical treatment, DWA (2009) indicates 1–2 Wh/m^3 for sedimentation (with or without flocculation) and 1–20 Wh/m^3 for sieving (with or without flocculation). Aerobic biological treatment exhibits the highest energy demand. For German conditions, LfU (1998) indicates 0.17 to 0.61 kWh/m^3, depending on the adopted treatment and the plant size (Table 2.2). For disinfection, for example with ozonation, UV irradiation or membrane filtration, additional 0.035–0.4 kWh/m^3 must be calculated (Haberkern *et al.* 2008). Nevertheless, the energy consumption on German wastewater treatment plants indicated still higher values as shown in Table 2.1. This implies that only the holistic treatment procedure has to be considered to estimate a realistic energy consumption.

Table 2.2 Range of energy demand for some selected biological treatment options (according to LfU, 1998).

Treatment process	Plant size	
	Small[1] kWh/m^3	Large[1] kWh/m^3
Aerated wastewater ponds	0.34	0.29
Biological contactor	0.23	0.25
Trickling filter	0.31	0.17
Activated sludge with aerobic sludge stabilization	0.61	0.31[2]
Activated sludge with nutrient elimination	0.48	0.30[2]

[1]Small stands for < 1,000 PE; large for < 5,000 PE
[2]Large for 10,000 to 100,000 PE

Bieker *et al.* (2009) show that for the complete treatment of greywater, 0.6–1.2 kWh/m^3 can be estimated for reaching non-potable reuse water quality for intra-urban use (disinfection not included). The lower value of 0.6 kWh/m^3 corresponds to conventional activated sludge or biological aerated filters, the upper value of 1.2 kWh/m^3 is when using membrane bioreactors for biological treatment.

In case an additional desalination step is used downstream, for example in order to use the recycled water for groundwater replenishment or for direct or indirect potable reuse, the energy demand has to be assessed at the top end. Keller (2008) states an energy demand of 0.9–1.2 kWh/m^3 alone for the further treatment of already biologically treated wastewater to become so-called purified recycled water, via micro-filtration, reverse osmosis, UV/H_2O_2 disinfection and chlorination. The recycled water was treated to meet the highest standard of drinking quality through a seven barrier treatment system (Keller, 2008; Western Corridor, 2009).

The operator of the new Goreangab water reclamation plant in Windhoek, Namibia, which, since 2002, daily produces approximately 21,000 m^3/d drinking water from a mixture of secondary effluent and reservoir water, quotes an energy demand of 1.34 kWh/m^3. This amount includes the whole multi-barrier process chain, consisting of powdered activated carbon, preozonation, coagulation, flocculation, dissolved air flotation, dual media filtration, ozonation, biological activated carbon, two-stage granular activated carbon, ultra-filtration, chlorination and stabilization including the raw water and high-lift pumps. Of course, these figures only reflect the energy for operation and cannot replace thorough life cycle assessments.

These examples point out the quality-dependent range of the energy demand and place special emphasis on the question of required quality. Thus, consequent energy management also means to supply the quality required for the respective use, that is fit for purpose, water for garden irrigation, toilet flushing or street cleaning does not necessarily be of potable water quality.

2.4 CONCLUSIONS

Wastewater is a multi-substance mixture containing urine and feces, a multitude of hygienically questionable germs, potential pathogens and helminth eggs, personal care products, bleaching agents, pharmaceuticals and endocrine disrupting compounds, inorganic pollutants such as heavy metals and salts. Yet, wastewater also contains potential nutrients such as nitrogen and phosphorus, as well as organic constituents which are potential sources of energy. The most important resource, however, can be the water itself, which in domestic wastewater accounts for more than 99.5%.

Quantitatively, agricultural reuse of adequately treated water represents by far the largest potential for water reuse. In some countries, for example Israel, more than 75% of municipal wastewater are reused for agricultural purposes. Depending on local conditions, nutrients contained in the wastewater can also be utilized. Hereby, care is to be recommended to prevent overfertilization. However, one of the real challenges lies in dealing with water supply and demand for irrigation not coinciding in terms of time and/or location, requiring long distance transport and storage of recycled water. Further challenges are the salt content of the wastewater, as salt can accumulate in soil, as well as the hygienic quality of recycled water with respect to farmers, farm workers and consumers. With climate changes, on the one hand, and the increasing cultivation of energy crops on the other hand, one has to expect the demand for irrigation water to increase as well. With the latter, the acceptance of recycled water will be high, as these plants are not used for food production.

Another large field of application is intra-urban reuse as non-potable water. Increasing urbanization and still growing megacities aggravate the gap between water demand and availability (regional/local). Via intra-urban water reuse, implemented in new housing or development areas, the specific freshwater demand can be reduced by 30% up to 50%. Intra-urban reuse fosters nodal, semicentralized supply and treatment, in order to minimize costly and energy-intensive multiple transport. As intra-urban reuse is mostly non-consumptive, multiple water use is possible, for example treated greywater for toilet flushing and, subsequently, treated blackwater for irrigation and nutrient utilization.

Reuse of water can contribute to the preservation of valuable freshwater resources. At the same time, water reuse might contribute in saving electric power. In particular, this is the case when freshwater has to be transported over long distances or elaborate water treatment is required, for example processing of potable water from brackish or seawater. The generation of recycled water only requires a fraction of the energy needed for the desalination of seawater.

Water reuse will become an integral part of integrated water resource management in most places of the world, especially where no infrastructure has to be planned and built. Water recycling fosters decentralized, nodal structures. Moreover, water supply and treatment will need to be linked more and more to energy demand and energy recovery.

REFERENCES

Allan T. (1997). Virtual Water: A Long-term Solution for Water Short Middle Eastern Economies? Speech at British Association Festival of Science, University of Leeds, UK, 9 September 1997.

Asano T., Burton F. L., Leverenz H. L., Tsuchihashi R. and Tchobanoglous G. (2007). Water Reuse Issues, Technologies, and Applications. McGraw Hill, New York.

BDEW (2008). Endenergieverbrauch in Deutschland 2007. *Bundesverband der Energie- und Wasserwirtschaft* (Final Energy Consumption in Germany 2007. Federal Association of the German Energy and Water Industries (BDEW)), Berlin, Germany, (in German).

Bieker S., Cornel P. and Wagner M. (2009). Semicentralized supply and treatment systems – integrated infrastructure solutions for fast growing urban areas. *Conference Proceedings of the 6th IWA Leading Edge Conference on Water and Waste Water Technology*, Singapore, 22–26 June 2009.

CEC (2005). Integrated Energy Policy Report 2005. California Energy Commission (CEC). http://www.energy.ca.gov/2005publications/CEC-100-2005-007/CEC-100-2005-007-CMF.PDF (accessed February 2011).

Cornel P. and Meda A. (2008). Wasserwiederverwendung zur landwirtschaftlichen Bewässerung (Water reuse for agricultural irrigation). *GWF Wasser Abwasser* 18–24, **149**(13), (Gas water application–water/wastewater) (in German).

Cornel P., Meda A. and Bieker S. (2011). Wastewater as a source of energy, nutrients and service water. In: Treatise on Water Science, P. Wilderer (ed.), Elsevier Verlag, Oxford, pp. 337–375.

DWA (2008). *Bewertung von Verfahrensstufen zur Abwasseraufbereitung für die Wiederverwendung. Deutsche Vereinigung für Wasserwirtschaft, Abwasser und Abfall* (Assessment of Process Steps for the Treatment of Wastewater for Reuse. DWA German Association for Water, Wastewater and Waste), GFA, Hennef, Germany, (in German).

DWA (2009). *Energie aus Abwasser – Wärme- und Lageenergie. Deutsche Vereinigung für Wasserwirtschaft, Abwasser und Abfall* (Energy from Water – Heat and Potential Energy. DWA – M114, DWA German Association for Water, Wastewater and Waste), GFA, Hennef, Germany, (in German).

Federal Statistical Office, Germany (2009). *Umwelt – Öffentliche Wasserversorgung und Abwasserbeseitigung. Fachserie 19 Reihe 2.1. Statistisches Bundesamt* (Environment – Public Water Supply and Wastewater Disposal, Technical series 19 Series 2.1, Federal Statistical Office (destatis)), Wiesbaden, Germany, (in German).

Gerbens-Leenes W., Hoekstra A. Y. and van der Meer T. H. (2009). The water footprint of bioenergy. *Proceedings of the National Academy of Science of the United States of America (PNAS)*, **106**(25), 10219–10223.

Gleick P. H. (1994). Water and energy. *Annual Review of Energy and Environment, Annual Reviews, Inc. Palo Alto, California*, **19**, 267–299.

Haberkern B., Maier W. and Schneider U. (2008). *Steigerung der Energieeffizienz auf kommunalen Kläranlagen. Umweltbundesamt* (Increasing the Energy Efficiency at Municipal Wastewater Treatment Plants. Federal Environmental Agency), Dessau-Roßlau, Germany, (in German).

Jiménez B. and Asano T. (2008). Water Reuse an International Survey of Current Practice, Issues and Needs. Scientific and Technical Report No. 20, IWA Publishing, London, UK.

Keicher K., Krampe J. and Steinmetz H. (2008). *Eigenenergieversorgung von Kläranlagen* (Energy self-supply of wastewater treatment plants). *Korrespondenz Abwasser - Abfall* (Correspondence wastewater–waste), **55**(6), 644–650 (in German).

Keller J. (2008). From Microbial Fuel Cells to Bio-electrochemical Systems How to Convert Organic Pollutants to Electric Energy and More. Speech at the IWA World Water Congress. Vienna, Austria, 8–12 September 2008.

Lazarova V., Levine B., Sack J., Cirelli G., Jeffrey P., Muntau H., Salgot M. and Brissaud F. (2001). Role of water reuse for enhancing integrated water management in Europe and Mediterranean countries. *Water Science & Technology*, **43**(10), 23–33.

LfU (1998). Stromverbrauch auf Kommunalen Kläranlagen, Handbuch Wasser 4, 13. *Landesanstalt für Umweltschutz Baden Württemberg* (Power Consumption at Municipal Wastewater Treatment Plants, Handbook Water 4, **13**. Regional Office for Environmental Protection) (LfU), Karlsruhe, Germany, (in German).

Mantovani P., Asano T., Chang A. and Okun D. A. (2001). Management Practices for Nonpotable Water Reuse. WERF, Project Report 97-IRM-6.

Plath M., Wichmann K. and Ludwig G. (2010). *Handbuch Energieeffizienz/Energieeinsparung in der Wasserversorgung* (Handbook Energy Efficiency/Energy Savings in the Water Supply Sector. DVGW-Information Wasser Nr. 77), Juli 2010, wvgw Wirtschafts – und Verlagsgesellschaft Gas und Wasser mbH, Bonn, Germany (in German).

Roedl & Partner (2006). *Verbändemodell Kennzahlenvergleich Wasserversorgung Baden-Württemberg, Ergebnisbericht für das Erhebungsjahr 2005* (Association Model – Benchmarking-Comparison of the Water Supply in Baden-Württemberg, Evaluation Report of the Survey-Year 2005). http://www.roedl-benchmarking.de/downloads/berichtbw2005.pdf (accessed February 2011) (in German).

Roedl & Partner (2007). *Effizienz- und Qualitätsuntersuchung der kommunalen Wasserversorgung in Bayern. 3. Abschlussbericht* (Efficieny and Quality Analysis of the Municipal Water Supply Sector in Bavaria. 3. Final Report). http://www.roedl-benchmarking.de/downloads/BerichtBY2007.pdf (accessed February 2011) (in German).

Schaum C., Schröder L., Lux J., Fehrenbach H., Reinhardt J., Cornel P., Kristeller W., Schmid S., Götz R., Himmelein D., Scholl B., Stegmayer K., Wagner G., Maurer M., Mauritz A., Hein A., Berchtenbreiter C. and Blotenberg U. (2010). *Behandlungskonzept der Zukunft?–Ergebnisse einer Grundsatzstudie zum Stand der Klärschlammbehandlung* (Treatment concept of the future?–results of the baseline study regarding the state of the sewage sludge treatment). Hennef, Germany. *Korrespondenz Abwasser–Abfall* (Correspondence wastewater–waste), **3**, 262–268 (in German).

Schmidt R., Lang F. and Thews G. (2005). Human Physiology with Pathophysiology. Springer Verlag, Heidelberg, Germany, (in German).

UN-HABITAT (2006). State of the World's Cities 2006/7. Nairobi, Kenya.

United Nations (2003). Water for People. Water for Life. The United Nations World Water Development Report, UNESCO Publishing, Barcelona, Spain.

USEPA, US Environmental Protection Agency (2004). Guidelines for Water Reuse. EPA/625/R-04/108.

Western Corridor (2009). http://www.westerncorridor.com.au/sitedocument.aspx?docId=86 (accessed April 2009).

WHO (2006). Guidelines for the Safe Use of Wastewater, Excreta and Grey-Water. Published in four volumes, World Health Organization (WHO), Geneva, Switzerland.

Zehnder A. J. B. (1997). Is water the first resource to control demographic development? *Food & Water, a Question of Survival*. 85–98. Forum Engelberg and vdf Hochschulverlag an der ETH Zürich. Zurich, Switzerland.

Zehnder A. J. B. (2001). *Wasserressourcen und Bevölkerungsentwicklung* (Water resources and population development). *Wasser – essentielle Ressource und Lebensraum*. Speech at Deutsche Akademie der Naturforscher Leopoldina, Halle-Saale 2001/046 in Halle, B. Parthier (ed.), Barth, Heidelberg, Germany (in German).

Zehnder A. J. B. (2003). Lecture at 27th International Exhibition-Congress on Chemical Engineering, Environmental Protection and Biotechnology. Frankfurt am Main, Germany, 21 May 2003.

ZVBWV (2009). *Geschäftsbericht 2009. Zweckverband Bodenseewasserversorgung* (Annual Report 2009. Administration Union Lake Constance Water Supply (ZVBWV)). http://www.zvbwv.de/upload/mediapool/BWV-Geschaftsbericht_2009.pdf (accessed February 2011) (in German).

Chapter 3

Water and energy link in the cities of the future – achieving net zero carbon and pollution emissions footprint

Vladimir Novotny

3.1 INTRODUCTION

The Cities of the Future or Ecocities represent a major paradigm shift in the way new cities will be built or older ones retrofitted to achieve a change from the current unsustainable status to sustainability, meet the net zero green house gas (GHG) emission targets, reuse and recycle water, and recover resources, including nutrients (Register, 1987; Novotny *et al.* 2010).

Current scientific research quoted in the National Science and Technology Council report (NSTC, 2008) indicates 60 to 70% of energy reductions in buildings in cities can be achieved with passive heating and cooling incorporated in the architecture of the building, more efficient appliances such as better water and space heaters, heat pumps, significant reduction of water demand by water conservation and use of rain and stormwater, organic solids management for energy and resources recovery, and other improvements. NSTC also estimated that 30 to 40% of energy can be produced by renewable sources, including heat recovery from wastewater (used water) and/or extracted from the ground and groundwater.

3.2 URBAN METABOLISM

3.2.1 Urban metabolism – reclaim, reuse and recycle

Cities are complex systems which accept, transform, use, and attenuate inputs and produce and emit outputs. This is called urban metabolism. Current urban systems have been usually linear in terms of their urban metabolism (Figure 3.1). Daigger (2009), Novotny (2008) and others agree the current "linear" approach, sometimes called the "take, make, waste approach" in the literature has become unsustainable and cannot continue. The linear system usually discourages reuse because the source of reclaimed water is far downstream from the city and reusing water in the city would require another pipe and pumping. Also the current economic benefit/cost or minimum cost evaluations do not consider important social and, in many cases, environmental costs and benefits of reuse traditionally considered as intangible.

Furthermore, in our understanding of the water use in a city we cannot just limit ourselves to considering water and its direct use and economics. Providing materials, energy, and food to the city requires additional water at the points of production, manufacturing, retail and fuel for transportation. This represents virtual water demand that could exceed the direct water use by an order of magnitude (Hoekstra & Chapagain,

2008). Linear systems result in far more pollution output and need more resources (water, air, soil and landscape) for dilution and assimilation of residuals than a semi closed system based on the three R's – reclaim, reuse and recycle.

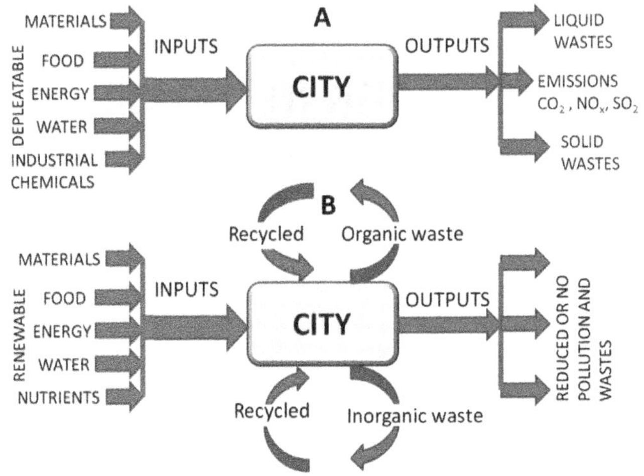

Figure 3.1 Linear (A) and circular (B) urban metabolism systems (from Novotny *et al.* 2010)

3.2.2 Water footprint – direct use of water

The per capita water use in cities is a local footprint which usually has regional significance. In the US, domestic indoor water use is relatively constant among the major urban areas (Heaney *et al.* 2000), averaging 242 L/cap · d for a household without water conservation but could be reduced to 136 L/cap · d in a household practicing water conservation. However, the total per capita water use is magnified by outdoor irrigation (using potable water), pipeline leaks, or swimming pools and in the US reaches almost 650 L/cap · d, the highest in the world, as compared to Europe which is around 150 to 200 L/cap · d or less. For example, Drbohlav and Jankovsky (2010) reported billed domestic water use in Prague (Czech Republic) as 140 L/cap · d. The high water use in the Southwest US, Middle East, and arid and semi arid areas in affluent countries puts enormous stresses on the water resources and availability. However, water shortages have also occurred in communities located in the more humid eastern US such as Atlanta (GA) or Florida and are endemic in megalopolises of developing world.

3.2.3 Water-energy nexus

In the US, based on national averages, buildings consume 40% of the energy of which 22% is residential and 18% commercial, respectively. Industries consume 32% and transportation 28%, respectively (NSTC, 2008). Providing treated water and disposal of wastewater in the US represents on average about 3% of the energy use but can be as high (California) as 20%. However, within buildings, 8% of the additional energy use is for water related processes such as cooking, wet cleaning, and water heating. A percent or more is needed to pump and transport water and wastewater.

The Energy Information Administration (2009) documented the total energy production in the US in 2007 to be 4,157 TWh[1] (4,157 × 10^9 kWh) which represented about 2.516 billons tons of CO_2 emitted.

[1] T – tera, equivalent to 10^{12}

In 2007, 55 billions m^3 of water was used by the population of 301.3 million in the US. Using the US EPA estimate of 3% energy use for water would result in the specific energy use of 2.26 kWh/m^3 attributed to water. Corresponding carbon emission is of 1.37 kg CO_2/m^3. Using the 3% estimate for providing and treating water, "water share" of the energy use is 124.7 TWh and 75.5 million tons of CO_2 is emitted as a result of providing clean and disposing polluted water, plus an additional 200 million tons of CO_2 for hot water heating, cooking, boiling, and wet cleaning.

The US Department of Energy (2000) published estimates of carbon equivalent of energy produced by fossil fuel power plants as:

- 0.96 kg of CO_2/kWh produced by coal fired power plants
- 0.89 kg of CO_2/kWh produced by oil fired power plants
- 0.60 kg of CO_2/kWh produced by natural gas power plants

Because 30% of energy is produced by processes that do not emit substantial quantities of GHG (nuclear, hydropower and other renewables), a weighted average of the CO_2 will be considered in this analysis as:

- 0.61 kg of CO_2 emitted per kWh of energy produced

In contrast, in France, Belgium, Austria and other EU countries, the GHG equivalent of energy is smaller because of much higher reliance on nuclear power (France) or hydropower (Austria, Switzerland). Vestraete *et al.* (2010) used the GHG equivalent 0.21 kg of CO_2 emitted per 1 kWh energy used. Growing use of wind and solar power in Germany, Spain, and Czech Republic is further decreasing the GHG equivalent of one kWh.

Figure 3.2 presents the possible relationship of water demand reduction leading to a closed urban water cycle and energy. Novotny (2010) suggests a hypothesis that there is a minimum inflection point beyond which further reduction of water use will increase energy demand and urban water metabolisms because of increased use of chemicals, energy and infrastructure (materials). A relationship can be developed for relating the cost of providing water to the magnitude of the water demand. The water – energy nexus relationship has three phases (Novotny, 2010; Novotny *et al.* 2010):

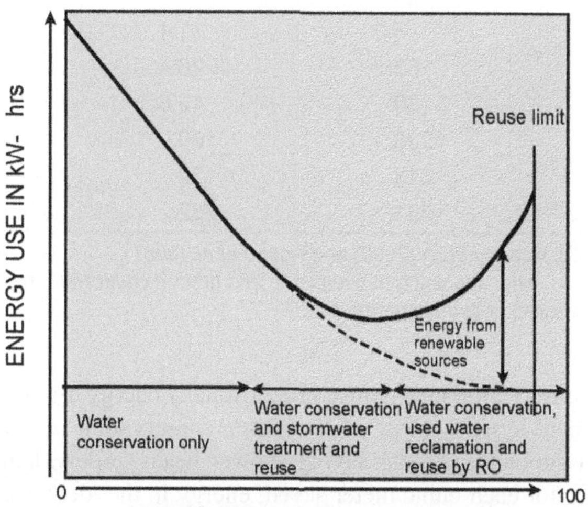

Figure 3.2 Three phases of the water-energy nexus (without energy recovery)

(1) The water conservation phase in which energy and GHG emission reduction is proportional to the reduction of the high water use.
(2) The inflection phase in which additional and substitute sources of water demanding more energy are brought in, treated and used.
(3) Rising energy (cost) phase in which energy use is increasing while water demand of the development is reduced by water recycling and multiple reuses.

Phase I – Water conservation – linear reduction

Table 3.1 shows the per capita volumes and proportions of the daily water use in a typical US single family home. The left part of the table is based on the AWWA RF (1999) study as reported by Heaney *et al.* (2000). On the right side are the estimates of water savings used by the AWWA RF study and by the Pacific Institute (Gleick *et al.* 2003) study for California. After implementing mostly common sense water conservation measures (for details see Novotny *et al.* 2010), the US domestic use can be reduced from the high 550 L/cap · d to less than 136 L/cap · d which is comparable to the current use in Europe. Because of the prevalence of single houses on more land in the US, the realistic conservation total use is 200 L/cap · d.

Table 3.1 Indoor and outdoor water use in a single family home in 12 monitored cities in North America.

Water use	Without water conservation*		With water conservation	
	L/cap · d	%	L/cap · d	%
Faucets	35	14.7	35	25.7
Drinking water and cooling	3.6	1.2	2.0	1.5
Showers	42	17.8	21	15.4
Bath and hot tubs	6.8	2.0	6.0	4.4
Laundry	54	22.6	40	29.4
Dish washers	3.0	1.4	3.0	2.2
Toilets	63	26.4	19	14.0
Leaks	30	12.6	10	7.4
Total Indoor	238	100	136	100
Outdoor	313	132	60**	44
Total	551	232	196	144

*Adapted from AWWA RF (1999), Heaney *et al.* (2000) and Asano *et al.* (2007)
**Reflects converting from lawn to xeriscape using native plants and ground covers with no irrigation. Water use is for swimming pools, watering flowers and vegetable gardens.

Reducing water use by conservation may reduce proportionally energy use. It works best if it is done in a distributed urban management system. Most of the water conservation reduction in Table 3.1 can be achieved by more efficient appliances (water saving shower heads, toilets, laundry wash machines, and so on) and xeriscape. Hence for each cubic meter saved, energy in the ideal average household would be reduced by the same proportions. This is the linear Phase I of Figure 3.2. The water saving potential shown in Table 3.1 is 65% reduction.

Phase II – inflection

In the inflection phase, a city is looking for additional sources of water or brings in sources that have worse quality such as more salty deeper groundwater. This will require more treatment, sometimes by energy demanding reverse osmosis, and/or water will have to be pumped from long distances or from deep geological layers. Many cities in the southwest US cannot meet the water demand using relatively inexpensive sources of fresh water and/or may be located on receiving water bodies that require a high degree of treatment. For example, pumping a volume $V = 1\ m^3$ of water from a depth of $H = 500$ m with a pump that has an overall efficiency of 80% will require work of $W = \gamma\ V\ H = 9{,}819 \times 1 \times 5000/0.8 = 6{,}131{,}125\ J = 1.7$ kWh (γ = specific density of water in N/m^3) and will result in 1 kg of additional CO_2 emissions. Many water short communities are pumping higher salinity water from depths as deep as 1000 m, for example El Paso, Texas. Water sources with low energy demand are rainwater harvesting (negligible energy needs) and stormwater (some pumping and treatment).

Phase III – increasing energy demand and CO_2 emissions

In the increasing phase, tapping higher salinity water sources (brackish sea or groundwater) is supplemented with water reuse that requires a two or three step high efficiency treatment (Figure 3.3). Table 3.2 presents energy and CO_2 emissions.

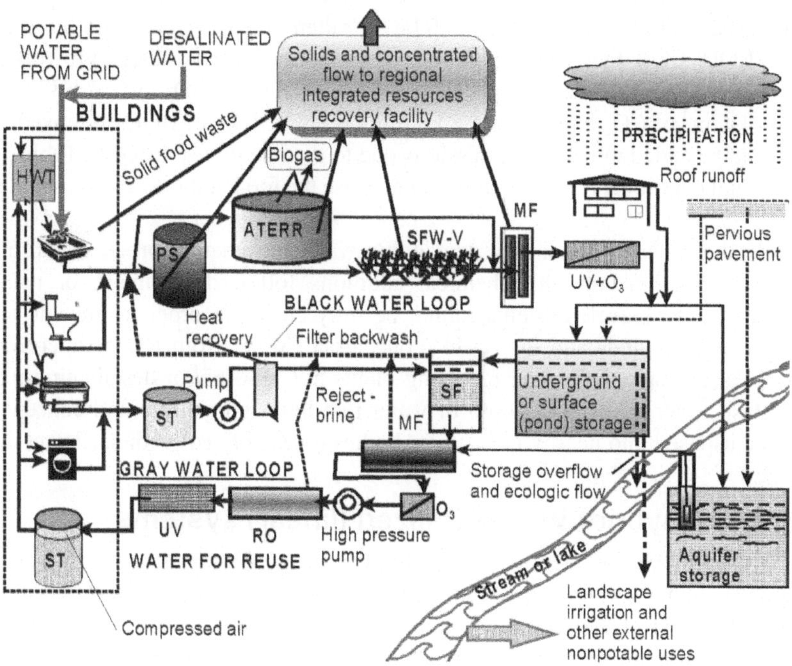

Figure 3.3 A double loop hybrid cluster (ecoblock) system that recycles grey water and some black water base on Qingdao Ecoblock (Fraker, 2008). From Novotny *et al.* (2010). Legend: ATERR – a generic anaerobic treatment and energy (methane) recovery reactor; HWT – hot water tank; MF – Membrane filter; NF – nanofilter; O_3 – ozone addition; PS – primary settler with solids removal; RO – reverse osmosis; SF – sand filter; SFW – subsurface flow multicell wetland; ST – storage; UV = ultra violet disinfestion; SFW-V – subsurface flow wetland with vertical flow

Table 3.2 Energy use of treated volume of municipal wastewater and corresponding CO_2 emissions.

Treatment process	Energy use kWh/m³ (CO_2 emissions kg/m³) Daily flow volume of treated wastewater (m³/d)		
	10,000	**25,000**	**>50,000**
Activated sludge without nitrification and filtration	0.55 (0.33)	0.38 (0.23)	0.28 (0.17)
Membrane bioreactor with nitrification	0.83 (0.51)	0.72 (0.44)	0.64 (0.37)
Reverse osmosis desalination			
Brackish water (TDS 1 – 2.5 g/L)		1.5 (0.91)–2.5 (1.52)	
Sea water		5 (3.05)–15 (9.15)	
Ozonization (ozone produced from air)			
Filtered nitrified effluent		0.24 (0.15)–0.4 (0.24)	
Desalination by evaporation		~25 (15.25)	

Source: Adapted from Asano *et al.* (2007) and from Novotny *et al.* (2010)

In activated sludge processes, for each mole of oxygen consumed in the aeration process, one mole of carbon dioxide is emitted.

Hence CO_2 emitted $= (12 + 2 \times 16)/(2 \times 16) = 1.37\ O_2$ consumed. For example, if the BOD_5 concentration in used water is $300\ mg/L = 0.3\ kg/m^3$ then the CO_2 emission in the aeration unit removing 95% of BOD_5 (80% by oxidizing carbonaceous matter, 15% by sludge growth) will be:

CO_2 emitted $(kg/m^3) = 1.4\ (BOD_{ultimate}/BOD_5) \times 0.80 \times 0.3\ (kg\ BOD_5/m^3) \times 1.37\ (CO_2\ emitted/O_2\ consumed) = 0.44\ kg/m^3$ of CO_2 emitted or 1.8 kg of CO_2 emitted/kg of BOD_5 removed.

This value should be added to the CO_2 emissions due to the energy use listed in Table 3.2 in the GHG balance, but it does not contribute to global warming because the food and feed crops assimilated CO_2 from the atmosphere.

To further close the water cycle, energy demanding water reclamation processes are needed such as highly efficient tertiary treatment with chemical additions followed by micro or nanofiltration and reverse osmosis (RO). The recycle systems cannot be fully closed in order to prevent accumulation of harmful conservative compounds and the reject (concentrate) water from RO or microfiltration usually cannot be reused. For example, RO water recycling plants in California or desalination plants in Florida using sea or brackish water reject 20 to 40% of water that has high salinity and concentrated pollutant content. It must be disposed into nearby large water bodies (ocean or sea) with sufficient dilution flow.

3.2.4 Distributed (hybrid) *vs.* centralized (linear) systems

A distributed Resource Management Cluster (RMC) or ecoblock is a semi-autonomous water management/drainage unit that receives water, implements water conservation inside the structural components throughout the cluster, captures and stores rainfall and stormwater, reclaims sewage for reuse, such as toilet flushing, irrigation, street washing and provides ecological flow to restored existing or daylighted streams, recovers heat energy from wastewater, and possibly recovers biogas from organic solids. Clusters may range from a large high-rise building, larger shopping center, large hotel, to a subdivision or a portion of a city (Furumai, 2007; Lucey & Barraclough, 2007; Heaney, 2007; Novotny *et al.* 2010; Daigger, 2009) or an entire (smaller) city or urban watershed.

The treatment level within the RMC is "fit for reuse". If, for example, in an RMC recycled water is used for landscape irrigation, removing nutrients does not make sense because the nutrients eliminated from reclaimed water would have to be replaced by industrial fertilizers, which would defy the purpose of reclamation and reuse. For all practical reasons, toilet flushing may require reduction of turbidity, disinfection, primarily to control bacterial growth in the toilets and urinals, and adding some colour. On the other hand, if recycled water is used for providing ecological flow to lakes or streams, nutrient removal may be necessary but nutrients should be recovered and not just removed (e.g. as sludge put in a landfill).

On a local ecoblock scale, aquifer recharge will be accomplished by infiltration of captured stormwater by best management practices which are foundation blocks of the Low Impact Development (LID) concept. LID practices include enhanced rainwater infiltration (rain gardens), pervious pavements, and infiltration ponds (Novotny *et al.* 2010).

Asano *et al.* (2007) suggested alternatives for retrofitting decentralized water management into existing urban environments. By starting with building satellite treatment in upstream portions of the urban drainage area, wastewater (used water from the local collection system) can be intercepted and treated to a high degree required for various reuse alternatives such as toilet flushing in houses, landscape irrigation, and ecological flow enhancement. This concept not requiring dual or triple plumbing and water separation into black and grey water was implemented at the Solaire Battery Park multi unit residential complex in New York City where recycled water is used for toilet flushing, irrigation and cooling. The residual effluent with solids from the cluster water reclamation facility is then conveyed to a central (regional) treatment plant and discharged into the environment. At this time, there are no guidelines that would establish the size of the ecoblock.

3.2.5 A double loop hybrid system – source separation

Source separation of wastewater is gaining popularity in some countries, for example, Sweden. Planners of the water frugal ecocities in Qingdao (China) and Masdar (UAE) considered a partially closed (hybrid) system similar to that shown on Figure 3.3. The Qingdao three hectare ecoblock would have contained 1500–2000 inhabitants out of the total 40,000 planned to live in the (future) ecocity (Fraker, 2008). Masdar in United Arab Emirates, operating as a single water/wastewater management cluster, will have a population of 50,000 permanent residents and 40,000 commuters. Population density is a more important parameter determining the cluster size than the total population the cluster may serve. The Qingdao ecocity cluster inhabitants were to live in several high-rise and medium height buildings. These ecoblocks are then combined to form the entire development (Fraker, 2008). The water management shown on Figure 3.3 reaches a maximum limit of reuse and would be applicable to regions with severe shortage of fresh water resources (Middle East, northeast China, inland Australia, some southwest US states, and many future cities in developing countries with burgeoning population and meagre or unusable water resources).

The Qingdao double loop (Fraker, 2008) was modified from the original one to avoid the possibility of direct potable reuse and includes restored surface water bodies and underground storage (Figure 3.3). The water reclamation and reuse is carried in black and grey water double loops. Black water flow includes water from toilets, kitchen sinks and dishwashers. In the original Qingdao system proposal sequencing batch reactors or septic tanks were proposed (Fraker, 2008). Verstraete *et al.* (2009, 2010) suggested an anaerobic upflow sludge blanket reactor (UASB) combined with a septic tank. In this application, the PS reactor, for example, an Imhoff primary settler (Novotny *et al.* 1989) with suppressed digestion achieved by a shorter sludge residence time is optional. The subsurface flow wetland treatment may emit small quantities of GHG carbon dioxide, methane and nitric oxide but vegetation growth in the wetland will assimilate CO_2, by converting it to alkalinity, biomass will be harvested for energy production, and nitrification will be minimal because of reducing conditions in the wetland and all nitrogen entering the wetland being in a form of Total Kjeldahl Nitrogen (TKN).

The system accomplishes several objectives: (1) it treats the grey water to a quality comparable to potable water for several in-house uses although direct potable use is not contemplated; (2) concentrates black water, including filter backwash from the grey water loop, and sends a part of the black water with solids to a regional integrated resource recovery facility (see next section); a part of the black water is further treated on site to supplement the grey water flow lost in reject and filter backwash flow; (3) in addition to providing water to inhabitants, the double loop system also provides some ecological flow to the surface water bodies within the ecocity and garden irrigation; and (4) it recovers some energy in a form of biogas and heat.

The water cycle balance of the double loop is presented on Figure 3.4. The figure shows the total water use in the cluster as 126 L/cap · d, which is similar to the conservation alternative from Table 3.1 without the leaks and outdoor use, but the municipal water supply grid provides only 50 L/cap · d. It can be seen that 50 L/cap · d water input from the municipal grid is not sufficient to sustain the total demand of 126 L/cap · d during dry weather. Rainwater harvesting and stormwater capture and infiltration (via pervious pavements and infiltration rain gardens) is needed to supplement the dry weather flow (Novotny & Novotny, 2009). Consequently, surface and/or subsurface storage is needed. Surface ponds, wetlands and architectural water features (Dreiseitl & Grau, 2009) could store collected surface runoff and a portion of the highly treated black water; subsurface storage should collect infiltration, excess clean water from dewatering basements and cooling condensate, and harvested rainwater.

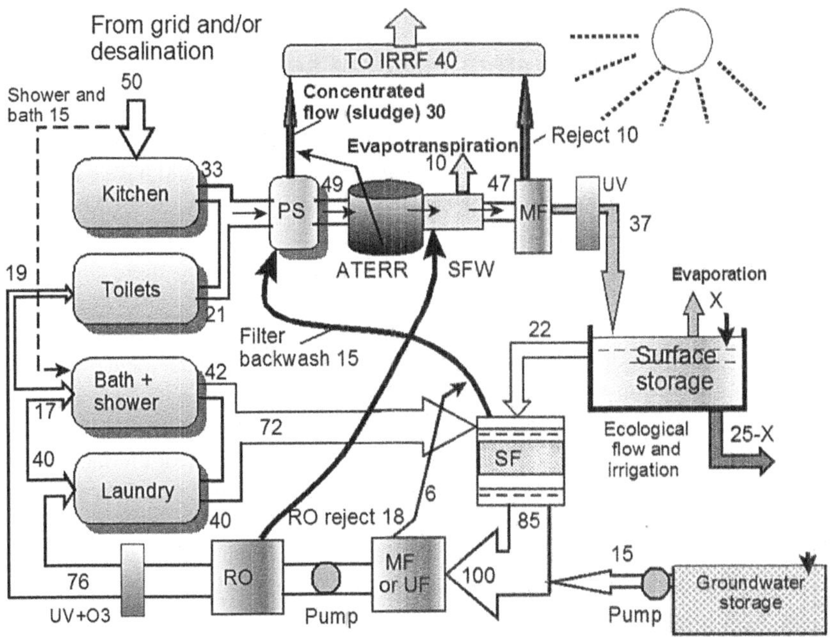

Figure 3.4 Mass balance of flows of the double loop hybrid water cycle system on a dry day. The numbers represent water flows and uses in L/cap · d; IRRF – integrated resource recovery facility; X – water lost by evaporation. From Novotny *et al.* (2010)

The Qingdao ecoblock also saves energy by passive heating and cooling and producing energy by solar panels, voltaics, and wind turbines. It could also produce biogas from digested sludge and provide organic solids harvested from the wetland, fallen leaves and gardens that can be transported to IRRF for processing to recover more efficiently energy and nutrients. The black water loop contains most nutrients and "fit for reuse" nutrients can be left in the treated black water effluent flow which may eliminate the need for industrial fertilizer if the reclaimed effluent is used for irrigation. In the overall scheme, the ecoblock should have net zero carbon emission and no pollution footprints.

The Qingdao ecoblock concept is now being implemented in Tianjin (China) Ecocity 150 km southwest of Beijing (Fraker, 2011) built by a China-Singapore consortium.

Heat recovery from wastewater

Water heating is one of the largest domestic energy uses. Total used water leaving the house has a temperature of about 27°C. Grey water is warmer; about 38–40°C because of warm showers, baths, and hot laundry (Roest *et al.* 2010). Therefore, it may be more efficient to retrieve heat from grey water after separation in the cluster. Actually, simultaneous warming and cooling can be accomplished in a 'water to water' heat pump. Water to air heat pump will provide space heating and cooling.

The net energy extracted from water is then (Novotny *et al.* 2010)

$$\Delta E = \mathrm{E}_{ac} - \mathrm{E}_{ap} = \mathrm{V}\rho\Delta\mathrm{TC}_p(1 - 1/\mathrm{COP})$$

where V is volume of water, ΔT is the temperature difference in the heat pump between incoming and leaving water, C_p is specific heat of water and COP is the coefficient of performance of the heat pump which is a ratio of energy acquired, E_{ac}, to energy applied, E_{ap}, to the compressor in the heat pump. COP ranges from 2 to 5, depending on temperature. The above equation works for both heating and cooling.

For the volume of 1 m^3 of grey water in the cycle the energy recovered, assuming COP = 4 and ΔT = 25°C, becomes:

$$\begin{aligned}\Delta E &= 1[\mathrm{m}^3] \times 10^6[\mathrm{g/m}^3] \times 25[^\circ\mathrm{C}] \times 4.2[\mathrm{J/g}^\circ\mathrm{C}]\{1 - 1/4\} \\ &= 78.75 \times 10^6\,\mathrm{J} = 78.75\ \mathrm{MJ} \\ &= 78.75\ \mathrm{MJ}/(3.6\ \mathrm{MJ/kWh}) = 21.87\ \mathrm{kWh/m}^3\end{aligned}$$

which is a very significant energy component.

Recovering heat from black water flow is counterproductive. It would significantly reduce treatment efficiency and, if anaerobic treatment is selected, black water would have to be reheated in the reactor. Recovering heat from wastewater provides energy for water heating. For the above calculation of energy recovery from 1 m^3 the carbon emission reduction would be:

$$21.87\ \mathrm{kWh/m}^3 \times 0.61\ \mathrm{kg\ CO_2/kWh} = 13.3\ \mathrm{kg\ CO_2/m}^3$$

This is a significant energy gain and GHG emission reduction.

3.2.6 Energy (CO_2) balance for a city switching to sustainable water management

Case study

In this illustrative analysis the starting reference point of water and energy use is a community with no water conservation, an open linear water management system and no reuse. The city and the subdivision type cluster is located in southwestern US. The available fresh groundwater source is unsustainably mined and surface source is ephemeral and uneconomical for development.

In this analysis the water heating and cooling is not included. The illustrative assumptions are:

- Total population in the cluster 10,000
- Water demand without conservation 500 L/cap · d
- Sustainable water available from freshwater source 100 L/cap · d
- Sustainable rainwater and stormwater reclamation with storage 20 L/cap · d
- Sustainable brackish groundwater (TDS ~1500 mg/L) 30 L/ cap · d
- Maximum water conservation limit (Table 3.1) 200 L/cap · d

Because the sustainable water from surface and freshwater groundwater sources can only provide 150 L/cap · d, water use must be reduced by water conservation and reuse. Wastewater treatment includes the activated sludge process with nitrification. Reuse will be done by filtration of the effluent, followed by reverse osmosis and ozonization. Reused water will not be available for potable use.

Calculations

A marginal and total water/energy nexus chart presented on Figure 3.5 has been prepared for carbon emissions. Marginal carbon is the CO_2 emission per one extra m^3 of water.

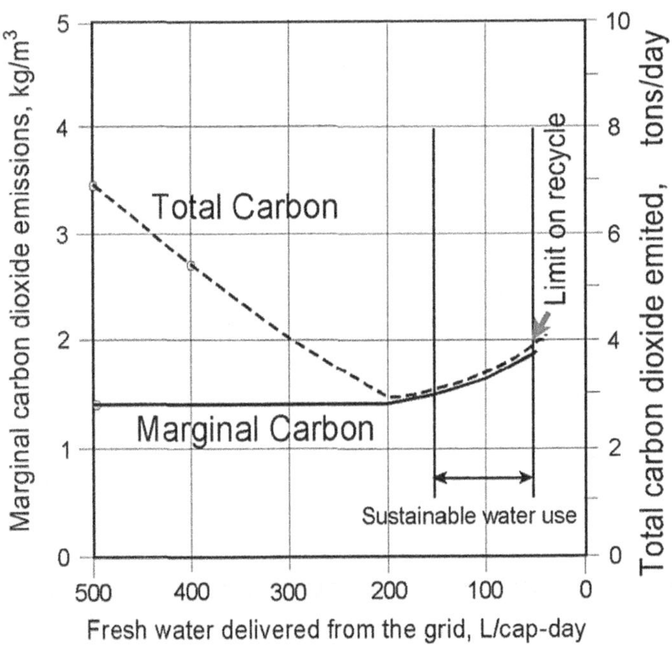

Figure 3.5 Water energy nexus chart that includes total and marginal carbon emissions related to water demand reductions by water conservation, additional sources and water recycling without heat energy recovery

Current unsustainable situation:

- Water use 0.5 $m^3/cap \cdot d$ – Total water use $0.5 \times 10{,}000 = 5{,}000\ m^3/d$
- Marginal energy use – $2.26\ kWh/m^3 \times 0.5\ m^3/cap \cdot d = 1.16\ kWh/cap \cdot d$
- Carbon emissions – 0.61 (kg of CO_2/kWh) $\times$ 2.26 = 1.37 kg of CO_2/m^3
- Total carbon emissions – 5,000 (m^3) $\times$ 1.37 (kg of CO_2/m^3) = 6,936 kg CO_2/d.

Reduction to 200 L/cap · d (60% reduction) or 2,000 m^3/d can be achieved solely by water conservation but the water use is still unsustainable and the available sources cannot provide enough water. At 100 L/cap · d of water available from the fresh treated water supply grid, additional water will originate from stored rainwater/stormwater (20), sustain-able brackish water (30) and reuse (50) to provide 200 L/cap · d of water. Rainwater/stormwater use will require storage, pumping and filtration which will result in estimated carbon emissions of 0.1 kg of CO_2/m^3. Brackish water has to be pumped (1.6 kWh/m^3 = 1 kg CO_2/m^3 if pumping depth is 500 m) and treated by reverse osmosis and UV/ozonization (1.7 kg CO_2/m^3). Reuse will approximately emit 2.0 kg of CO_2/m^3.

At 100 L/cap · d of fresh water from the grid the marginal kg CO_2/m^3 emissions become:

- $[\text{freshwater}] \times 1.37 + 0.02[\text{rain}] \times 0.1 + 0.03[\text{brackish}] \times 2.7 + 0.05[\text{reuse}] \times 2)/0.2$

$$= 1.6\ kg\ CO_2/m^3$$

The total carbon emissions at 200 L/cap · d demand and 100 L/cap · d fresh water availability from the grid will be 1.6 kg $CO_2/m^3 \times 2{,}000\ m^3$ = 3,200 kg CO_2/d. The marginal kilograms of CO_2/m^3 and the total CO_2 emissions based on additional calculations are plotted on Figure 3.5.

Additional energy saving and carbon emission reductions can be achieved if recycled water is treated to the fit for use level and by recovering heat. For example, water for irrigation outdoor use, 60 L/cap · d, does not have to be treated by reverse osmosis and nutrients can remain in the recycled water. This would require dual piping but this is already common in some US cities located in arid zones (e.g., Salt Lake City, Utah).

3.3 INTEGRATED RESOURCE RECOVERY FACILITY (IRRF)

A completely distributed water/stormwater/wastewater management system with independently operated clusters fully reclaiming and recycling all water is unrealistic. The major reasons are: (a) the cycle needs make-up water to prevent accumulation of salts and of "new" conservative contaminants (pharmaceuticals, nanopollutants, endocrine disrupting compounds) in the system, and (b) the need for safe disposal of reject water (brines) from RO systems. Also sludge management and biogas recovery may not be advisable in small cluster size reclamation facilities and may be objectionable to the citizens living in the cluster. An IRRF will be a new complete water/energy recovery and management facility which accepts organic solids, concentrated excess wastewater and recovers water, nutrients, solids, electric energy, heat in much greater quantities than it is possible in the traditional "water recycling plants".

In the proposed double loop distributed system, a large volume of water and some energy are reclaimed/recovered and reused on site in the cluster/ecoblock. Consequently, the main objectives of the IRRF could be:

(1) Treating and reclaiming water for:
   (a) Ecological flow of the receiving water body
   (b) Beneficial downstream uses for:

(i) Irrigation
(ii) Water supply from alluvial deposit
(iii) Recreation

(c) Providing water and nutrients to algal aquaculture producing biomass and energy

(2) Recovering phosphorus and removing nitrogen
(3) Recovering and producing heat energy for heating the anaerobic treatment and fermentation units as well as the buildings in the facility and nearby
(4) Producing biogas that may include methane or syngas (a mixture of carbon monoxide and hydrogen) (O'Riordan *et al.* 2008) and/or hydrogen
(5) Producing organic solids for soil conditioning
(6) Providing water and nutrients to algal aquaculture producing biomass for biogas and oil production and, in the future, hydrogen
(7) Converting biogas and hydrogen into electricity
(8) Deriving all energy needs from on-site energy recovery, additional renewable sources (solar) and sequestering carbon

Such facilities will generate no pollution, produce excess electricity and will be net sequesters of carbon. Laboratory and field tested technologies that enable to propose this revolutionary resource recovery system, summarized in Novotny *et al.* (2010), include:

- new developments of the more than century old anaerobic treatment and digestion of organic solids and sludge in upflow anaerobic sludge blanket (UASB) reactors (Lettinga and Hulshoff-Pol, 1991; Verstraete *et al.* 2009, McCarty *et al.* 2011);
- microbial fuel cells that convert organic matter to hydrogen or electricity (Rabaey & Verstraete, 2005; Logan, 2004, 2008; Call & Logan, 2008; Wagner *et al.* 2009);
- hydrogen fuel cells converting biogas to hydrogen and electricity (US Department of Energy, 2010);
- heat recovery from water by heat pumps and other heat reclamation devices;
- production of struvite (ammonium magnesium phosphate) fertilizer from wastewater effluents and digester supernatants (Barnard, 2007; Parsons *et al.* 2001);
- improved production of nutrient rich solids from sludge (Verstraete *et al.* 2010);
- more efficient biogas and biofuel production;
- production of algal biomass and subsequently hydrogen (James *et al.* 2009);
- more efficient capture of renewable solar energy by concentrated solar panels and photovoltaics.

Figure 3.6 is a schematic of the system composed of ecoblocks interconnected with IRRF. Note the important role of surface water bodies in the system.

3.3.1 A concept of a future IRRF

A future possible IRRF alternative was conceptually presented in Novotny (2010) and Novotny *et al.* (2010) and is shown on Figure 3.7.

This facility would accept both concentrated liquid wastewater flows from the clusters and organic solids (vegetation residues, food waste, and manure in developing countries). The produced biogas could be converted to electricity by a combustion engine and generator, or, in a more distant future, biogas and hydrogen would be generated and converted to electricity in a hydrogen fuel cell. Energy can be recovered in a form of biogas, syngas, heat, or hydrogen. It is anticipated that hydrogen based systems will be the future because such systems can recover the maximum energy output and produce the smallest or no GHG emissions.

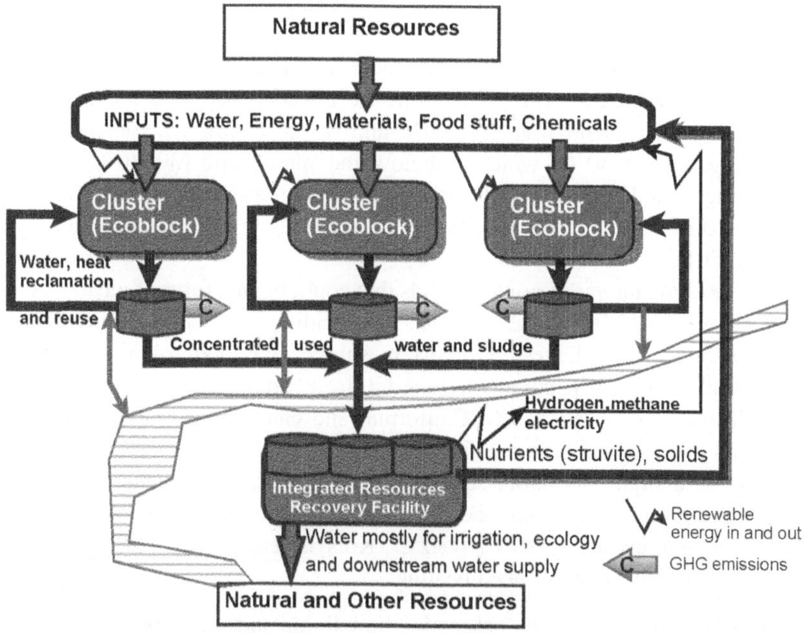

Figure 3.6 Distributed urban water/stormwater/wastewater management system with IRRF

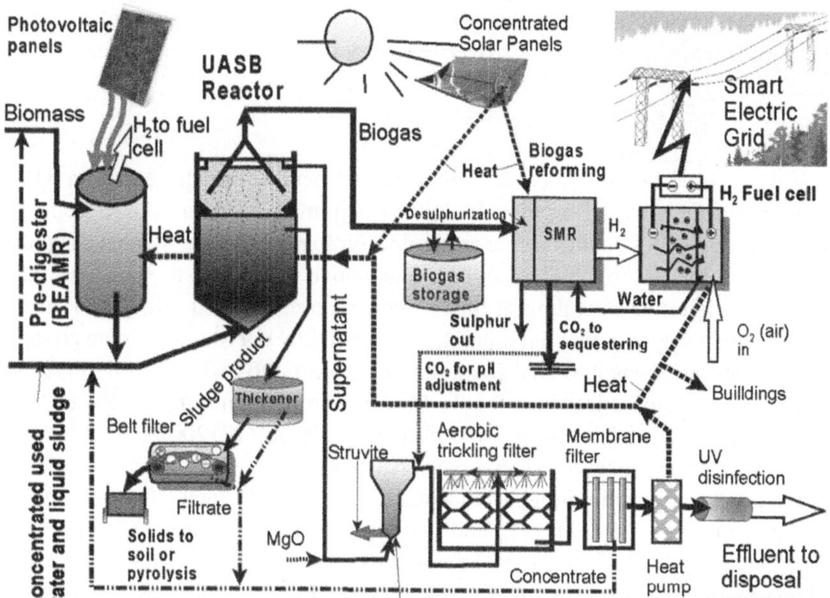

Figure 3.7 A vision of the future – hydrogen/energy and resources recovering IRRF for concentrated wastewater and organic solid waste. Credits Novotny (2010) and Novotny *et al.* (2010). BEARM – Bioelectrically assisted microbial reactor (Logan, 2008); SMR – stream methane reforming

The two key energy producing anaerobic reactors in this proposed IRRF are:

- the upflow anaerobic sludge blanket (UASB) reactor for liquid concentrated wastewater containing mostly black water and backwash solids from the cluster water and reclamation plants, legacy leachate from abandoned solid waste landfills, and so on. UASB reactors work efficiently in the temperature range of 20–30°C, which is below the mesophilic (30–40°C) range of traditional anaerobic digesters, and require much smaller hydraulic residence time (HRT) than traditional digesters. In a combination with membrane filtration they can provide high quality effluents (McCarty *et al.* 2011).
- a pre-digester for fermentation of organic solids that could be in a form of a Bio-electrically Assisted Microbial Reactor (BEAMR) (Logan, 2008) or a traditional (black) fermentation digester with suppressed methanogenesis. Digesters digesting sludge and food solids and producing methane have been installed in many countries ranging from small installations in India and China, municipal applications in Sweden, US (California) and Germany, to large digesters for manure in The Netherlands and Denmark. In the first ten years of the 21st century, BEAMR digesters have been built only on a small laboratory scale (Logan, 2004; Call & Logan, 2008; Cheng & Logan, 2011). The pre-digester producing acetates and/or hydrogen may work satisfactorily in the psychophilic temperature range of < 10–30°C (Kotsyurbenko *et al.* 1993; Nozhevnikova *et al.* 2001), hence, this reactor may not need heating.

This dual symbiotic arrangement allows co-processing both liquid and solid organic wastes, eliminating the need for landfills and recovering the resources in one facility. Furthermore, the heat recovered in the conversion of organic matter into energy will exceed the need for heating the reactors and converting water into steam for methane conversion into hydrogen in the SMR process. Additional heating energy can be provided by the concentrated solar panels installed on the premises (roof) of the IRRF. Hence, the proposed IRRF could operate with an excess overall energy output even in colder climatic conditions.

UASB reactors

The UASB reactor proposed by Lettinga *et al.* (1980) is a suspended growth reactor in which the primary COD removal mechanism is by absorption, biological action of bacteria and fermentation of organic suspended and dissolved solids in the active anaerobic biological sludge blanket. Typical HRT at the temperature range between 20 to 26°C is 7–10 h and above 26°C is 6 h, which is far less that that required for anaerobic digesters operating at higher temperatures and requiring HRT in weeks. UASBs do not remove nutrients beyond that incorporated into sludge nor do they convert ammonium into nitrate. Input COD concentrations into UASB should be between 0.6 to 15 g/L and maximum input total suspended solids (TSS) should be less than 8 g/L. Methane production ranges between 0.2 to 0.4 m^3/kg of COD removed (Lettinga & Hulshoff-Pol, 1991; Tchobanoglous *et al.* 2003).

Typical COD removal efficiencies of UASB reactors (75 to 85%) are somewhat smaller than those of a typical aerobic activated sludge unit but it has to be realized that the influent COD and TSS concentrations will be much higher, more than 1000 mg/L, and in a typical aerobic activated sludge process with pure oxygen aeration, treatment would be costly. The smaller COD (BOD) removals could also be caused by dissolved methane in the effluent. Tandukar *et al.* (2007) proposed and used for post treatment after UASB a downflow hanging sponge (DHS) filter, similar to a trickling filter with plastic media. A sequence of UASB and DHS reactors provided removals of 94% of BOD and 90% of COD, respectively. Solids separation by membrane filters should provide effluent quality commensurate to a well functioning activated sludge membrane reactor.

Because the treated effluent from IRRF will be relatively warm (>20°C), heat can be extracted by a heat pump and, after disinfection, the effluent should be ready for disposal into the environment or non potable reuse.

Pre-digester or bio-electrically assisted microbial reactor (BEAMR)

A pre-digester is a (black) fermentation digester accepting organic solids, food waste, vegetation residues, and so on. Its role is wetting and pre-processing the solids using a portion of the influent or effluent from the UASB reactor which would provide moisture and nutrients. The traditional anaerobic digestion process progresses in three stages:

(1) Hydrolysis and fermentation of organic solids into organic acids;
(2) Fermentation of organic acids into acetates and hydrogen;
(3) Methanogenesis in which acetates are broken down and hydrogen is scavenged by methanogenic microorganisms to form methane and carbon dioxide.

The third step, methanogenesis, is endothermic and requires heating and longer HRT otherwise acetates with small amount of hydrogen are the "dead end" products of fermentation (Call & Logan, 2008). Without assistance of electricity providing electrons, the hydrogen yield is relatively small, a few percent of unrecoverable H_2. Dark fermentation products (organic acids) can be utilized by specific species of bacteria in a photo fermentation process to develop additional H_2 under nitrogen deficient and illuminated conditions (Hallenbeck & Benemann, 2002; Lu *et al.* 2010). The rates of hydrolysis and fermentation in the pre-digester are much greater than those for methane generation; therefore, the particle breakdown and formation of acetates and hydrogen can occur in much shorter HRT than in a typical digester and, after a shorter HRT, the pre-processed mixture full of acetates can be sent into the UASB influent.

A future alternative to a pre-digester is BEAMR which is an enhanced anaerobic fermentation reactor capable of producing far more hydrogen (Logan, 2004, 2008; Call & Logan, 2008) directly from organic matter and concentrated wastewater (Wagner *et al.* 2009). Logan and co-workers and other researchers discovered that if electricity is added to the reactor by DC current, the hydrogen production by electrogenesis in a microbial fuel cell can be much greater than that in the fermentation process without electricity assistance. Wagner *et al.* (2009) found that by adding electricity providing electrons, hydrogen was produced at a high rate. Furthermore, COD removals comparable to conventional digesters were achieved. The energy equivalent of the produced hydrogen is also much higher than the used electricity which, to a large degree, could be provided by solar photovoltaics with a storage battery. The produced gas in swine wastewater, BEAMR treatment was 87% pure hydrogen. Hydrogen production from acetates per reactor volume was about 2.5 to $3\ m^3H_2/m^3 \cdot d$ (Call & Logan, 2008). However, by decreasing electrodes spacing, Cheng & Logan (2011) achieved production rates as high as $17\ m^3H_2/m^3 \cdot d$ in a laboratory setting with applied voltage of 1 V. The production rate of $10\ m^3H_2/m^3 \cdot d$ would be very attractive for full scale applications and seems to be achievable (Cheng & Logan, 2011).

The needed DC electricity of 0.3 to 0.8 Volts is much smaller than the electricity needed in the electrolysis of water without microorganisms. For the swine wastewater in Wagner *et al.* (2009) research, energy used in BEAMR per COD removed was about equivalent to the energy used for COD removal in a traditional active sludge plan, which is 0.2 – 2 kWh/kg COD (Tchobanoglous *et al.* 2003), but the net energy in hydrogen produced by BEAMR was more than twice the energy used.

Nutrient recovery

Nitrogen in the atmosphere is abundant and can be converted to fertilizer forms of nitrate and ammonium by the energy demanding Haber-Bosh process. However, the world is running out of phosphate resources and it is expected that in less than 50 years with the present rate of consumption high grade ore sources will be exhausted (Barnard, 2007) which could be devastating to the world agriculture feeding a substantially larger population (approximately 10 billion) than today. Hence, the main task of the future IRRFs is to recover phosphorus. If nitrogen recovery can be accomplished simultaneously, it would save money and reduce GHG emissions of the Haber-Bush process (Barnard, 2007). Traditional oxic/anoxic nitrogen removal by Bardenpho process requires energy.

A process simultaneously removing both N and P without energy is available by struvite precipitation from wastewater and digester supernatant rich in nutrients (Barnard, 2007; Cecchi *et al.* 2003). Struvite is chemically an ammonium magnesium phosphate which grows in sewers carrying flows high in magnesium (hardness). Magnesium is added to the treatment/recovery process as magnesium hydroxide or magnesium chloride. Struvite precipitates at pH greater than 9; hence, if magnesium chloride is added, pH has to be increased by sodium hydroxide which adds another chemical to the process. After precipitation in the reactor and struvite removal, pH has to be adjusted back to neutral which can be done by carbon dioxide produced in the treatment process. Struvite is recovered in fluidized bed or pellet reactors.

Methane conversion to hydrogen and electricity

Hydrogen is a highly valuable commodity which also has the highest energy content. Also, using hydrogen for producing electricity is more efficient than electricity production by combustion of methane. Furthermore electricity production from hydrogen is clean energy.

Hydrogen has a very high heating value per unit of mass, as seen in the Table 3.3 below, and has a high economic value estimated as US \$6/kg which far exceeds that of methane given as US \$0.47/kg of CH_4 (Logan, 2004). Hydrogen, if needed, can be stored compressed or liquefied.

Table 3.3 Heating value of common fuels common in energy recovery from used water and organic solids.

Fuel	Heating value		
	MJ/kg	MJ/m^3	kWh/kg
Hydrogen	141.8	13.00	39.4
Methane	55.5	39.80	15.42
Gasoline	47.3		13.14
Natural gas (US)	37.9		18.53
Carbon monoxide	10.11		2.80

Source: www.engineeringtoolbox.com

A hydrogen fuel cell (HFC) produces clean electricity from hydrogen and oxygen, which react in the presence of an electrolyte by combining hydrogen and oxygen into water. Water and heat are the only by-products. Hence the hydrogen fuel cell produces no polluting emissions (Sammers, 2006).

Hydrogen fuelling the cell is produced from methane by the steam methane reforming (SMR) process. In the first step of SMR, biogas is cleaned by removing sulphides, which produces a purified mixture of methane and carbon dioxide (CO_2) (Sammers, 2006). Altogether, four molecules of H_2 are produced from one molecule of CH_4 or, theoretically, one kg of CH_4 produces 0.5 kg of hydrogen. Carbon dioxide can be sequestered by injecting underground or used to (1) neutralize reclaimed water after production of struvite, and/or (2) grow algae in algal ponds and reactors to produce more energy biomass. SMR is the most common method for producing commercial bulk hydrogen.

About 30% of energy in incoming methane would be lost for heating the SMR unit. This can be remedied by a solar heat powered SMR unit (Bakos, 2005). Since the SMR unit uses less water than produced in the HFC unit, the joint twin system produces clean water which can be reused. Each molecule of CH_4 produces net 2 molecules of water, or for each kg of CH_4 reformed to hydrogen, 1.8 kg of water is produced.

Additional process units of the IRRF

Additional process units of the IRRF are:

- disinfection by ultraviolet radiation,
- post treatment and polishing units such as aerobic trickling filter and membrane filters for solids separation,
- thickener and belt filter for residual solids dewatering,
- pyrolysis (produces syngas and charcoal from dewatered residual solids),
- algal farm producing biomass from carbon dioxide and nutrients generated by IRRF,
- heating system and its computerized real time control, including energy from renewable sources,
- heat pump to recover heat from the warm effluent.

3.4 COMPARISON OF ALTERNATIVES

Three alternative solutions have been analyzed to illustrate the water and energy use, production and associated GHG emissions. Water demand and its division into domestic (kitchen, bathroom, dishwasher, laundry, and so on) and outdoor (irrigation) components for average US households without water conservation and after installing water conserving appliances and water conserving practices were based on the study of the American Water Works Association Research Foundation reported in Table 3.1. Methods and parameters for the calculation and original reference sources are included in Novotny *et al.* (2010).

3.4.1 Description of three alternatives

Alternative I is an average end-of-the-century US household with a sprinkler lawn irrigation water demand, practicing no water and energy conservation and discharging its wastewater into a conventional sewer system connected to a conventional activated sludge treatment plant with nitrification that deposits residual sludge on land or in a landfill. Water demand is 550 L/cap · d of which 313 L/cap · d is for outdoor irrigation. Heated water daily volume is 106 L/cap · d and there is no heat energy recovery.

Alternative II is a household practicing water conservation indoor and xeriscape outdoor planting with minimal irrigation. Water conservation reduces the total water demand to 166 L/cap · d, similar to a typical demand in Europe and Japan. This house is located in a cluster which has a capability to reclaim some water and reuse it for toilet flushing (20 L/cap · d) and about 50% of the irrigation needs of the estimated outdoor use of 30 L/cap · d. One half of outdoor use would be from captured rainwater. Per Table 3.1 the heated water volume is reduced to 71 L/cap · d by installing water saving shower heads and other fixtures. On the cluster level, a portion of heating energy can be recovered by heat pumps. Recycled water for toilet flushing would be treated by microfiltration and ozonization. The rest of wastewater would be delivered to a regional activated sludge treatment plant with nitrification that would produce methane from sludge for heating of the digesters and buildings. This alternative needs separate piping, storage and a pump with a pressure tank for delivering recycled water to the toilets.

Alternative III is a hybrid distributed system that on the cluster (ecoblock) level separates water into black and grey water cycles as shown on Figures 3.4 and 3.5. The black water cycle (BWC) includes solids separation and treatment of a portion of the BWC flow for the local supplement of the grey water cycle (GWC), which needs make-up water to replace water lost in backwash and reject water (concentrates) of the filtration (including RO) units of the GWC and for irrigation and ecological outdoor flow. The daily volume of fresh water from the grid provided by freshwater sources is 50 L/cap · d.

A part of the BWC with all solids separated in BWC and GWC is conveyed to the regional IRRF shown on Figure 3.7. Admittedly, this is a visionary concept still lacking prototype testing and parameter derivations. Nevertheless, the system units have been developed and tested. The IRRF also accepts yard and food solid waste. The IRRF includes UASB reactor that accepts concentrated liquid flows and a BEAMR reactor in which solid waste (sludge, yard and food waste) is diluted by a portion of the BW flow to provide optimum moisture for fermentation. Because of the uncertainty with the future hydrogen yields in the BEAMR, the assumed yield is 5 $m^3/m^3 \cdot d$ (Cheng & Logan, 2011) and HRT 5 days.

The energy needed to extract, treat and deliver potable water from the grid is based on the US average of 2.26 kWh/m^3 and the GHG CO_2 emissions average equivalent from the power generating plants in the US is 0.61 kg CO_2/kWh. If the grid water is provided by desalination, energy use by the system may increase by about 0.15 kWh/cap · d. Table 3.4 presents the water and energy balances for the three alternative systems.

Table 3.4 Water and energy balance of three alternative water/wastewater management.

Parameter		Alternative I Traditional linear system with no conservation	Alternative II Mostly linear system with water conservation and small reuse	Alternative III Hybrid system with energy recovery and conversion to H_2
Water flow from the grid	L/cap · d	551	166	50
Energy to deliver and use water	kWh/cap · d	1.245	0.375	0.113
Water used for irrigation from grid	L/cap · d	313	30[1]	0
Energy use for irrigation[2]	kWh/cap · d	0.169	0.016	0
Total heating water flow	L/cap · d	106	71	71
Energy use for heating	kWh/cap · d	3.876	2.60	2.60
Total wastewater (WW) flow[3]	L/cap · d	297	116	NA
Pumping WW in the sewers [4]	kWh/cap · d	0.030	0.012	<0.01
COD content of used water	g/cap · d	95	95	95
Energy used to treat WW[5]	kWh/cap · d	0.125	0.072	0
Methane recovery from sludge	kWh/cap · d	0	−0.05	0
Grey water (GW) recycle	L/cap · d	0	20[6]	76
Energy to treat recycle	kWh/cap · d	0	0.015[7]	0.160[8]

(*Continued*)

Table 3.4 Water and energy balance of three alternative water/wastewater management (*Continued*).

Parameter		**Alternative I Traditional linear system with no conservation**	**Alternative II Mostly linear system with water conservation and small reuse**	**Alternative III Hybrid system with energy recovery and conversion to H_2**
Heat recovery from GW	kWh/cap · d	NA	NA	−1.00
Concentrated BW flow to IRRF	L/cap · d	NA	NA	69
Pumping BW to IRRF,	kWh/cap · d	NA	NA	0.007
Methane recovery from UASB	kg/cap · d	NA	NA	−0.02
H_2 from UASB methane conversion by SMR	kg/cap · d	NA	NA	−0.035
H_2 from BEAMR fermenting solids[9]	kg/cap · d	NA	NA	−0.02
Total energy from hydrogen	kWh/cap · d	NA	NA	−1.50
Heat recovery from effluent	kWh/cap · d	0	1.78[10]	−1.20[10]
Total energy expenditure (production)	kWh/cap · d	**5.45**	**1.26**	**(−0.83)**
Carbon GHG emissions (credit)	kg CO_2/cap · yr	**1263**	**281**	**(−185)**
Area of a solar concentrated panel to provide 50% energy for water heating (line 6)	m^2/cap	NA	1.2	1.2
GHG credit	kg CO_2/cap · yr	NA	**(−13.5)**	**(−473)**

[1]Water use for xeriscape and other outside uses assuming 50% irrigation demand satisfied by captured rainwater
[2]Includes sprinkler flow pumping energy and lawn mower energy estimate
[3]Includes indoor water use +25% increase of sewer flow by infiltration/inflow into sewer
[4]Pumping in lift stations for 30 m head loss to keep minimum velocities to prevent solids settling and providing hydraulic head in the treatment plant
[5]Assuming extended aeration and nitrification
[6]Water recycle only for toilet flushing treated by filtration and ozonization
[7]Water recycle treated by microfiltration and ozonization
[8]Grey water recycle treated by microfiltration, reverse osmosis and ozonization
[9]Per US EPA (2010) food and yard organic waste is 0.68 kg/cap · d and the recovery is 60%
[10]Total effluent for Alternative II, IRRF effluent for Alternative III

3.4.2 Discussion of the alternatives

Both water heating energy and heat energy recovery (less than the former) from wastewater were included in the balance. Heating energy is responsible for the largest portion of the overall energy use and corresponding energy gain when heat is recovered.

Alternative I is a baseline against which the other two, or any other improving alternative, are measured. This currently common linear management alternative would result in 1.26 ton of CO_2 emissions per capita per year, which represents about 12% of the average US per capita emissions, including also traffic and home space heating. Its average water use is unsustainable and cannot be repeated in most countries, especially in arid and semi-arid regions. Achieving the net zero carbon emissions goal is impossible and the alternative may also result in widespread water shortages even in more humid regions.

Alternative II could be a near future (less than ten years) goal. It incorporates reasonable water and energy saving measures. Methodology, parameters and pertinent references are included in Novotny *et al.* (2010). By means of water saving measures with a reuse and rain water reclamation, the per capita water use can be reduced in the US to the levels common in some European countries and Japan. By these measures, the CO_2 equivalent emissions can be reduced by 75%. Furthermore, the net zero carbon emissions goal would be achievable if about 50% or more of water heating energy is derived from renewable sources, for example, by installing concentrated heat panels that are already common in many environmentally conscious communities in California, Europe, Australia, China, and elsewhere. The size of the panel for a community in the southwestern US, which has approximately 320 insolation days, average insolation rate of 4 kWh/m^2 and heat recovery efficiency of 30% (a near future estimate) would be about:

- $A = 1.26[kWh/cap \cdot d] \times 365[days]/[4(kWh/m^2) \times 0.3 \times 320(days)] = 1.2\ m^2/capita$

Alternative III represents a more distant future (>15 years). It has a double loop separating grey water and black water flows and on site cluster/ecoblock treatment. The water demand of this alternative is very low, 50 L/cap · d, which is deemed to be the lowest limit for urban water delivery needed for providing adequate water supply and water based sanitation and some irrigation and ecologic flow. However, there is not enough energy in the used water (wastewater) to satisfy the needs for the high recycle without co-digestion of organic solids and/or using renewable (solar) energy for heating water. The energy deficiency without these auxiliary energy sources would be about 0.3 kWh/cap · d.

Alternative III with co-digestion is a very attractive and efficient solution with better than net zero GHG emissions. It results in excess energy produced by the system and commercially attractive production of hydrogen, electricity and commercial grade nutrients. Furthermore, the process can accomplish some carbon sequestering. Implementing solar power similarly to Alternative II for heating would further improve the energy balance. For a community of 10,000 inhabitants, the IRRF energy production would then be approximately equivalent to a 1 MW power plant that would provide, based on average 2010 electricity consumption (12 146 kWh/cap), electric power to 725 people, which could be improved in the future.

It is also interesting to compare the base line (current) Alternative I with the more sustainable Alternatives II and III. Implementing Alternative II would eliminate almost 1 ton CO_2/cap · yr from the current US emissions. This is even more impressive with Alternative III that would eliminate almost 1.5 ton of CO_2/cap · yr or about 14% from the current total household emissions (including car use and space heating), while producing commercially valuable products (water, fertilizers, hydrogen, biogas, electricity) and save money on dumping fees for organic solid wastes and water delivery fees.

3.5 CONCLUSIONS

Water and energy uses are intertwined and represent a significant portion of the total carbon emissions reaching the environment. Based on US EPA and IPCC (2007) statistics about 3–7% of the total energy use and the equivalent portion of GHG emissions are attributed to water and wastewater delivery, treatment and disposal.

In high water use countries such as US water conservation is the best alternative solution to a water availability problem because it reduces proportionally energy use and carbon emissions. Furthermore, energy can be extracted from wastewater by heat pumps for a carbon credit which can be used to warm water in the buildings or generate energy. A water to water heat pump provides 4–5 times more energy than it uses.

If the water use reduction goals cannot be met by water conservation only, a distributed water management system is the best solution whereby highly treated effluents, after heat energy is extracted, provide irrigation, ecological flow to receiving waters and water for downstream uses. Such a system has been implemented in Hammarby Sjőstad in Stockholm (Novotny & Novotny, 2009). Total urban water management is best carried out in a hybrid system whereby in local clusters used black and grey waters are separated, grey water and a portion of black water are reused and concentrated black water with other solids is sent to a regional integrated resource recovery facility for further resource and energy recovery.

Wastewater has recoverable energy content larger than that needed to satisfy energy needs of the water conveyance, reclamation and reuse. However, heating energy requirements, typically not accounted for in the water-energy nexus analyses, would result in the overall energy balance in energy deficit and GHG emissions. This deficit can be more than compensated by co-processing food and yard waste in the IRRF. The co-digestion and hydrogen production would produce energy in excess of the energy consumption in the hybrid urban water/used water cycle. The produced energy could be commercially sold along with nutrients and other resources produced by the IRRF and count as credit.

Reuse with high efficiency solids and pollutant removals (e.g. microfiltration and reverse osmosis) in a closed cycle (e.g. Masdar in UAE or Orange County in the US) relying on desalination for water supply requires more energy because of the energy requirement of desalination employed in the treatment processes and multiple cycle reuse (i.e. the water is reclaimed and treated more than once). This leads to a higher marginal carbon emission rate and total energy use. In order to stay sustainable, the extra energy has to be provided by renewable energy sources as it is done in Masdar or proposed for Qingdao in China. Methane production in the treatment and recycle process, if burned to produce energy, is carbon neutral.

The regional hydrogen based integrated resource recovery facility (H_2-IRRF) would:

- treat concentrated wastewater and solids to a level commensurate for safe disposal into the environment and safe regional reuse,
- produce excess energy in a form of biogas, commercial hydrogen, and electricity,
- recover nutrients in the form of commercial grade struvite,
- sequester carbon and provide carbon dioxide for a possible algal biomass production that can be used in the IRRF to produce more energy or commercially processed for biofuel,
- produce nutrient rich organic solids for farms or for production of syngas and charcoal by pyrolysis.

For all practical purposes, the H_2-IRRF would not use outside energy and the entire system could be a carbon sequester and energy producer. The net produced energy could be counted towards credit for a net zero GHG emission goal. The H_2-IRRF is a vision and implementation challenge. All components have been tested in laboratories and practice for some time, some for decades. The conception of H_2-IRRF should now be followed by a translational research, pilot testing and full scale implementations.

REFERENCES

AWWA R. F. (1999). Residential End Use of Water. American Water Works Association Research Foundation, Denver, CO.

Asano T., Burton F. L., Leverenz H. L., Tsuchihashi R. and Tchobanoglous G. (2007). Water Reuse – Issues, Technologies, and Applications. Metcalf & Eddy/WECOM, McGraw–Hill, New York.

Bakos J. (2005). Renewable Solar Hydrogen Production, Environmental Science and Engineering. May 2005, http://www.esemag.com/0505/solar.html (accessed October 2010).

Barnard J. (2007). Elimination of Eutrophication Through Resources Recovery. The 2007 Clarke Lecture, National Water Research Institute, Fountain Valley, California.

Call D. and Logan B. E. (2008). Hydrogen production in a simple chamber microbial electrolysis cell lacking a membrane. *Environmental Science and Technology*, **42**, 3401–3406.

Cecchi F., Battistoni P. and Boccadoro R. (2003). Phosphate Crystallisation for P Recovery Applied at Treviso Municipal Wastewater Treatment Plant. University of Verona, Italy, http://www.nhm.ac.uk/research-curation/research/projects/phosphate-recovery/Treviso2003.pdf (accessed June 2010).

Cheng S. and Logan B. E. (2011). High hydrogen production rate of microbial electrolysis cell (MEC) with reduced electrode spacing. *Bioresource Technology*, **102**, 3571–3574.

Daigger G. (2009). Evolving urban water and residuals management paradigms: water reclamation and reuse, decentralization, resource recovery. *Water Environment Research*, **81**(8), 809–823.

Drbohlav J. and Jankovský J. (2010). Vývoj potřeby vody v Praze (Water demand development in Prague- in Czech), Hydroprojekt. www.smv.cz/res/data/014/001642.pdf (accessed February 2011).

Dreiseitl H. and Grau D. (2009). Recent Waterscapes: Planning, Building and Designing with Water. 3rd edn, Birkhäuser Verlag, Basel, Boston, p. 176, ISBN: 978-3-7643-8984-0.

Energy Information Administration (2009). Net Generation by Energy Sources: Total (All Sectors). Report released September 11, 2009, http://www.eia.doe.gov/cneaf/electricity/epm/table1_1.html (accessed January 2010).

Fraker H. Jr. (2008). The Ecoblock-China Sustainable Neighbourhood Project. Power point presentation – Connected Urban Development Conference, September 24, 2008, Amsterdam, http://bie.berkeley.edu/ecoblocks

Fraker H. (2011). Personal communication. University of California, Berkeley (accessed February 2012).

Furumai H. (2007). Reclaimed stormwater and wastewater and factors affecting their reuse. In: Cities of the Future: Towards Integrated Sustainable Water and Landscape Management, V. Novotny and P. Brown (eds), IWA Publishing, London, UK, pp. 218–235.

Gleick P. H., Haasz D., Henges-Jeck C., Srinivasan V., Wolf G., Kaocushing K. and Mann A. (2003). Waste Not, Want Not: The Potential for Urban Water Conservation in California. Pacific Institute for Studies in Development, Environment and Security, Oakland, CA, http://www.pacinst.org/reports/urban_usage/waste_not_want_not_full_report.pdf (accessed December 2009)

Hallenbeck P. C. and Benemann J. R. (2002). Biological hydrogen production; fundamentals and limiting processes. *Int. J. Hydrogen Energy*, **27**, 1185–1193.

Heaney J. P. (2007). Centralized and decentralized urban water, wastewater, & storm water systems. In: Cities of the Future-Towards Integrated Sustainable Water and Landscape Management, V. Novotny and P. Brown (eds), IWA Publishing, London, pp. 236–250.

Heaney J. P., Wright L. and Sample D. (2000). Sustainable urban water management. In: Innovative Urban Wet-Weather Flow Management Systems, R. Field, J. P. Heaney and R. Pitt (eds), TECHNOMIC Publ. Comp., Lancaster, PA.

Hoekstra A. Y. and Chapagain A. K. (2008). Globalization of Water: Sharing the Planet's Freshwater Resources. Blackwell Publishing, Oxford.

IPCC (2007). Summary for Policy Makers, Climate Change 2007: The Physical Scientific Basis, Fourth Assessment Report, Intergovernmental Panel on Climatic Change, Working Group (WG 1), Geneva.

James B. D., Baum G. N., Perez J. and Baum K. N. (2009). Technoeconomic Boundary Analysis of Biological Pathways to Hydrogen Production, Final Report NREL, Directed Technologies, Arlington, VA, http://www.directedtechnologies.com/publications/fuel_options/BioH2_Boundary_Analysis.pdf (accessed 23 May 2011).

Kotsyurbenko O. R., Nozhevnikova A. N. and Zavarin G. A. (1993). Methanogenic degradation of organic matter by anaerobic bacteria at low temperatures. *Chemosphere*, **27**(9), 1745–1761.

Lettinga G. and Hulshoff Pol L. W. (1991). UASB-process design for various types of wastewater. *Water Sci. & Technology*, **24**(8), 201A–208A.

Lettinga G., Van Velsen A. F. M., Hobma S. W., de Zeeuw W. and Klapwijk A. (1980). Use of upflow sludge blanket (USB) reactor for biological wastewater treatment. *Biotechnol. and Bioeng.*, **22**, 699–734.

Logan B. E. (2004). Harvesting energy from wastewater treatment. http://www.engr.psu.edu/ce/enve/logan/bioenergy/presentations/Logan-BuschAward-10-5-04.pdf (accessed 23 May 2011).

Logan B. E. (2008). Microbial Fuel Cells. Wiley–Interscience, Hoboken, NJ.

Lu L., Xing D., Xie T., Ren N. and Logan B. E. (2010). Hydrogen production from proteins via electrohydro-genesis in microbial electrolysis cells. *Boiosensors and Bioelectronics*, **25**, 2960–2695.

Lucey P. and Barraclough C. (2007). Accelerating adoption of integrated planning & design: a water-centric approach, green value and restoration economy. Power point presentation at Northeastern University, AquaTex Victoria, BC.

McCarty P. L., Bao J. and Kim J. (2011). Domestic wastewater treatment as a net energy producer – can this be achieved? *Environmental Sci. & Technol.*, **45**, 7100–7106.

National Science & Technology Council (2008). Federal Research and Development Agenda for Net-Zero Energy, High Performance Green Buildings. Committee on Technology, Office of the President of the United States, Washington, DC.

Novotny V. (2007). Effluent dominated water bodies – their reclamation and reuse to achieve sustainability. In: Cities of the Future- Towards Integrated Sustainable Water and Landscape Management, V. Novotny and P. Brown (eds), IWA Publ. Co., London, UK, pp. 191–214.

Novotny V. (2008). Sustainable urban water management. In: Water & Urban Development Paradigms, J. Feyen, K. Shannon and M. Neville (eds), CRC Press, Boca Raton, FL, pp. 19–31.

Novotny V. (2010). Water energy nexus – towards zero pollution and GHG emissions of future (eco) cities. In: Water Infrastructure for Sustainable Communities, X. Hao, V. Novotny and V. Nelson (eds), IWA Publishing, London, pp. 35–58.

Novotny V. and Novotny E. V. (2009). Water centric ecocities – Towards macroscale assessment of sustainability. *Water Practice and Technology*, **4**(4), pp. 1–9, ISSN Online: 1751-231X, doi:2166/wpt.2009.062.

Novotny V., Ahern J. F. and Brown P. R. (2010). Water Centric Sustainable Communities: Planning, Retrofitting and Constructing the Next Urban Environments. J. Wiley & Sons, Hoboken, NJ.

Novotny V., Imhoff K. R., Olthof M. and Krenkel P. A. (1989). Handbook of Urban Drainage and Wastewater Disposal. J. Wiley Publishers, New York, p. 500.

Nozhevnikova A. N., Simakov M. V., Parchina S. N. and Kotsyurbenko O. K. (2001), Temperature characteristics of methanogenic archaea and acetogenetic bacteria isolated from cold environment. *Wat. Sci. Tech.*, **44**(8), 41–48.

O'Riordan K., Lucey W. P., Baraclough C. L. and Corps C. G. (2008). Resources from waste, an integrated approach to managing municipal water and waste systems. *Industrial Technology*, **4**(3), 238–245.

Parsons S. A., Wall F., Doyle J., Oldeing K. and Chuerley J. (2001). Assessing the potential for struvite recovery at sewage treatment works. *Environmental Biotechnology*, **22**, 1279–1286.

Rabaey K. and Verstraete W. (2005). Microbial fuel cells: novel biotechnology for energy generation. *Trends in Biotechnology*, **23**(6), 291–298.

Register R. (1987). Ecocity Berkeley: Building Cities for a Healthy Future. North Atlantic Books, Berkeley, CA. ISBN: 1556430094.

Roest K., Hofman J. and Loosdrecht M. (2010). The Dutch water cycle can produce energy (in Dutch). *H_2O*, **43**(25/26), 47–51.

Sammers N. (2006). Fuel Cell Technology – Reaching Towards Commercialization. Springer Verlag, London, UK.

Tandukar M., Ohashi A. and Harada H. (2007). Performance comparison of a pilot-scale UASB and DHS system and activated sludge process for the treatment of municipal wastewater. *Water Research*, **41**, 2697–2705.

Tchobanoglous G., Burton F. and Steusel D. (2003). Wastewater Engineering: Treatment and Reuse. 4th edn, Metcalf and Eddy, McGraw–Hill, New York.

US Department of Energy (2000). Carbon Dioxide Emissions from the Generation of Plants in the United States. Also published by US EPA, Washington, DC.

US Environmental Protection Agency (2010). Municipal Solid Waste Generation, Recycling, and Disposal in the United States: Facts and Figures for 2009. http://www.epa.gov/osw/nonhaz/municipal/pubs/msw2009-fs.pdf (accessed 20 June 2011).

Verstraete W., Van de Caveye P. and Diamantis V. (2009). Maximum use of resources present in domestic 'used water'. *Bioresource Technology*, **100**, 5537–5545.

Verstraete W., Bundervolt B. and Eggermont B. (2010). Zero Waste Water: Short-cycling of Wastewater Resources for Sustainable Cities of the Future. 2nd Xiamen International Forum on Urban Environment, Lab. of Microbial Ecology and Technology, (LabMET), University of Ghent, Belgium, http://www.LabMET.UGent.be (accessed June 2010).

Wagner R. C., Regan J. M., Oh S. E., Zuo Y. and Logan B. E. (2009). Hydrogen and methane production from swine wastewater using microbial electrolysis cells. *Water Research*, **43**(4), 1480–1488.

Chapter 4

Embodied energy in the water cycle

Michael J. Wilson

4.1 INTRODUCTION

The traditional linear approach to water resource management is increasingly proving to be unsustainable due to water stress being placed on urban water management (Daigger, 2009). The water qualities of various reclaimed wastewaters and the energy consumed by the unit processes have not been taken into consideration when evaluating and making decisions on water portfolios by the industry as a whole. This inequality in the full life cycle cost analysis has grown out of traditional thinking and has lead to an inefficient and compartmentalized approach to water resource management. For example, a U.S. Department of Energy Report to Congress in 2006 on Energy Demands on Water Resources did not even discuss the energy associated with recycled water nor was there a comparison to show the energy benefits associated with the embodied energy within recycled water (U.S. Department of Energy, 2006). Surprisingly this concept is only beginning to be grasped and taken into account in decision making. Consequently it becomes evident that energy considerations be included in water resource planning and water production.

This chapter will provide a methodology to include embodied energy in a full life cycle cost analysis within the water cycle by incrementally adding the energy consumed by each unit process to obtain a water quality product in terms of kilowatt-hours per cubic meters and million gallons (kWh/m^3 and kWh/MG). The chapter includes discussion on how embodied energy and energy intensity impacts water quality through a value chain. Additionally, a sustainability principle is presented for recycled water that is derived from energy usage, greenhouse gas emissions (GHG) and full life cycle costs to show that water reuse is a key element in water resource management.

4.2 ENERGY CONTENT OF WASTEWATER

The internal energy content of municipal wastewater has been measured through a calorimeter and combustion process and is defined as energy of combustion at a constant volume $-\Delta Uc$ on units of kWh/kg dry (kWh/lb). This is the energy that may be recovered through biological and physical-chemical processes that accounts for the theoretical energy within raw wastewater or solids. The calorimeter measures only the organic fraction of the wastewater, and therefore, the units are on a dry weight basis.

Typical measured energy content values for municipal raw wastewater and sludge samples are shown in Table 4.1 (Shizas & Bagley, 2004). The values from the table below are in contrast to the embodied energy which is imposed externally through the energy applied at each unit process and not internal or available for direct energy recovery from the process.

Table 4.1 Energy content of wastewater.

Parameter	Raw wastewater	Primary sludge	Secondary sludge	Anaerobically digested sludge
$-\Delta Uc$, kWh/kg dry (kWh/lb)	0.18 (0.40)	0.91 (2.00)	0.71 (1.56)	0.73 (1.60)
TSS (mg/L)	1,980	30,500	3,160	39,100
VS (%)	12%	67%	60%	51%

Adapted from Shizas & Bagley, 2004

As may be seen from the table above, the energy content of raw wastewater is much lower because, on a dry weight basis, the percent volatile solids is 12% for raw wastewater in comparison to 51% for anaerobically digested sludge. Consequently, most of the energy content of wastewater is contained in the volatile solids.

An energy balance may be prepared for a wastewater treatment plant based on the energy content of the unit processes. The energy balance around a typical municipal wastewater treatment plant with anaerobic digestion is depicted in Figure 4.1.

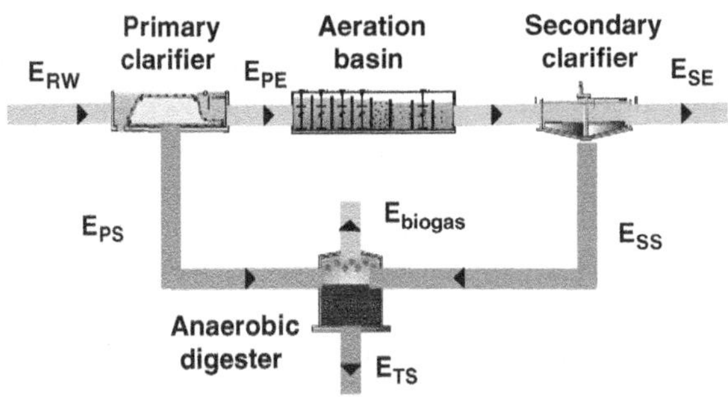

Figure 4.1 Energy balance diagram

Internal energy has not been included or accounted for in the following definition of embodied energy. The net potential energy would be reduced by the energy required for process treatment and distribution as described below.

4.3 EMBODIED ENERGY IN THE WATER CYCLE

Embodied energy in the water cycle has been defined as the energy consumed by all of the processes associated with the production of water and treated wastewater, from raw water pumping, treatment and distribution, through raw wastewater, pumping and conveyance, treatment and discharge. For example, the embodied energy and emissions of the production of wastewater would include the entire value chain of water production due to the direct relationship between potable water use and wastewater production.

The consumption of fossil fuels and the energy consumed in the production and delivery of water products has often been overlooked by decision-makers due to accounting principles and in-ability to account for energy consumed by each water and wastewater treatment process. The connection of unit processes or value chain approach is a new paradigm that provides an alternative view of the entire water product and energy relationship. Electricity use is typically quantifiable for treatment, but the combined energy costs across the water cycle are not known and vary dramatically from each water source, both regionally and by treatment process.

Figure 4.2 depicts the typical energy usage range by water and wastewater treatment unit processes (Wilkinson, 2000; Focus on Energy, 2006; Turgeon & Burgess, 2009). This graph summarizes several databases and several hundred data points by unit process for treatment plants in the United States including the Environmental Protection Agency (EPA) Wastewater Energy Benchmarking Tool Database and unpublished sources. The data was either originally presented in kWh/MG (converted to kWh/m^3) or was converted directly from the raw data.

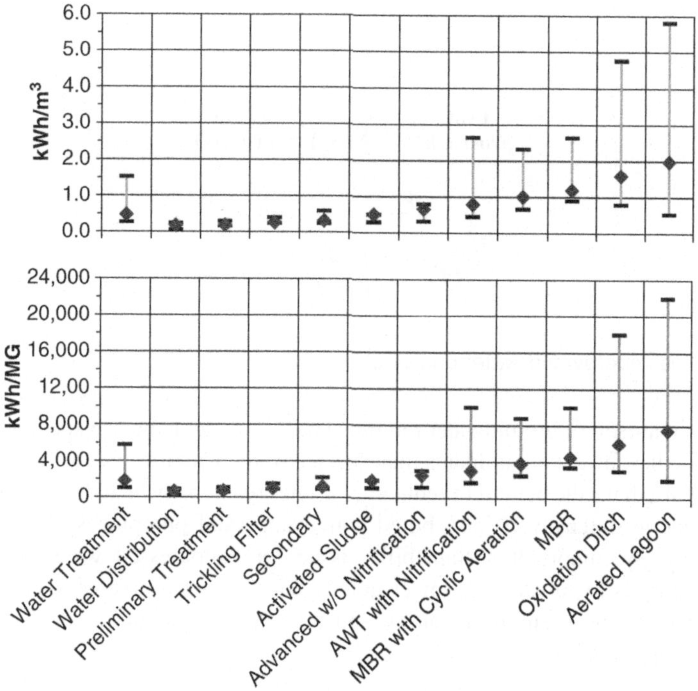

Figure 4.2 Energy usage range by water and wastewater process

4.4 WATER QUALITY AND ENERGY INTENSITY

Water quality is a function of increasing energy input. An increase in water quality is observed for each unit of energy, although the relationship is often non-linear or proportional. Increasing levels of water quality can be characterized by lower levels of turbidity, suspended solids, biochemical oxygen demand, bacterial count, colour, total dissolved solids and taste. Water products that have higher purity are either naturally pristine due to environment or have been transformed from lower quality to higher quality from a man-made treatment system or series of unit processes.

Energy intensity is an energy-time based relationship and can occur in both natural treatment systems and physical-chemical (man-made) treatment systems. The main difference being the energy intensity in a natural treatment system is typically much lower than a physical-chemical system. Therefore, the cycle time needed to obtain a specific water quality is longer in the natural system.

Figure 4.3 depicts a typical unit process energy and water quality cycle. Water quality varies depending on the location within the cycle. The figure shows that each unit process is additive in that as you progress around the cycle, energy consumed is cumulative. For example, if surface a water treatment plant (Unit process i) was pumped from sea level through a distributed conveyance system (Unit process i + 1) to a reclamation plant (Unit process n) at elevation 152 m (500 feet), the pumping energy would be added at the beginning the subsequent unit process for a cumulative unit process energy consumption. The water quality of the surface water would likely degrade within the distributed conveyance and pumping system and following reclamation, the water quality would increase. The energy consumed is location specific and includes many factors such as the quality of the source water, types of unit processes used to obtain the desired water quality, topography and size of the service area.

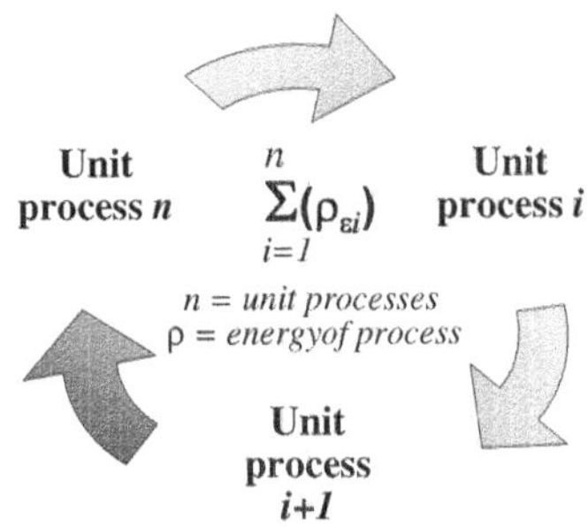

Figure 4.3 Unit process energy and water quality cycle

For example, the treatment of groundwater requires less energy to achieve a specific water quality since groundwater normally has fewer contaminants than surface water. Capture and treatment of stormwater and rainwater harvesting also have the benefit of reducing the energy intensity to achieve a specific water quality without the added energy intensity of distributed conveyance and pumping systems. Optimizing water quality is an increasingly important sustainability issue for overstressed watersheds. One method to reduce embodied energy is to substitute a lower quality source with a pristine source, however, pristine sources are often located near environmentally sensitive areas. Therefore, decision makers need to balance energy and water quality.

4.5 WASTEWATER RECLAMATION AS PRODUCT

Water reuse, in terms of wastewater reclamation, is a method to optimize the embodied energy within the water cycle since recycling a reclaimed (higher) water quality product instead of discharging the reclaimed water into a lower quality water through a single cycle use. Life cycle assessment, as defined by the US Green Building Council, includes a "cradle-to-discharge" approach (Scheuer & Keoleian, 2002). When wastewater is reclaimed in lieu of using treated raw source water it has the potential to reduce embodied energy because the energy required to treat wastewater is less than the energy required to treat raw water. Based on this method there is energy value in reclaimed wastewater that is not typically accounted for in a full life cycle analysis.

Wastewater is a product in the same way as manufacturing cement. Wastewater is only wastewater when we choose to waste it, that is typically the energy consumed in producing this product is wasted to a water body or into the ground in the same fashion that one would dispose of a plastic bottle at a landfill. Recycling is typically not an option because the municipal infrastructure does not exist or there are other economic and social impediments for wastewater to be recycled.

Figure 4.3 depicts the incremental energy steps in the water cycle. Three examples of embodied energy within the water cycle are presented in Figure 4.4. These examples do not account for the chemical energy within the water and wastewater nor do the examples include any contribution of energy added within the residence or industry. Table 4.2 summarizes the energy usage for the three systems on the East Coast of the United States presented also in Figure 4.4.

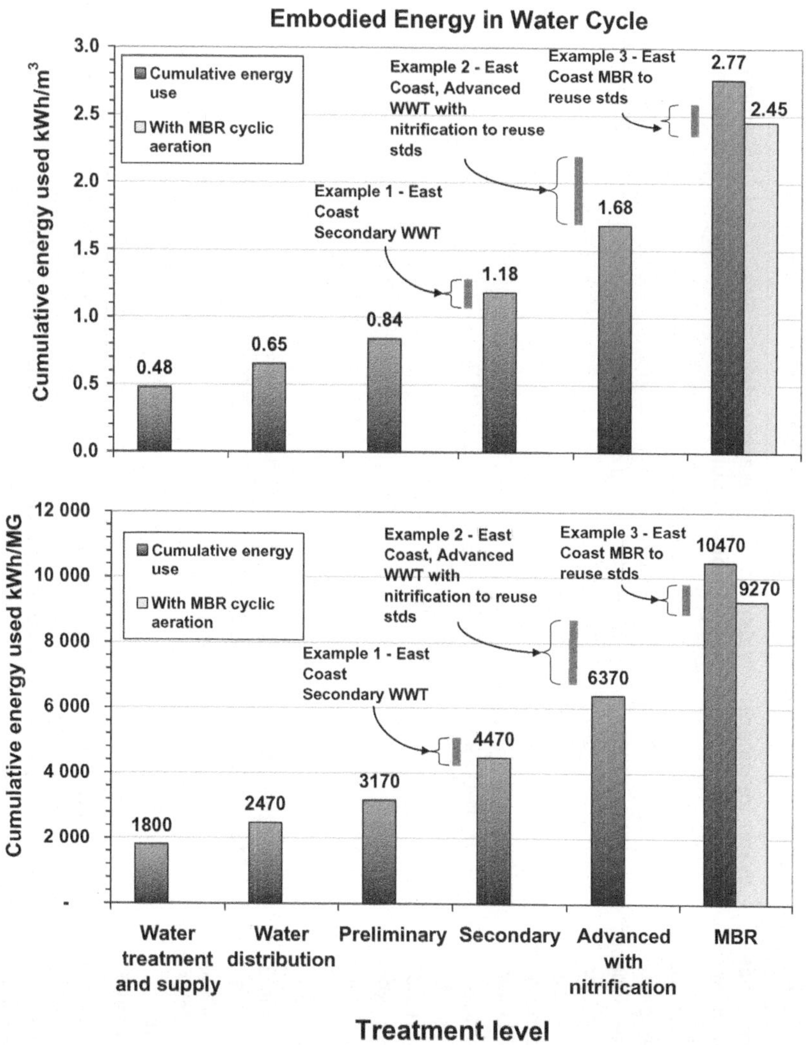

Figure 4.4 Incremental energy steps in water cycle

Table 4.2 Energy Usage by Process expressed in kWh/m^3 (kWh/MG).

Final process treatment level	Water treatment & supply	Water distribution	Wastewater pumping & conveyance	Pre-treatment	Secondary treatment	Advanced wastewater & nitrification	Membrane bioreactor (MBR)
Example 1-East Coast secondary wastewater treatment with nitrification	0.26–0.37 (1000–1400)	0.26–0.29 (1000–1100)	0.05–0.07 (200–280)	–	0.50–0.53 (1900–2000)	–	–
Example 2-East Coast advanced wastewater treatment to reuse standards	0.26–0.29 (1000–1100)	0.24–0.29 (900–1100)	0.21 (790)	–	–	1.00–1.69 (3800–6400)	–
Example 3-East Coast membrane bioreactor to reuse standards	0.26–0.32 (1000–1200)	0.26–0.32 (1000–1200)	0.05–0.08 (200–300)	–	–	–	1.82–1.93 (6900–7300)

All three examples have gravity water treatment and supply systems and pump from one reservoir to water distribution. All three example wastewater systems have gravity discharges for plant effluent.
All data obtained from plant utility metering records and supervisory control and data acquisition systems in 2008.
1 Million Gallons, MG = 3785.4 m^3

The data for the cumulative energy by unit process on Figure 4.4 graphs was obtained from three East Coast water and wastewater systems. There are three levels of wastewater unit processes depicted including a secondary treatment process, a secondary treatment process with advanced filtration to tertiary reuse standards and a reclamation plant with a membrane bioreactor process (MBR). The cumulative energy used in the membrane bioreactor differs significantly between a standard MBR process and one that uses cyclic aeration. The water quality increases as you progress from preliminary treatment to secondary, and to advanced treatment with nitrification to MBR.

The data presented in Table 4.2 were obtained directly from energy usage by unit process from the values collected by the water utility and wastewater utility at three locations on the East Coast. These values were compiled from either sub-metering data or supervisory control and data acquisition systems and cumulative energy usage was calculated based on the unit process data attributable to each unit process from the potable water source to reclamation as applicable. Each example combines the energy data across the specific water treatment, wastewater collection and treatment system within a utility district or watershed.

4.6 SUSTAINABILITY PRINCIPLE IN WASTEWATER RECLAMATION

Sustainability in water resources requires that we consider the principles of embodied energy when we are making decisions regarding the water cycle value chain. The value of recycled water includes the embodied energy that is reclaimed from the water cycle. Figure 4.5 depicts the principle of sustainability in water reuse (wastewater reclamation). This figure depicts how the cumulative energy at the point of water reuse shortens the natural water cycle to reduce the embodied energy through a recycle loop that circumvents and substitutes the energy intensity of the potable water treatment and distribution system.

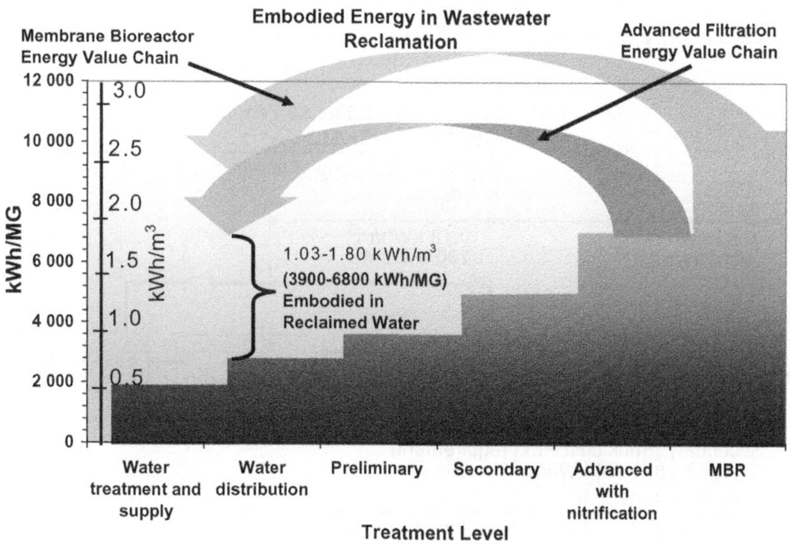

Figure 4.5 Sustainability Principle in Wastewater Reclamation

The energy value chain associated with the embodied energy in the water recycling process is cumulative and significant. The methodology for accounting for the amount of energy expended within the supply, treatment and conveyance is extremely variable and is primarily based on lifting the weight of the water and filtering pressure during processing. Figure 4.6 and 4.7 depict the embodied energy value chain with and without water reuse. These figures provide the differences in energy consumption between a system with a water recycling loop and one without water reuse. They show the differences between each value chain and the challenges in calculating the energy consumed by the unit processes. The value chain can never be a zero sum due to regulatory permit compliance and environmental regulations. Consequently, the true value chain of the reclaimed product is difficult to measure. However, the value added through source substitution is evident with the ability to short-circuit the complete water cycle.

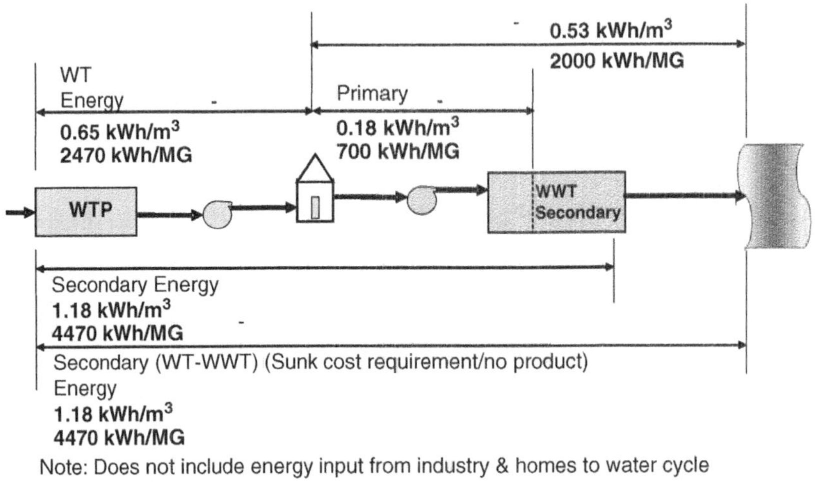

Figure 4.6 Embodied energy value chain without water reuse

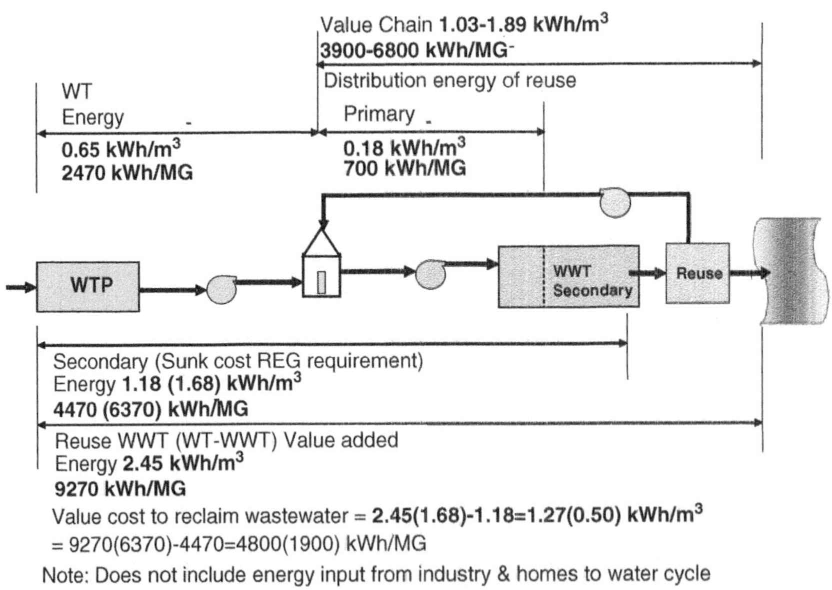

Figure 4.7 Embodied energy value chain with water reuse

4.7 LIFE CYCLE ANALYSIS OF WATER REUSE

A life cycle analysis at each step in the production process reveals that the embodied energy value chain is an additive process, whereby at each step more energy is consumed in creating the water product. The energy envelope can be generalized when you consider the major differences between raw water pumping and treated water distribution between the East Coast and West Coast of the United States. Water is pumped and treated and then re-pumped in many cases several times in Western States where elevation changes can be significant (Wilkinson, 2000). In contrast, when we look at the relative pumping and distribution costs for Eastern States, the amount of energy consumed in delivery of the product can be an order of magnitude less due to the predominance of more gravity systems and less pumping energy.

The life of the piping infrastructure and structural components of most water treatment and wastewater treatment systems is 50 years or more. The process equipment, chemical systems, instrumentation and electrical systems within the facilities used in creating the water product typically has a life of 20 years.

A full life cycle analysis on a 50 year term shows the relative difference between the embodied energy in operational costs of water and wastewater treatment processes and the energy embodied in the material construction of the treatment plant (Figure 4.8). The embodied energy associated with the operation for the creation and distribution of the water product is about 30% of the operational life cycle cost through the life of a facility (Reiner *et al.* 2007). The relative cost of recycled water is similar to the life cycle cost of advanced treatment with nitrification. This is due to the value of the embodied energy in the life cycle costs of the recycled water that is applied and recovered from the system.

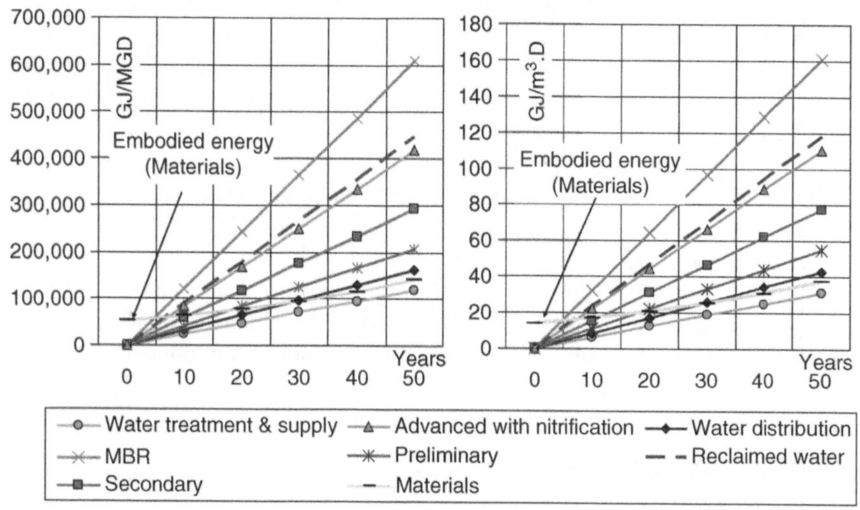

Figure 4.8 Life cycle energy usage envelope expressed as cumulative energy in GJ/MGD and $GJ/m^3 \cdot d$

Figure 4.8 shows the Life Cycle Energy Usage Envelope for each process in terms of giga-joules per million gallons per day (GJ/MGD) and cubic meters per day ($GJ/m^3 \cdot d$). It was compiled from secondary research (Scheuer & Keoleian, 2002) and the data sources were obtained from a life cycle materials and energy balance on a membrane bioreactor recycling plant from several data sources (SAIC, 2006; Wilson, 2009). This graphic depicts an extraordinary amount of energy. As a point of reference one giga-joule is equal to 277 kWh, and on average, 98 kg (216 lbs) of carbon-dioxide (CO_2) is produced per giga-joule.

When embodied energy in recycled water is accounted for over a 50 year life cycle term, at 12 cents per kWh, and not including interest, it has a value of US\$ 3925/m^3 per cubic meter or US\$14,857,000/MG. The capital cost embodied in the construction materials for the recycled water infrastructure is diminimus when compared with the overwhelming operational energy recovery savings from the substitution of recycled water for a more pristine potable water source in the value chain. The analysis shows that the life cycle of a membrane bioreactor reclaimed water process is commensurate with the energy of an advanced treatment process when the natural water cycle of water treatment, supply and distribution is able to be substituted by the recycled water. In other words, recycled water value is artificially increased by the regulatory permitting requirements to meet secondary treatment standards and the energy usage sunk into secondary treatment.

4.8 CARBON FOOTPRINT AND GREENHOUSE GASES

Fossil fuels are the predominate fuel used in the production of energy for water and wastewater reclamation processes. The amount of carbon dioxide (CO_2) produced is based on an emissions formula and is a combination of converting the purchased electricity by sub-region into kilograms per kilowatt-hour per cubic meter (kg/kWh/m^3) or pounds per kilowatt-hour per million gallons (lbs/kWh/MG). This can then in turn be converted to kilograms of carbon dioxide per cubic meters or pounds of carbon dioxide per million gallons (kgCO_2/m^3 or lbsCO_2/MG) when cumulative energy usage is applied to the value chain for wastewater reclamation.

Table 4.3 provides a summary of the CO_2 emission factors for each sub-region and the respective amount of CO_2 that is used to produce one cubic meter or one million gallons of potable water assuming an average energy consumption of 0,65 kWh/m^3 (2,470 kWh/MG). Since there is a regulatory requirement to meet certain minimum standards for water treatment and wastewater effluent quality, a portion of the amount of the CO_2 generated in the production of the water is a sunk cost, and reducing GHG emissions further is difficult due to increasing treatment requirements.

However if the wastewater is reclaimed and value is added to reduce water consumption through source substitution (potable water), it is possible that a reduction in greenhouse gases occurs through the exchange of water recycling with water supply and distribution. This would predominately occur in the Western United States where the energy to produce and distribute potable water can be between 1.32–3.96 kWh/m^3 (5,000–15,000 kWh/MG; Wilkinson, 2000) and is in contrast with the 0.48–0.66 kWh/m^3 (1,800–2,500 kWh/MG, U.S. Department of Energy, 2006) typical for potable water treatment and distribution in the Eastern United States.

4.9 ENERGY OPTIMIZATION OF SYSTEMS

There are a wide range of water treatment processes that are available to improve water use efficiency and reduce the amount of source water demand from the watershed to meet urban water needs including natural treatment systems. These processes include many water treatment, stormwater management, rainwater harvesting and wastewater reclamation processes that capture and treat these water sources locally. The benefit is the reduced energy required for a specific water quality and then using the water locally. The energy required that achieves a commensurate level of water quality through use of stormwater or rainwater is typically less than the energy required for water recycling since wastewater reclamation requires the removal of more contaminants.

One approach that would increase synergy and increase potential for a broad based regional water resources strategy to minimize the "information silos" typically found in water resources is the concept

Table 4.3 Carbon footprint reduction due to water recycling for Northeast Coast.

Subregion Name – Abbr.	CO2 emission factor kg/MWh, (lbs/MWh)	Input for energy total to produce potable water (CO_2 reduction)[1] kWh/m^3 (kWh/MG)	Input for energy total to produce potable water (CO_2 reduction)[1] MWh/m^3 (MWh/MG)	CO_2 emissions for potable production kg/MWh/m^3 (lbs/MWh/MG)	CO_2 emissions for potable production kg/kWh/m^3 (lbs/kWh/MG)	Reduction, kgCO_2/m^3 (lbsCO_2/MG)	Reduction, tonsCO_2/m^3 (tonsCO_2/MG)
NEWE – NPCC New England	412.3 (908.9)	0.65 (2470)	0.00065 (2.47)	0.27 (2245)	0.00027 (2.24)	0.66 (5545)	0.00066 (2.52)
NYLI – NPCC Long Island	640.6 (1412.2)	0.65 (2470)	0.00065 (2.47)	0.42 (3488)	0.00042 (3.49)	1.03 (8616)	0.00103 (3.91)
NYUP – NPCC Upstate NY	371.8 (819.68)	0.65 (2470)	0.00065 (2.47)	0.24 (2025)	0.00024 (2.02)	0.60 (5001)	0.00060 (2.27)
RFCE – RFC East	496.9 (1095.53)	0.65 (2470)	0.00065 (2.47)	0.32 (2706)	0.00032 (2.71)	0.80 (6684)	0.00080 (3.03)

[1]Embodied Energy of Typical East Coast Value
Units: 1 pound (lbs) = 0,4536 kg; 1MG = 3785,4 m^3

of an intelligent network (Matsumoto *et al.* 2004) or smart water grid. An intelligent network that is based on a common control platform would allow water management decisions to be made through real time controls for system water demand levelling similar to the automatic switching that occurs between wireless and local area network (LAN) based servers in computer networks. Another important aspect to system design is the balance between centralized and decentralized systems and the security enhancement due to partial system redundancy (Daigger & Crawford, 2007).

Optimization of water quality is an important public health consideration and yet in many locations we do not have the required infrastructure to allow choices to be made including water reuse, that is to use non-potable water for irrigation or other domestic purposes. Obviously water conservation and leakage detection are also important elements that are to be integrated into a well planned sustainable water management program. Intelligent networks and smart water grids within a watershed or service area would be a management strategy that could promote and balance energy demand and water quality. These networks are being integrated into the existing distribution and conveyance systems by utilities to optimize the energy demand required for water distribution and water reuse (Jentgen *et al.* 2005).

4.10 WATER RECYCLING PURPLE PIPE INFRASTRUCTURE

Based on the life cycle analysis presented above there is absolutely no excuse for not building new water reuse systems. The capital cost is recovered in a fraction of the life of the system (Scheuer & Keoleian, 2002; Reiner *et al.* 2007). The operating costs are derived from user fees on a predetermined rate basis. The distribution system that allows recycled water to be used by the consumer and the installation of purple pipe is recovered in energy savings which may significantly offset the cost of the distribution system.

4.11 SUMMARY

Until recently, the hidden value of embodied energy in water reuse has not been well considered in a full life cycle analysis. The value of recycled water includes the embodied energy that is reclaimed from the water cycle. When embodied energy in recycled water is accounted for the capital and operating cost, the water reuse infrastructure is substantially offset with the overwhelming operational energy recovery savings.

Consequently, a cornerstone of sustainable water resources is water reuse. Reclamation of treated wastewater promotes environmental benefit by allowing for both high water quality and optimizing the energy intensity within the water cycle. Water conservation, while an important component of a sustainable water resources management plan, does not provide the same value chain impact in terms of reclaiming the energy already embodied in the water product.

REFERENCES

Daigger G. T. (2009). Evolving urban water and residuals paradigms: water reclamation and reuse, decentralization, resource recovery. *Water Environment Research*, **81**(8), 809–823.

Daigger G. T. and Crawford G. V. (2007). Enhanced water system security and sustainability by incorporating centralized and decentralized water reclamation and reuse into urban water management systems. *Journal of Environmental Engineering Management*, **17**(1), 1–10.

Focus on Energy (2006). Water and Wastewater Best Practice Guide Book. Science Applications International Corporation SAIC, State of Wisconsin, Department of Administration, Division of Energy, Madison, Wisconsin. Focus on Energy, SAIC, December 2006, 93.

GEMI (2002). Connecting the Drops Towards Creative Water Strategies a Water Sustainability Tool, June 2002.

Jentgen L., Conrad S. and Lee T. (2005). Optimizing system operations. *Journal American Water Works Association*, **97**(8), 58–65.

Matsumoto S., Imai K. and Samejima S. (2004). Intelligent control technology water pipe network simulation system supporting daily water processing business. *Journal Meiden Jiho*, **298**, 36–39.

Reiner M., Pitterle M. and Whitaker M. (2007). Embodied Energy Considerations in Existing LEED Credits. Symbiotic Engineering, Boulder, Colorado. September 2007.

Scheuer C. W. and Keoleian G. A. (2002). Evaluation of LEED Using Life Cycle Assessment Methods, NIST GCR 02–836, September 2002.

Shizas I. and Bagley D. M. (2004). Experimental determination of energy content of unknown organics in municipal wastewater streams. *Journal of Energy Engineering*, **130**(2), 45–53.

Towards A Sustainable America (1999). The President's Council on Sustainable Development, May 1999.

Turgeon J. and Burgess J. (2009). Environmental Protection Agency Wastewater Benchmarking Database Tool.

U.S. Department of Energy (2006). Energy Demands on Water Resources. Report to Congress on the Interdependency of Energy and Water, December 2006.

Wilkinson R. (2000). Methodology for analysis of the energy intensity of California's water systems, and an assessment of multiple potential benefits through integrated water-energy efficiency measures. Environmental Sciences. California Institute for Energy Efficiency, January 2000, 89.

Wilson M. (2009). Embodied Energy in the Water Cycle. Water Environment Association, WEFTEC Session 83, Life Cycle Model for Analysis of Embodied Energy, October 2009.

Chapter 5

Microbial electrochemical technologies for energy sustainability of the water infrastructure

Bruce E. Logan

5.1 INTRODUCTION

This century will place great strains on our planet and lifestyles. The release of large amounts of carbon dioxide and other greenhouse gases into the atmosphere is causing changes in our climate that are altering global patterns of temperature, rainfall, and runoff, and thus water availability. Along with these changes are also increases in the global demand for energy and water. The challenge to the water community is how to achieve energy sustainability of our water infrastructure through the use of only sustainable and carbon neutral methods of energy production. We can do this partly by exploiting an unused energy resource: the organic matter present in various domestic and industrial wastewaters. Rather than expend energy to remove organic matter in wastewater, we should be improving on existing methods and developing new methods for energy recovery from wastewater, effectively converting our wastewater treatment plants into power plants.

New methods for obtaining energy from wastewater have being investigated that are based on microbial fuel cell (MFC) technologies. These systems use bacteria to directly produce an electrical current which can be used to power the wastewater treatment plant, and additionally to provide electricity for potable water treatment. The current produced in this bioelectrochemical system (BES) can alternatively be captured directly into gaseous fuels such as hydrogen or methane gas. Modification of these devices allows direct water desalination, coupled to wastewater treatment and electrical energy production, providing for the first time a biomass-based process for water desalination that does not require an external electrical energy input. Further developments have recently resulted in systems that boost power generation through the capture of energy from salinity gradients (Kim & Logan, 2011a,b). While MFC technologies are at an early stage of development, they show great promise as one approach for ensuring the energy sustainability of the water infrastructure.

5.2 THE WATER-ENERGY NEXUS

The energy needed to maintain the water infrastructure in the USA is already quite large, and without changes we are currently on a path to adopt more energy demanding treatment systems. Therefore, unless new technologies are developed, our water infrastructure will continue to require non-sustainable amounts of energy. The amount of energy needed for our water infrastructure has been estimated to consume 4 to 5% of the electricity produced in this country, with about 1.5% of this used for wastewater treatment alone. In California, it was recently estimated that 19% of the electricity used in the state was for water related services, including agricultural operations (Stokes & Horvath, 2009). This high energy demand poses a problem for the US in the future, and it particularly affects the ability of developing countries to achieve sufficient supplies of water and acceptable levels of sanitation if they use the same technologies.

Over 1 billion people lack reliable supplies of potable water, and 2.4 billion people lack adequate sanitation. Not only would the cost of this energy be prohibitive for developing countries, but producing this much electricity from fossil fuels for the water infrastructure would add considerably to the release of carbon dioxide and other greenhouse gases to the atmosphere.

Climate change and the large amount of energy needed for the water infrastructure together create a large obstacle for providing water and adequate sanitation to the global population of over 6 billion people. The average temperatures around the world are changing, with projected increases of 3–5°C, for example, in Southern California. With these temperature changes, it is expected that there will be a decrease in water runoff, and thus a reduction in needed water flow into rivers and critical water storage systems. These changes in runoff will be in addition to natural variations and cycles in rainfall that already are challenging water supplies in some areas of the nation. Lake Mead in Arizona, for example, is already exhibiting critical decreases in water levels as a result of reduced flow into this major reservoir of water for many southwestern US cities. Water conservation and changes in water uses will allow us to respond to some of these changing patterns, but energy costs for pumping and moving water remain a challenge.

Energy used for producing potable water is increasing as more advanced treatment technologies are used, such as membrane-based reverse osmosis (RO) compared to conventional coagulation and flocculation systems. Decreases in freshwater availability could also result in an increased reliance on water desalination. Various desalination technologies are being used around the world, and energy requirements for these methods vary substantially, although overall energy requirements have decreased in recent years with development of RO systems. Desalination techniques and their associated energy requirements normalized to a cubic meter of water produced include: single stage evaporation, 650 kWh/m^3; multistage flash, 55–80 kWh/m^3; multi-effect distillation, 40–65 kWh/m^3; and RO, which requires only 3.7 kWh/m^3 with seawater (Semiat, 2008). If all the water used in California were to be provided by water desalination, however, 52% of the electricity used in the state would be needed to meet water demands (Stokes & Horvath, 2009).

The amount of energy used for wastewater treatment varies by over a factor of ten, depending on the technology used (Table 5.1). For example, trickling filters use 0.12 kWh/m^3 of wastewater treated, although this process needs additional systems such as solids contact to meet discharge requirements. The energy demand for activated sludge is higher (0.28 to 0.71 kWh/m^3) but energy requirements are decreased with biological nutrient removal to 0.23 kWh/m^3. The use of membrane bioreactors (MBRs) substantially increases the energy needed for treatment, with an average power use of 2.4 kWh/m^3. MBRs are the only treatment technology that use more energy than that in a typical domestic wastewater in the US, which is approximately 1.2 kWh/m^3 (based on 300 mg/L of chemical oxygen demand, COD).

The energy used at domestic wastewater treatment plants for treatment is being reduced through online and dynamic control of aeration systems to better match oxygen utilization, through energy recovery processes and the replacement of energy utilizing processes such as aerated sludge thickeners with stirred systems, and other processes. The largest gains in energy recovery are being made by recovering biogas (natural gas) from anaerobic digesters. This gas can be used to generate electricity to power the plant. Some plants can even become completely self-sufficient in terms of energy needs using the natural gas produced (Nowak *et al.* 2011). In cases where the organic matter concentration in industrial wastewaters is on the order of several grams per liter, it is usually feasible to use anaerobic processes to directly treat the wastewater and produce methane. These types of plants are used more frequently in Europe than in the US (see Chapter 3).

Table 5.1 Energy usage for different wastewater treatment processes (from Cooper *et al.* 2007).

Wastewater treatment process	Energy used, kWh/m^3
Oxidation pond	0.047–0.12
Trickling filter (TF)	0.12
Activated sludge (AS)	0.28–0.71
Activated sludge (AS) + nutrient control	0.23
Membrane bioreactor (MBR)	2.4

Substantial funding is needed for our aging water infrastructure in the US, and so now is the time to put in place new technologies that have low energy demands or that can produce net energy. It was estimated that $2 trillion was needed in the US for the next 20 yrs for building, operating and maintaining wastewater and drinking water facilities (WIN, 2001). About $45 billion was needed for wastewater alone, in addition to the estimated annual expenditure of $25 billion. The costs for operating these treatment plants, and the energy used by these plants, could be greatly reduced by using the energy in the wastewater. The energy in the organic matter in domestic, animal and food wastewaters is estimated to be about equal to the electrical energy currently used for our water infrastructure (Logan, 2008). Capturing this energy should be our next main goal in designing new wastewater treatment systems.

5.3 MICROBIAL FUEL CELL BASED TECHNOLOGIES FOR ENERGY RECOVERY

While energy conservation and advancement of existing technologies can reduce energy consumption at water and wastewater plants, the next generation of treatment systems that can produce a large excess of energy will require substantially different approaches. One of the most promising approaches is using microbial fuel cells (MFCs) to simultaneously treat wastewater while generating electricity (Liu *et al.* 2004; Logan *et al.* 2006). Through modifications of MFCs, it is possible to develop systems called microbial electrolysis cells (MECs) that can be used for other purposes, such as producing gaseous fuels like hydrogen and methane (Logan *et al.* 2008). Further modification of these MFC-based technologies can achieve electricity generation and water desalination (microbial desalination cell, MDC) (Cao *et al.* 2009; Mehanna *et al.* 2010b), or hydrogen production and water desalination (microbial electrodialysis desalination cell, MEDC) (Mehanna *et al.* 2010a). In the rest of this chapter, these technologies are described, with a focus on their potential applications for wastewater treatment and eventual energy sustainability.

5.3.1 Microbial fuel cells for bioelectricity generation

The basic concept of energy capture in an MFC is similar to that used by many organisms, ranging from bacteria to humans, to generate and capture energy: food is consumed and broken down, and energy is captured during respiration. When we eat food, our bodies break down complex organic matter into simple molecules that are then oxidized, releasing electrons to an energy carrier in our cells. When these electrons flow through a series of respiratory enzymes, our cells capture and store that energy (in ATP), but then the electrons must then be discarded to continue the cycle. We get rid of these electrons by breathing air, so that the electrons combine with oxygen and form water. Thus we see that production of energy requires two steps: chemical oxidation (removal of electrons from organic molecules), and oxygen reduction (oxygen accepting electrons). These two oxidation and reduction processes are used to generate electricity in MFCs, but the key capturing the electrical current from the bacteria is to keep them separated from a source of oxygen.

An MFC consists of two electrodes, and anode and a cathode (Figure 5.1). Bacteria oxidize the organic matter and release electrons to the anode and protons into the water. Electrons flow through the circuit to the cathode where they react with protons and oxygen, forming water. It is possible in MFCs to use electron acceptors other than oxygen, but their use so far has not been considered to be practical or sustainable. The maximum amount of energy that can be captured is the chemical potential between the electron carrier in the bacterium and oxygen, which is about 1 volt (V). In an MFC, the potential of the anode is between these potentials, so that bacteria capture the part of the energy between that of the carrier and the potential of the anode (a few tenths of a volt). The remaining energy (about half a volt), which is the difference in the anode potential and the cathode potential (based on the chemical reduction of oxygen), is captured in the MFC, with the remaining electrical energy lost as heat.

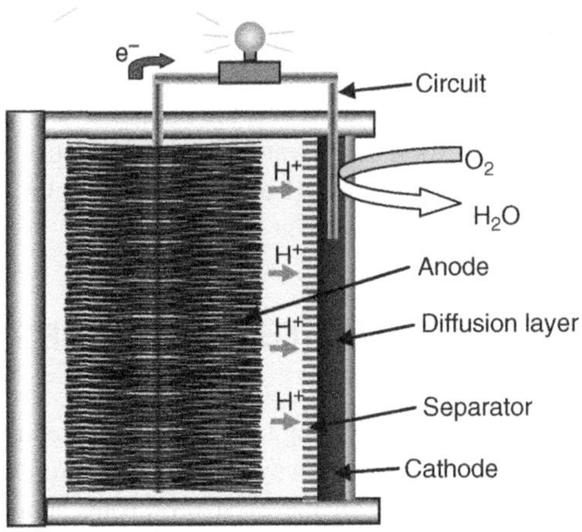

Figure 5.1 Example of a microbial fuel cell where the anode is made of an electrically conductive graphite fiber brush and the cathode is a piece of carbon cloth containing a catalyst on the water facing side to catalyze oxygen reduction (not shown) and a diffusion layer on the air side (to prevent water loss and reduce oxygen diffusion into the chamber). This MFC design contains a separator that is placed between the electrodes to insulate the cathode against touching to anode, while allowing for charge transfer (Modified from Logan, 2009)

The outcome of organic matter degradation in an MFC that is fueled by wastewater is that the organic matter in the wastewater is removed, so the wastewater is cleaned, and useful power is produced. This step of the treatment process requires no net energy input as the whole process is thermodynamically spontaneous. As long as oxygen reacts at the cathode in a passive manner, which is typical of "air-cathode" MFCs, there is no need to aerate the wastewater. Because the bacteria derive much less energy than they would if they directly used oxygen, there is less biomass produced in an MFC than in an aerobic treatment process such as activated sludge.

This approach results in three advantages of MFCs compared to conventional aerobic treatment:

(1) a useful product (electricity) is produced;
(2) no aeration is used so electricity for that is not needed;
(3) bacterial solids (sludge) production is greatly reduced compared to an aerobic process.

The lack of a need for aeration is a significant contribution to the savings of the process as wastewater aeration can consume 30–50% of the electricity used at a wastewater treatment plant. Solids reduction is another saving, as solids handling can account for 20–50% of operating costs at a treatment plant.

The power densities produced in MFCs has substantially increased in the past decade. Systems initially developed by Kim and co-workers produced only 0.001 to 0.01 mW/m^2 of surface area of the anode (projected) (Kim *et al.* 1999; Logan & Regan, 2006). This levels were rapidly increased to about 50 mW/m^2 in 2002, to levels of 2770 mW/m^2 (2.77 W/m^2) in recent years when both the anode and the cathode have the same size (Logan, 2009; Xing *et al.* 2008). It is possible to increase this power density further to ~7 W/m^2 (relative to the anode size) by using much larger cathodes than anodes, but systems with these very large cathodes are unlikely to be practical (Fan *et al.* 2008). It is estimated that substrate diffusion to biofilms of bacteria would not limit power generation until ~15–17 W/m^2 (Logan, 2008, 2009). Thus, it is currently thought that the MFC architecture is limiting power generation, not bacteria.

Work today continues to focus on increasing power densities based on volume of the reactor (volumetric power densities), and reducing the cost of the materials. Typical MFC designs with an air cathode produce ~10–100 W/m^3. Decreasing the space between the electrodes can increase power, but if the electrodes get too close to each other then oxygen contaminates the anode and power decreases (Cheng *et al.* 2006a). In one recent design, where a cloth separator was placed between the anode and cathode to reduce oxygen intrusion and the electrodes were closely spaced together, power reached 2.7 kW/m^3 (Fan *et al.* 2007). Such designs may not be practical for wastewater treatment, however, due to the high cost of the materials used in that system. High performance anodes and cathodes can cost US$1000/$m^2$, so 100 m^2/m^3 of projected area would cost US$100,000 per cubic meter. Obviously these are too high to make systems with these materials practical.

MFCs are being developed that use much less expensive materials. Graphite fiber brush anodes, which consist of electrically-conductive graphite fibers held in a twisted non-corrosive metal core (this looks like a common bottle brush), can have very high surface areas (10,000 to 15,000 m^2/m^3) and low costs (US $0.58/$m^2$) (Logan *et al.* 2007). In tests with these electrodes, power densities of 73 W/m^3 were achieved (Logan *et al.* 2007). It is essential that we reduce the cost of the cathode, and develop cathodes that do not use precious metals, to produce MFC designs that are practical and economical for scale up. The cathode in an MFC typically contains a platinum catalyst for oxygen reduction. Alternative catalysts that use non-precious metals have been developed (Cheng *et al.* 2006b), and bacteria can catalyze oxygen reduction (Clauwaert *et al.* 2007). Recent tests with activated carbon as the catalyst show great promise for low cost cathodes in MFCs (Zhang *et al.* 2009).

5.3.2 Microbial electrolysis cells for hydrogen or methane gas production

With modification of the MFC to exclude oxygen, it is possible to create a system that produces hydrogen or methane gas based on the same principle used to produce hydrogen gas in a water electrolysis system. A hydrogen fuel cell captures the chemical energy from hydrogen gas oxidation and oxygen reduction as electrical energy. A water electrolyzer is essentially the opposite of a fuel cell: electrical energy is put into the system and water is electrolyzed at the anode to produce electrons that go into the circuit, protons that diffuse to the cathode, and oxygen. When two electrons and two protons combine at the cathode, they form hydrogen gas which then bubbles out of the water. Thus, electrical energy is converted to hydrogen gas and oxygen gas at an applied voltage of about 1.8 V. It is important to keep these two gases separated (by a membrane) so that they do not (explosively!) re-combine.

Bacteria on the anode of an MFC break down organic matter to produce electrons and protons, effectively "electrolyzing" organic matter into electrons, protons and carbon dioxide. However, the voltage produced by the anode is insufficient to allow these electrons and protons recombine at the cathode to form hydrogen gas. By adding a small voltage to that produced by the bacteria (>0.2 V), however, it is possible to produce hydrogen gas at the cathode (Figure 5.2). This voltage is only about 1/10th that needed for water electrolysis, and thus hydrogen can be produced in this microbial electrolysis cell (MEC) using much less electrical energy than that needed for water electrolysis.

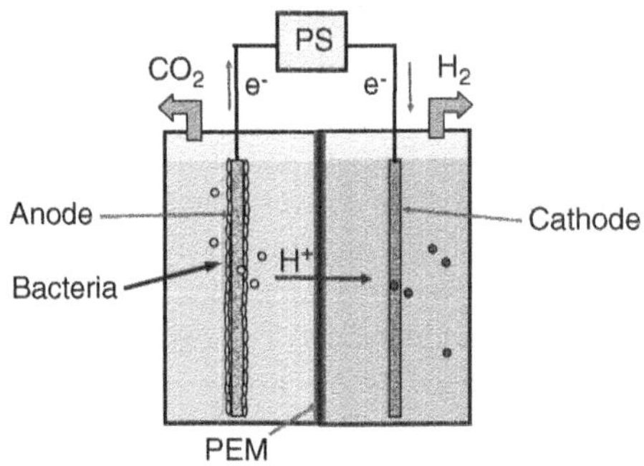

Figure 5.2 A microbial electrolysis cell (MEC) that produces hydrogen gas. An MFC is modified by removing oxygen from the cathode chamber, and adding additional voltage to that produced by bacteria on the anode. The result is that electrons and protons combine on the cathode to form hydrogen gas, that bubbles out of the water (Adapted from Liu *et al.* 2005)

The development of improved MECs has followed a path similar to that of MFCs. Over time, the current densities and efficiencies have been increasing as we gain a better understanding of the factors that affect current generation. Recent MEC designs have increased hydrogen production rates to 3.1 to 6.3 liters per day, per liter of reactor volume (L/L-d), using acetic acid (an end product produced by glucose fermentation), with precious metal catalysts such as Pt on the cathode (Call & Logan, 2008; Tartakovsky *et al.* 2008). While membranes can be placed between the anode and cathode (Figure 5.2) to produce a high purity hydrogen gas at the cathode, these membranes are not essential as the main gases produced (hydrogen and carbon dioxide) do not chemically react with each other. Avoiding the use of precious metals in these systems is essential. It was recently demonstrated that a non-precious metal cathode made with a high surface area stainless steel brush cathode (810 m^2/m^3) could be used with a graphite fiber brush electrode to produce hydrogen at the same rate (based on applied voltages and reactor volume) as that achieved with a similar system that used platinum on the cathode (Call *et al.* 2009). This finding provides a route for generating hydrogen without the need for precious metal catalysts.

The energy recoveries in MECs can be very high. In the MEC with a stainless steel cathode, for example, the energy efficiency relative to the electrical energy input was 221% at an applied voltage of 0.6 V. This efficiency is greater than 100% because the energy in the substrate (acetic acid) is not included in this calculation. When both the electrical and chemical energy are included, the overall energy efficiency was 78%, which is still a very high energy recovery. The electrical energy efficiencies can be even higher, and need to be improved through the use of other cathode catalysts. When Pt is used on the cathode, electrical energy efficiencies have reached 400%. Efficiencies in theory could exceed 900%, allowing an even more effective method capturing energy in organic matter (Logan *et al.* 2008).

The production of methane instead of hydrogen gas in an MEC represents another new opportunity for capturing electrical energy as a gaseous fuel. The conversion of acetate or hydrogen gas into methane is a thermodynamically favourable reaction, and in nature both acetate and hydrogen can be used by methanogenic microorganisms to make methane. It was recently discovered that methanogens can directly accept electrons from an electrode (and thus presumably from other microorganisms) so that an organic fuel (e.g. acetic acid) or hydrogen gas does not need to be produced to make methane (Cheng *et al.* 2009). This ability of methanogens to convert electrical current into methane means that renewable sources of electricity, for example from a windmill or solar panel, could be converted into methane. This fuel could then be used, stored for later use, or transported in trucks or conventional natural gas pipelines. The production of methane using MECs, is at a very early stage compared to MFCs or MECs for hydrogen gas production.

5.3.3 Microbial desalination cells

While the energy generated by MFCs or MECs can be used for water desalination, modification of these reactor designs allows for a more direct method of water desalination based on the concept of electrodialysis (ED). In a conventional ED a high voltage is applied to a stack of cells containing both anion and cation exchange membranes. Water is desalinated by the motion of the ions through these membranes in response to the applied charge and current flow. This concept can be modified to achieve water desalination using the voltage produced by bacteria (Cao *et al.* 2009).

In a microbial desalination cell (MDC), the saline water is placed between an anion (AEM) and cation exchange membrane (CEM) as shown in Figure 5.3. The AEM is placed next to the anode, and the CEM next to the cathode. Bacteria grow on the anode using organic matter (for example in a wastewater) and produce electrical current and release protons in to the water. The protons cannot move to the cathode, however, because they cannot diffuse through the AEM as only negatively charge ions can pass through this membrane. In order for charge balance to occur, anions (*i.e.* Cl^-) in the middle desalination chamber flow into the anode. At the cathode, protons are removed from water, and so Na^+ ions in the desalination chamber move into the cathode chamber to balance charge. As a result of this process, the NaCl salt in the middle chamber between the AEM and CEM electrodes is removed, and thus the water is desalinated.

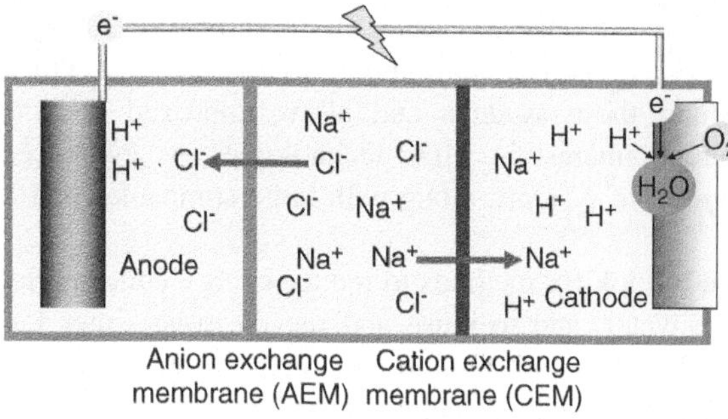

Figure 5.3 Schematic of a three-chamber microbial desalination cell (MDC) that can produce electricity and desalinate water in the middle chamber. Bacteria growing on the anode produce electrons and protons, causing a chloride ion to move into the anode chamber to balance charge. The reduction of oxygen at the cathode removes a proton, so a sodium ion moves into the cathode chamber to balance charge. This process results in water desalination in the middle chamber (from Mehanna *et al.* 2010b)

There are multiple advantages for desalinating water using an MDC. First, water is desalinated without the need for electricity. Second, wastewater or any source of biodegradable organic matter can be used as a source of renewable energy for desalination. Third, electrical energy is generated, which makes it possible to gain additional energy for water pumping and other processes. The fourth advantage of the MDC has to do with improving the treatability of the wastewater. High solution conductivities (*i.e.* saltier water) can improve power generation in an MFC because it improves charge transfer in the water and reduces the internal resistance of the system. The electrical conductivity of many wastewaters is quite low (1 mS/cm or lower), and thus power generation is limited using real wastewaters compared to laboratory systems that operate with solutions with much higher conductivity (5 to 20 mS/cm). As the water is desalinated, the solution conductivity of the wastewater increases, allowing higher current densities. Thus, this may help more effectively treat the wastewater.

The idea of water desalination using an MDC has just been discovered (Cao *et al.* 2009), so it is too early to know how cost effective this approach to water desalination will be compared to others. However, this technology is rapidly advancing as shown by development of stacked MDCs (Kim & Logan, 2011c).

5.3.4 New MFC-based technologies

Technologies that are based on electrical current generated by exoelectrogenic bacteria continue to be invented and refined. For example, integration of a reverse electrodialysis (RED) stack of membranes into the MFC between the anode and cathode allows current and voltage to be increased well beyond that possible using only an MFC. This new MFC-RED technology, called a microbial reverse electrodialysis cell (MRC), increased power production to 4.2 W/m^2, and produced a voltage of 1.2 V (Kim & Logan, 2011b).

Other technologies being developed include production of value added products, such as hydrogen peroxide (Rozendal *et al.* 2009), caustic (Rabaey *et al.* 2010) and acetate (Nevin *et al.* 2010).

5.4 CONCLUSIONS AND OUTLOOK

The next level of development of these technologies requires pilot-scale testing. There have been tests of a larger scale (1000 L) MFC in Australia (Logan, 2010), and an MEC at a winery in Napa County, California (Cusick *et al.* 2011). Hopefully, many more larger-scale tests will follow, allowing engineers to gain experience in running these systems and allow improved designs and performance. It is encouraging to see commercial interest in MFC technologies, as evidenced by several new startup companies working to design these systems, along with large companies that are global leaders in water infrastructure technologies.

It is important to continue to look for methods to reduce energy consumption at water and wastewater treatment plants, to conserve water, and to reuse and recycle wastewater for the maximum benefit to society. We must continue to innovate and explore new methods for conservation and treatment, such as using wastewater as a source and not sink of energy.

The MFC-based technologies that have been described here are just emerging from the laboratory but they show tremendous promise as a new approach for wastewater treatment. It is still too early to know when they will become efficient and economical methods for wastewater treatment and energy production, but it is important that we continue to develop these and other techniques through this new century, and beyond, to ensure an adequate supply of energy for our water infrastructure.

Acknowledgement

This chapter is a modified version of the written Athalie Richardson Irvine Clarke Prize lecture by the author, delivered on July 9, 2009, in Newport Beach, California. The Clarke Prize is given for lifetime achievements, and thus the acknowledgements given here are consistent those of the lecture. The work described here was made possible through the tremendous support of my family, colleagues and students. First and foremost, I thank the loving support of my wife Angela, and my children, Alex and Maggie. Second, my research has benefitted from my research collaborations over the years on a variety of topics, ranging from studies of particles in oceans to environmental remediation. I could not have worked in many different fields had it not been possible to collaborate with people having expertise in other scientific and engineering areas. I would like to especially thank three people for their wonderfully collaborative and team-based approach to research: Professor Alice Alldredge at UC Santa Barbara, for working with me when I was a young assistant professor and for collaborations in oceanography; Professor Robert Arnold, my colleague when I was at the University of Arizona for more than a decade, for his pioneering work on iron reducing bacteria as that helped to educate me on how bacteria could respire using solid substrates; and Professor John Regan at Penn State, who works with the MFC group and whose expertise in microbiology and molecular biology methods continues to help us advance our understanding of exoelectrogenic bacteria and microbial fuel cells. I am indebted to the support of several research sponsors for the research discussed here, most notably the National Science Foundation, Air Products and Chemicals, Inc., and the generous support from Award KUS-I1-003-13 by the King Abdullah University of Science and Technology (KAUST). I would especially like to thank Mrs. Joan Irvine Smith and the Irvine Family, and the National Water Research Institute for honouring my work with the Clarke Prize.

REFERENCES

Call D. and Logan B. E. (2008). Hydrogen production in a single chamber microbial electrolysis cell (MEC) lacking a membrane. *Environ. Sci. Technol.*, **42**(9), 3401–3406.

Call D., Merrill M. D. and Logan B. E. (2009). High surface area stainless steel brushes as cathodes in microbial electrolysis cells (MECs). *Environ. Sci. Technol.*, **43**(6), 2179–2183.

Cao X., Huang X., Liang P., Xiao K., Zhou Y., Zhang X. and Logan B. E. (2009). A new method for water desalination using microbial desalination cells. *Environ. Sci. Technol.*, **43**(18), 7148–7152.

Cheng S., Liu H. and Logan B. E. (2006a). Increased power generation in a continuous flow MFC with advective flow through the porous anode and reduced electrode spacing. *Environ. Sci. Technol.*, **40**, 2426–2432.

Cheng S., Liu H. and Logan B. E. (2006b). Power densities using different cathode catalysts (Pt and CoTMPP) and polymer binders (Nafion and PTFE) in single chamber microbial fuel cells. *Environ. Sci. Technol.*, **40**, 364–369.

Cheng S., Xing D., Call D. F. and Logan B. E. (2009). Direct biological conversion of electrons into methane by electromethanogenesis. *Environ. Sci. Technol.*, **43**(10), 3953–3958.

Clauwaert P., Van der Ha D., Boon N., Verbeken K., Verhaege M., Rabaey K. and Verstraete W. (2007). Open air biocathode enables effective electricity generation with microbial fuel cells. *Environ. Sci. Technol.*, **41**(21), 7564–7569.

Cooper N. B., Marshall J. W., Hunt K. and Reidy J. G. (2007). Less power, great performance. *Water Environ. Technol.*, **19**(2), 63–66.

Cusick R. D., Bryan B., Parker D. S., Merrill M., Mehanna M., Kiely P. D., Liu G. and Logan B. E. (2011). Performance of a pilot-scale continuous flow microbial electrolysis cell fed winery wastewater. *Appl. Microbiol. Biotechnol.*, **89** (6), 2053–2063.

Fan Y., Hu H. and Liu H. (2007). Enhanced coulombic efficiency and power density of air-cathode microbial fuel cells with an improved cell configuration. *J. Power Sources*, **171**(2), 348–354.

Fan Y., Sharbrough E. and Liu H. (2008). Quantification of the internal resistance distribution of microbial fuel cells. *Environ. Sci. Technol.*, **42**(21), 8101–8107.

Kim Y. and Logan B. E. (2011a). Hydrogen production from inexhaustible supplies of fresh and salt water using microbial reverse-electrodialysis electrolysis cells. *Proc. Natl. Acad. Sci. USA*, **108**(39), 16176–16181.

Kim Y. and Logan B. E. (2011b). Microbial reverse electrodialysis cells for synergistically enhanced power production. *Environ. Sci. Technol.*, **45**(13), 5834–5839.

Kim Y. and Logan B. E. (2011c). Series assembly of microbial desalination cells containing stacked electrodialysis cells for partial or complete seawater desalination. *Environ. Sci. Technol.*, **45**(13), 5840–5845.

Kim B. H., Kim H. J., Hyun M. S. and Park D. H. (1999). Direct electrode reaction of Fe(III)-reducing bacterium, Shewanella putrefaciens. *J. Microbiol. Biotechnol.*, **9**(2), 127–131.

Liu H., Ramnarayanan R. and Logan B. E. (2004). Production of electricity during wastewater treatment using a single chamber microbial fuel cell. *Environ. Sci. Technol.*, **38**(7), 2281–2285.

Logan B. E. (2008). Microbial Fuel Cells. John Wiley & Sons, Inc., Hoboken, NJ.

Logan B. E. (2009). Exoelectrogenic bacteria that power microbial fuel cells. *Nature Rev. Microbiol.*, **7**(5), 375–381.

Logan B. E. (2010). Scaling up microbial fuel cells and other bioelectrochemical systems. *Appl. Microbiol. Biotechnol.*, **85**(6), 1665–1671.

Logan B. E. and Regan J. M. (2006). Electricity-producing bacterial communities in microbial fuel cells. *Trends Microbiol.*, **14**(12), 512–518.

Logan B. E., Aelterman P., Hamelers B., Rozendal R., Schröder U., Keller J., Freguiac S., Verstraete W. and Rabaey K. (2006). Microbial fuel cells: methodology and technology. *Environ. Sci. Technol.*, **40**(17), 5181–5192.

Logan B. E., Cheng S., Watson V. and Estadt G. (2007). Graphite fiber brush anodes for increased power production in air-cathode microbial fuel cells. *Environ. Sci. Technol.*, **41**(9), 3341–3346.

Logan B. E., Call D., Cheng S., Hamelers H. V. M., Sleutels T. H. J. A., Jeremiasse A. W. and Rozendal R. A. (2008). Microbial electrolysis cells for high yield hydrogen gas production from organic matter. *Environ. Sci. Technol.*, **42** (23), 8630–8640.

Mehanna M., Kiely P. D., Call D. F. and Logan B. E. (2010a). A microbial electrodialysis cell for simultaneous water desalination and hydrogen gas production. *Environ. Sci. Technol.*, **44**(24), 9578–9583.

Mehanna M., Saito T., Jingling Y., Hickner M. A., Cao X., Huang X. and Logan B. E. (2010b). Using microbial desalination cells to reduce water salinity prior to reverse osmosis. *Energy & Environ. Sci.*, **3**(8), 1114–1120.

Nevin K. P., Woodard T. L., Franks A. E., Summers A. M. and Lovley D. R. (2010). Microbial electrosynthesis: feeding microbes electricity to convert carbon dioxide and water to multicarbon extracellular organic compounds. *mBio*, **1** (2), 1–4.

Nowak O., Keil S. and Fimml C. (2011). Examples of energy self-sufficient municipal nutrient removal plants. *Wat. Sci. Tech.*, **64**(1), 1–6.

Rabaey K., Butzer S., Brown S., Keller J. and Rozendal R. A. (2010). High current generation coupled to caustic production using a lamellar bioelectrochemical system. *Environ. Sci. Technol.*, **44**(11), 4315–4321.

Rozendal R. A., Leone E., Keller J. and Rabaey K. (2009). Efficient hydrogen peroxide generation from organic matter in a bioelectrochemical system. *Electrochem. Commun.*, **11**(9), 1752–1755.

Semiat R. (2008). Energy issues in desalination processes. *Environ. Sci. Technol.*, **42**(22), 8193–8201.

Stokes J. R. and Horvath A. (2009). Energy and air emission effects of water supply. *Environ. Sci. Technol.*, **43**(8), 2680–2687.

Tartakovsky B., Manuel M. F., Wang H. and Guiot S. R. (2008). High rate membrane-less microbial electrolysis cell for continuous hydrogen production. *Int. J. Hydrogen Energy*, **34**, 672–677.

WIN (2001). Water infrastructure now: recommendations for clean and safe water in the 21st century.

Xing D., Zuo Y., Cheng S., Regan J. M. and Logan B. E. (2008) Electricity generation by Rhodopseudomonas palustris DX-1. *Environ. Sci. Technol.*, **42**(11), 4146–4151.

Zhang F., Cheng S., Pant D., Bogaert G. V. and Logan B. E. (2009). Power generation using an activated carbon and metal mesh cathode in a microbial fuel cell. *Electrochem. Commun.*, **11**(11), 2177–2179.

Part 2

Energy Footprint of Wastewater Treatment

By The Editors

FOREWORD

Historically, the objective of wastewater treatment has been to protect human health by the implementation of centralized sewerage systems and large treatment works. Nowadays, these objectives have been reviewed and wastewater is being considered as an alternative water resource, as well as a source of energy and nutrient recovery. As a consequence, new generations of wastewater treatment technologies not only have to consistently achieve water quality objectives at acceptable costs, but also integrate the new concerns of energy efficiency and reduced carbon footprint into their operations.

Part 2 of this book "Energy Footprint of Wastewater Treatment", (e.g. Chapters 6 to 10), gives some examples of how to reduce the carbon and environmental footprints of wastewater treatment, recovering heat and green energy and improving water quality for water recycling.

Achieving energy positive wastewater treatment, minimising thus greenhouse gas emissions, is becoming a reality not only for developed countries, mainly in Europe, but also in the developing world, as shown from the successful experience of the As Samra wastewater treatment plant in Jordan. Nowadays, a number of best practices and innovative technologies are enabling wastewater treatment processes to reduce energy consumption, recover chemically embodied energy to produce more biogas, recover potential and thermal energy, as well as produce electricity and biofuels in wastewater treatment facilities.

To meet the increasingly stringent regulations for effluent discharge and water reuse, membrane bioreactors (MBRs) have become more popular for the treatment of municipal and industrial wastewater. Despite the undeniable advantage of producing high quality water, free of suspended solids, with a small footprint, this relatively new technology is disadvantaged by its high energy consumption.

A typical MBR energy consumption is in the range of 0.5–1.9 kWh/m^3, which is up to 100% higher than conventional biological treatment. Several strategies for minimising the energy consumption of MBRs are presented, including best practices for operation, improved design and feedback from full-scale operations. Promising options for enhancement of energy efficiency include automatically attenuated aeration, membrane flux control, and effective membrane cleaning. The large variation of energy usage from one

MBR to another is primarily linked to hydraulic loads, especially in the case of combined sewers or plants with high flow variations.

In the context of global warming concerns, wastewater treatment facilities will increasingly be converted to energy recovery plants capable of harvesting the potential energy available in the organic carbon in wastewater. However, the initiatives to reduce energy consumption can impact the emission of greenhouse gasses other than CO_2, for example, N_2O and CH_4. Some of the short-term measures for energy saving and new treatment configurations are associated with a risk of increased N_2O emission from the activated sludge process or CH_4 emission from sludge handling. This indicates the need for the development and implementation of appropriate monitoring strategies for greenhouse gasses other than CO_2 to avoid any negative impacts of enhanced energy efficiency on the global environmental footprint of wastewater treatment.

Chapter 6

Toward energy self-sufficiency of wastewater treatment

Valentina Lazarova, Carlos Peregrina and Pascal Dauthuille

6.1 INTRODUCTION

Over recent years, the World Water Vision has shifted drastically. Certainly, the most significant milestone of the 20th century is the recognition of the multiple values of water in terms of sustainable development. The Dublin and Rio principles (1992) on sustainable use of water resources have been adopted internationally and completed by new initiatives proposed during the next World Water Forums. Another very important milestone is the recognition of sanitation as important element of sustainable human development, as stated in the recommended targets of the World Summit on Sustainable Development in Johannesburg (2002).

Historically, the objective of wastewater treatment has been to protect human health by the implementation of centralized sewerage systems and large treatment works. Nowadays, these objectives are being revisited and wastewater is considered not only a competitive alternative resource for water reuse and environmental protection (Asano *et al.* 2007), but also as a source of energy and nutrient recovery (GWRC, 2008; Keller, 2008; McCarty *et al.* 2011). Wastewater contains organic matter and nutrients, and also conveys heat and mechanical energy, all of which are renewable energy sources which can be produced in wastewater treatment plants (Meda & Cornel, 2010). Wastewater treatment plants (WWTPs) have the potential to become environmental platforms and to provide an energy source for tomorrow's eco-cities as part of a system characterised by the smallest possible ecological footprint (GWRC, 2011).

The growing interest in the energy embedded in wastewater is a consequence of the increasing pressure on energy supply caused by the fluctuation in energy costs (EIA, 2012) together with climate change concerns (ESSEA, 2010) (Figure 6.1). Fuel and related costs have increased significantly for the period 2001–2006 and fluctuated sharply over the period 2007–2008, with a US$100 difference per barrel, thus strongly affecting global energy, stock markets and industrial profitability. The need for alternatives to fossil fuels is also emphasised by the increasing energy demand. From an economic and social point of view, global energy demand has risen in such a manner that a failure in the petroleum demand/production ratio was revealed in 2007 (International Energy Agency, 2007). From an environmental point of view, fossil fuels are, by definition, limited resources, although they represent today 81% of the world energy production (35% for oil, 21% for gas and 25% for coal). Furthermore, the combustion of fossil fuels produces pollution with substantial environmental impacts and costs.

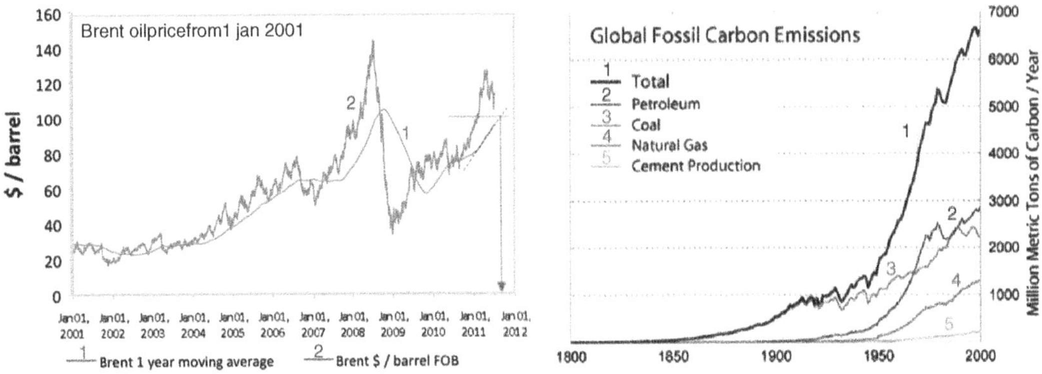

Figure 6.1 Crude oil prices chart and global fossil carbon emissions (*Sources*: ESSEA, 2010; EIA, 2012)

In 2008, the European Council approved a legislative package of measures for a renewed energy policy and global warming reduction strategy (Tilche & Galatola, 2007; European Commission, 2008). The Heads of State of the 27 EU countries took the deliberation of unilaterally cutting their greenhouse gases (GHG) emissions by 20% by 2020, with the mandate of negotiating with the other main World partners to reach even more ambitious objective of a 30% decrease by 2020. EU leaders also set a binding overall goal of 20% for renewable energy sources by 2020, compared to the present 6.5%. A binding minimum target of 10% for the share of biofuels in overall transport petrol and diesel consumption by 2020 was also set.

With the increasing concern of climate change and water scarcity, there has been a paradigm shift over the last few years, from treating wastewater effluent to recycling/reusing wastewater. The time has come to reveal the sustainable energy potential of wastewater treatment plants and networks, helping to reduce the ecological footprint of sanitation activities.

Wastewater contains organic matter that constitutes a potential energy source: a large portion of the calories from the inflow, energy from hydraulics and green energy from biomass are all recoverable during the treatment process.

It is also an important issue to optimise and rationalize the consumption of energy needed for wastewater treatment. At last, all of these sources of energy, or even green energy sources linked to wastewater treatment, can be combined with external renewable energy sources (wind, solar, geothermal) to achieve the following main goals:

- Minimizing the energy needed for wastewater treatment.
- Significantly reducing the carbon footprint associated with the wastewater treatment process.
- Making wastewater treatment self-sufficient.

6.1.1 Energy consumption of wastewater treatment plants

It's well documented that energy costs represent a primary expense for wastewater treatment operations, typically second only to labour costs (Lindtner *et al.* 2004, 2008; Lazarova & Kamisoulis, 2005). The energy costs to operate a wastewater treatment facility are highly variable according to local technical and economic contexts. Electricity consumption may range from 0.2 to 1.5 kWh/m^3 based upon the treatment process employed and the level of treatment required, with the general trend that unit power increases as effluent water quality requirements become more stringent(Lazarova & Kamisoulis, 2005;

Meda & Cornel, 2010; GWRC, 2011). For example, a facility employing process treatment solely for carbon removal will be at the lower end of this range whereas plants operating MBR technology to achieve carbon and nitrogen removal along with stringent limits on suspended solids and bacteriological quality will, on average, consume up to three times more energy.

Aeration, which is used to biologically remove soluble pollutants, namely, carbon, nitrogen and phosphorus, is the most energy consuming process in wastewater treatment plants accounting for more than 50% of the total energy requirements (Bischof *et al.* 1996). Additional energy is required to pump the sewage and remove settled solids. Only a minor fraction of the input pollution leaves the process within the treated water in compliance with the wastewater discharge or reuse standards.

Energy consumption is a major component of wastewater operating costs and depends on plant size, water treatment train and sludge processing. Large variations have been reported from 9% to 52% of the total operating costs, with typical value is $25 \pm 5\%$ for conventional activated sludge plants (Balmér, 2000; Lindtner *et al.* 2004, 2008; Lazarova & Kamisoulis, 2005; Jonasson, 2007). The type of sewer system (combined or separated) and the degree of plant utilization greatly influence the specific operation costs and energy consumption, especially in tourist areas, where the ratio between hydraulic loading in the summer and winter can vary with a factor of 7–10. For this reason, several performance indicators are used in different countries and for various purposes, such as kWh/p.e. (population equivalent calculated on the basis of the specific daily pollution of one person COD, BOD_5 or m^3, calculated as an average value for a given period of time), kWh/m^3 or kWh/kg BOD_5 removed (on yearly basis).

Thus for a given facility, energy requirements may range from 1 kWh/kg BOD_5 with a basic treatment removing only carbonaceous pollution and portion of the suspended solids to 5 kWh/kg BOD_5 for most sophisticated treatment such as that employed in membrane bioreactors, which produces high-quality treated water, almost completely disinfected and free of suspended solids (Figure 6.2).

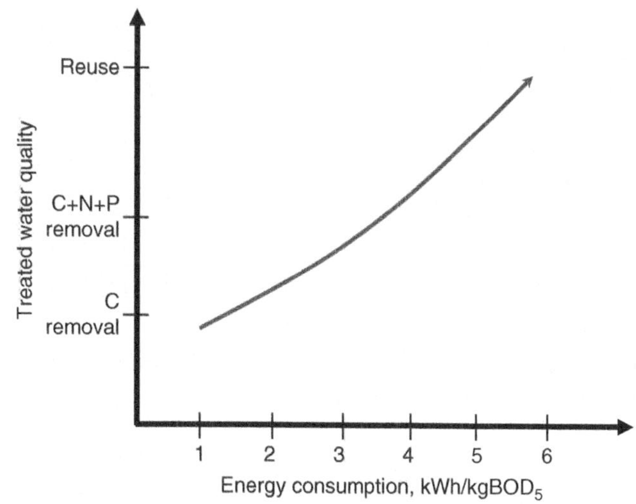

Figure 6.2 Energy consumption in WWTPs for different levels of treatment for WWTPs in France

Typical energy consumption levels of activated sludge plants in France are in the range of 2.0 ± 0.5 kWh/kg BOD_5 for carbon removal and 3.0 ± 0.5 kWh/kg BOD_5 activated sludge with nitrification/denitrification and capacity more than 50,000 p.e. Specific energy consumption per unit of water volume is an useful benchmark in comparing selected processes. For WWTPs, pumping head has a great importance and typical values vary in a wide range from 0.4 to 1.2 kWh/m^3. The benchmark used in several European countries (Austria, Germany, Sweden, Switzerland) is expressed as population equivalent (p.e.) corresponding to 110 gCOD/d or 60 $gBOD_5/d$ for one year period, ($pe_{110} \cdot yr$). Typical energy consumption of large WWTPs (over 100,000 p.e.) in Austria and Germany is 23 $kWh/pe_{110} \cdot yr$, while higher value of 42 $kWh/pe_{110} \cdot yr$ has been reported in Sweden, due probably to more diluted wastewater (Jonasson, 2007). Energy input rises with the increasing level of treatment and nutrient removal, as well as with the decreasing of the plant size. In addition to plant specificity, the difference in the applied methodology and wastewater characteristics for the estimate of energy consumption makes any comparison of the results very difficult.

Figure 6.3 illustrates such an example, comparing the results of on-site energy audits of several full scale WWTPs in Europe, with in-depth analysis of the plant's performance and measurement of the power use and power factor for all major electromechanical equipment such as motors and blowers (Martin & Aguilera, 2011). The majority of the facilities are activated sludge plants with combined sewers, designed for carbon and nitrogen removal with few tertiary treatment systems, and membrane bioreactors (MBR). Compared to design capacity, hydraulic loads vary from 25% to 125% and BOD_5 loads from 26 to 103%.

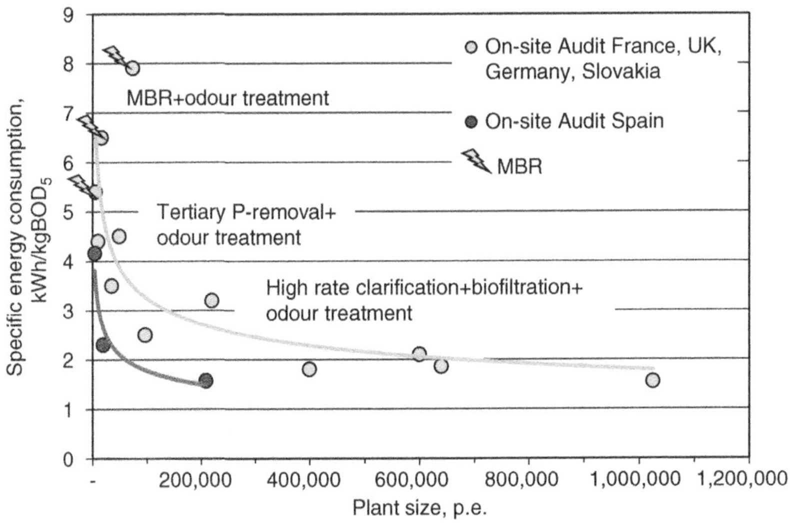

Figure 6.3 Distribution of specific energy consumption of different wastewater treatment plants (predominantly activated sludge) depending on plant capacity

The specific energy consumption of the WWTPs in Spain from the survey appeared lower than the values from the other on-site audits, 1.28 ± 0.63 kWh/kg BOD_5 and 2.84 ± 1.09 kWh/kg BOD_5, respectively (MBR plants not included). For this indicator, MBR facilities had significantly higher values of 5.4 to 7.9 kWh/kg BOD_5. The specific energy consumption by unit volume varied from 0.35 to 1.16 kWh/m^3, with values of 0.9, 1.1 and 2.5 kWh/m^3 for MBR facilities. As a rule, the addition of a tertiary treatment or odour removal accounted for an increase in energy consumption by 10–15%.

A major influence of sludge treatment was not observed, including plants with anaerobic digestion.

6.1.2 Energy recovery by anaerobic digestion

In the European context of sustainable development, anaerobic digestion (AD) is becoming a key method for both waste reduction and for recovery of a renewable fuel and other valuable co-products such as compost (EuroObserv'ER, 2010). In 2009, 25.2 TWh was produced from biogas, which is an increase of 17.9% as compared to 2008. The WWTP contribution was 9.38%. This increase in biogas production and the diversification of energy resources and technologies is supported by European policy. The cogeneration promotion (Directive 2004/8/CE) leads to significant subvention for energy valorisation of biomass (Belgium, France, the Netherlands, Spain, UK) and specific purchase prices (Germany, Italy). Hence, the increase of anaerobic digestion efficiencies (sludge reduction, biogas production) is an attractive way in terms of waste and wastewater treatment benefits, energy efficiency and environmental protection.

Nowadays, the economic factor is favouring the development of new methods for renewable energy production, in particular the improvement of biogas production from sludge by partly disintegrating the sludge prior to digestion (Table 6.1, Camacho *et al.* 2009a). Hence, a reconsideration of these different technologies is necessary to support technical, economic and environmental aspects and feedback from full operation.

Table 6.1 Summary of different available processes for the improvement of anaerobic digestion (Camacho *et al.* 2009a).

Pre treatment technology	Description
Thermal	High temperature to enhance hydrolysis
Mechanical	High pressure or pressure shift to generate cavitation
Lysing centrifuge	High shear force to disintegration
Ultra sound	Ultrasonic wave energy to generate cavitation
Ozonation	Oxidation by ozone to increase solubility and biodegradability
Biological	Enhance biological activity by optimizing reaction condition
Combined	Chemical-mechanical or biological-thermal

The increase of anaerobic digestion efficiency results not only in the reduction of disposed waste volume and weight, but also in the reduction of digester size, which decreases the threshold size of wastewater treatment plants (50,000 p.e.) for its implementation (Camacho *et al.* 2010b). On the other hand, biogas produced from AD is considered a renewable energy source. Therefore, an increase in biogas valorisation would not only bring a net energy producing process, but would also open a new renewable energy market for the wastewater sector. Furthermore, increasing AD efficiency would have an indirect positive impact on reduction of CO_2 and CH_4 emissions, odour control and generation of sanitised final product.

Energy recovery from wastewater for biogas production can be improved not only by anaerobic digestion of sludge, but also by anaerobic treatment of wastewater. Upflow anaerobic sludge blanket (UASB) reactors are widely used for wastewater treatment, mainly for industrial applications, in Europe and Brazil (Seghezzo *et al.* 1998; Verstraete *et al.* 2009). New technologies such as complete anaerobic secondary treatment combined with membrane filtration would allow the achievement of good treatment efficiency, lower cost and of the doubling of CH_4 production as compared to conventional aerobic treatment (McCarty *et al.* 2011). As demonstrated by recent pilot studies in Singapore, combined anaerobic (UASB) – aerobic treatment of wastewater allows to decrease oxygen demand by 35% increasing at the same time the CH_4 production by 18% (Cao, 2011).

The current state of wastewater treatment technology offers a challenging opportunity for significant energy savings, with the opportunity for production of renewable energy in order to reduce both dependence on fossil fuels and greenhouse gas (GHG) emissions. Using current best practices and new technological solutions, a strong improvement in energy efficiency can be expected.

6.1.3 The concept of positive energy plant

The ideal of self-sufficient or even energy producing wastewater treatment is becoming an ambitious objective and visionary target to reduce the carbon footprint of water industry (GWRC, 2008, 2010, 2011). The label "Positive energy plant" which refers to energy self-sufficient wastewater treatment plants should be considered as a series of "toolboxes" (Figure 6.4), including innovative technologies, tools and best practices. These "toolboxes" need to be developed and optimised by fundamental and applied research, as well as by pilot studies and feedback from full-scale operation. The choice of the most appropriate solution for a given wastewater treatment facility will depend on the target objectives, process characteristics and local conditions. This approach can be applied to the design and construction of new "Positive Energy Plants" or to enhance the energy efficiency of existing facilities.

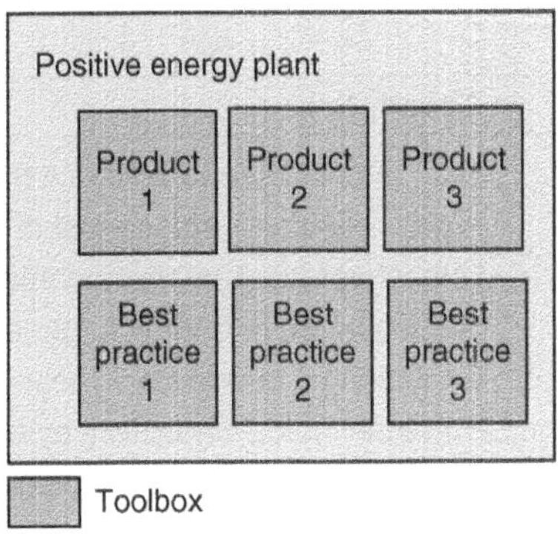

Figure 6.4 Positive energy plant label

The approach developed by Suez Environnement to achieve the challenge of the "Positive Energy Plant" consists of four main areas for the improvement of energy efficiency of WWTPs towards energy self-sufficiency, as illustrated on Figure 6.5. In fact, these zones of improvement can be considered as the major components of the "Positive Energy Plant", namely (1) Energy savings, (2) Energy from sewage flows, (3) Energy generation from sludge and (4) Renewable energy.

Energy recovery from sewage flows aims to valorise potential (kinetic) and thermal energy embodied in wastewater. However, energy yield is relatively limited and depends on local conditions as discussed in the next sections. Chemically bound energy from the organic matters in wastewater has the highest potential for recovery (Meda & Cornel, 2010).

Energy balance depends on organic concentration of wastewater, on country's specific water consumption per capita, type of sewers, percentage and type of industrial effluents and other local conditions. Figure 6.6 illustrates exemplary energy balances for a typical WWTP in France, compared with the exemplary energy balance provided by Meda *et al.* in Chapter 2. In both cases, the treatment processes are the same, including primary settling, nitrification/denitrification, anaerobic sludge stabilisation and biogas valorisation. The main assumptions for the French example are COD of 132 g/cap · d (per inhabitant (capita) and per day, on the basis of the typical COD/BOD_5 ratio), methane production of 0.35 L CH_4/gCOD and energy potential of methane of 10 kWh/m^3. Despite the slightly more concentrated wastewater in the French effluents, the energy balance is quite similar: 57% (96 kWh/cap · yr) are transferred to the digester and only 26% (43 kWh/cap · yr) are conveyed into methane. Using the same conversion efficiency of 32%, only 8.9% of the embodied chemical energy is recovered as electricity (15 kWh/cap · yr).

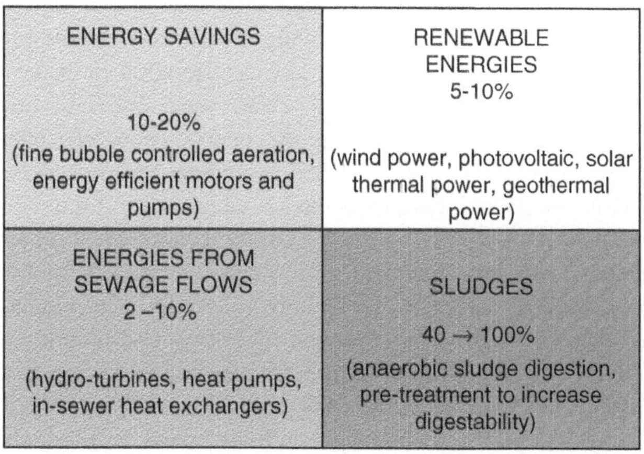

Figure 6.5 Major components of the "Positive Energy Plant", that is zones of improvement of energy efficiency of wastewater treatment

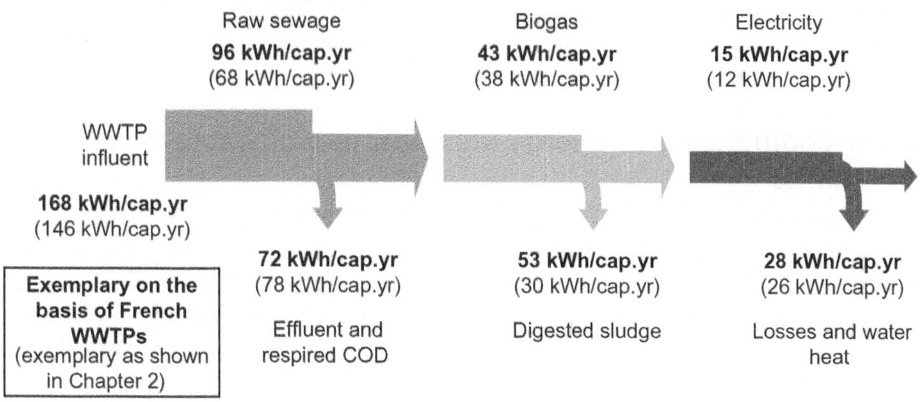

Figure 6.6 Schematic energy COD-balance for a typical WWTP in France and exemplary energy balance as shown by Meda *et al.* in Chapter 2

Energy optimization of a given WWTP will require a global analysis of the wastewater treatment train that is needed in respect to the current water quality requirements, as well as the best integration of the energy concerns. The aim is threefold: to improve energy efficiency, to maximize the use of sludge for energy production and to recover energy from internal or external sources. Achieving energy self sufficiency of wastewater treatment is a major challenge – how to achieve it will be discussed and illustrated in the following section of this chapter.

In addition to the more conventional options of energy savings and biogas production, new options for energy recovery and renewable energy sources have to be explored. In any case, the only valid solutions would be site-specific and would be tailored for each case.

6.2 ENERGY SAVING

In order to achieve the objectives defined in the European Union's Green Paper on reduction of energy consumption by 20% by 2020, a number of energy savings "toolboxes" would provide WWTP operators with effective technical solutions, and this, without any deterioration of water quality or the level of service provided to consumers (Lindtner *et al.* 2004, 2008; Wett, 2007; Lazarova *et al.* 2010; Novak *et al.* 2011; Cao, 2011). The main objective is to use energy more efficiently by means of better technologies and optimized performance of the installed equipment and processes.

Energy reduction is being widely pointed out as the key element for a more energy sustainable future because it can be implemented immediately by using existing technologies. As a rule, an on-site audit of energy consumption, plant's loads and treatment performance is needed for the selection of the most appropriate best practices, tools or products for energy reduction. The practical experience of Suez Environnement indicates that up to 10–20% reduction of energy consumption is achievable for many existing WWTPs at relatively low capital expenditure. Similar energy saving potential was identified in Sweden (Rydh & Åkesson, 2007), while up to 30% and 50% of energy saving potential hase been reported for wastewater treatment plants in Germany and Switzerland, respectively (Jonasson, 2007).

Two types of measures for improvement of energy efficiency can be implemented in existing WWTPs:

- Incremental corrective measures, including dissolved oxygen control, proper use and maintenance of pumps and main motors, implementation of VFD control of motors, peak demand management, etc.
- Renewal/replacement of old equipment, such as, for example the replacement of antiquated and, inefficient equipment with high-efficiency or premium efficiency motors, blowers, ventilation devices and/or motor drivers, lighting devices, and so on.

6.2.1 Major components of energy consumption of wastewater treatment plants

As mentioned previously, energy consumption varies considerably between treatment processes and wastewater facilities. Typically, the most energy consuming processes in WWTPs are aeration for biological treatment, and pumping, mechanical treatment and ventilation for odour control. Figure 6.7 illustrates a typical breakdown of energy consumption for French conventional wastewater treatment plants with nutrient removal (Martin & Aguilera, 2011). Tertiary treatment, in particular membrane processes, require more energy for production of recycled water, which rises with the increase of the level of treatment for removal of pathogens, and trace organics and salts.

Aeration of activated sludge is characterised with the highest energy consumption for conventional wastewater treatment facilities, typically accounting for $45 \pm 15\%$ of total energy needs depending on plant size, design and additional treatment processes. A similar value of 42.4% is reported for the Ulu Pandan water recycling plant in Singapore – a 361,000 m^3/d, facility which includes three different treatment lines, odour treatment facilities and a 25,000 m^3/d MBR (Cao, 2011). A higher typical value of 54.1% is reported in the US for conventional secondary treatment with chlorination, but without odour control (WERF, 2010b).

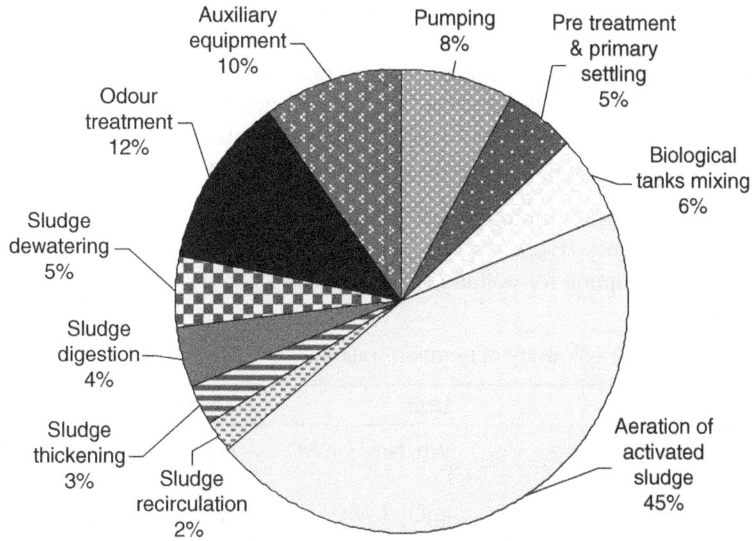

Figure 6.7 Typical energy consumption breakdown of a conventional wastewater treatment plant (100,000 p.e., France)

Even if advanced patented models allow to estimate the real aeration demand for endogenous respiration, carbon and nitrogen removal as a function of environmental conditions and operating parameters (mass loading, sludge age, sludge concentration, temperature), on-line control of dissolved oxygen is not enough for the optimisation of aeration efficiency. Other technological options have to also be considered, such as the use of energy efficient blowers, variable frequency drive (VFD) control of motors, control of diffusers' aging, and so on.

In recent years, membrane bioreactors have been increasingly implemented for wastewater treatment at small to large scale, due to their advantages of small footprint, high reliability of operation, automation and the production of almost completely disinfected effluent free of suspended solids. In MBR system, overall aeration demand is increased by the membrane air scouring, typically accounting for $65 \pm 10\%$ of WWTP energy consumption (Lazarova *et al.* 2009; Barillon *et al.* 2011).

In addition to the analysis of the energy consumption of the various treatment steps and equipment, another important element to be taken into account is the hydraulic and mass loading of the facility. As a rule, WWTPs are characterised by high load variations. At low loads, specific energy consumption can be up to 50–300% higher than the value under nominal loads.

6.2.2 Improvement of aeration system's efficiency

Significant energy savings are frequently realized through implementation of automated aeration control via integration of appropriate sensors, blowers with variable speed drives and appropriate control logic in order to match oxygen delivered with the regular variations in system requirements. Another key opportunity is to implement practices which permit operational flexibility – for example aeration tanks may be taken out of service during periods of low flow or loading or high temperature. Of course, such options are best validated via dynamic modelling simulation prior to implementation. Furthermore, enhanced understanding of process interactions is critical in order to optimise energy consumption while ensuring treatment performance and minimizing greenhouse gas production at the same time.

The aeration system of a typical wastewater treatment plant is divided into several units that can all be optimised. In France, a new methodology has been developed and implemented for improving the energetic efficiency of the aeration system of activated sludge plants (Carrand & Martin Ruel, 2006; Besnault *et al.* 2011), focussing on three specific ratios of energy consumption (Table 6.2):

- Specific energy consumption for air production (E_{blow}),
- Specific aeration efficiency (E_{aer}),
- Specific energy consumption for pollution removal (E_{BOD}).

Table 6.2 Indicators for energy efficiency of aeration (ratios).

Indicator	Unit	Target
Specific energy consumption for air production, E_{blow}	$Wh/Nm^3 \cdot mWC$	Blowers 4.5–5.0 for roots
Specific aeration efficiency, E_{aer}	kgO_2/kWh	Aeration devices 1.4–1.5 for new FBM*
Specific energy consumption for pollution removal E_{BOD}	$kWh/kgBOD_5$	Pollution removal 1.0–1.3 for new FBM

*FBM: Fine Bubble Membranes

These three ratios take into account the critical elements of the aeration system and can help the choice of the most efficient corrective actions for the improvement of the energy efficiency of aeration. A comparison with reference values of the three key ratios is a good way to evaluate the actual efficiency of a given aeration system and is the first step of the global methodology towards energy savings on aeration.

6.2.2.1 Optimisation of air production

The production of compressed air is the most energy consuming process of fine-bubble aeration systems (currently, the most common aeration system in WWTPs). If the specific energy consumption for air production ratio is high (>5 $Wh/Nm^3 \cdot mWC$ for roots), key parameters to take into account for energy reduction are the choice of blowers, as well as their control and operation strategy, start-up and maintenance. New premium efficiency motors, such as hybrid permanent magnet motors, allow a reduction of the energy consumption by up to 20%. Controlling blowers with a variable frequency converter or with an airflow measurement instead of a pressure set point are interesting energy saving actions that can be implemented. Reducing the number of start-ups of non-volumetric equipment can also allow additional energy savings. And finally, an optimal maintenance of the blowers is necessary to limit the energy consumption, including special care of the oil, filters, valves and belt's tension. The best practices for the reduction of specific energy consumption for air production are summarised in Figure 6.8.

It is important to underline the complexity of the overall aeration system. To optimise it, all the elements must be taken into account without forgetting that these elements are linked and have mutual influence and interactions. The global recommendations and best practices presented below should be followed not only to improve energy efficiency, but also to preserve the good operation of the system. The selection of the most appropriate strategy for improvement of aeration energy efficiency should be on a case-by-case basis.

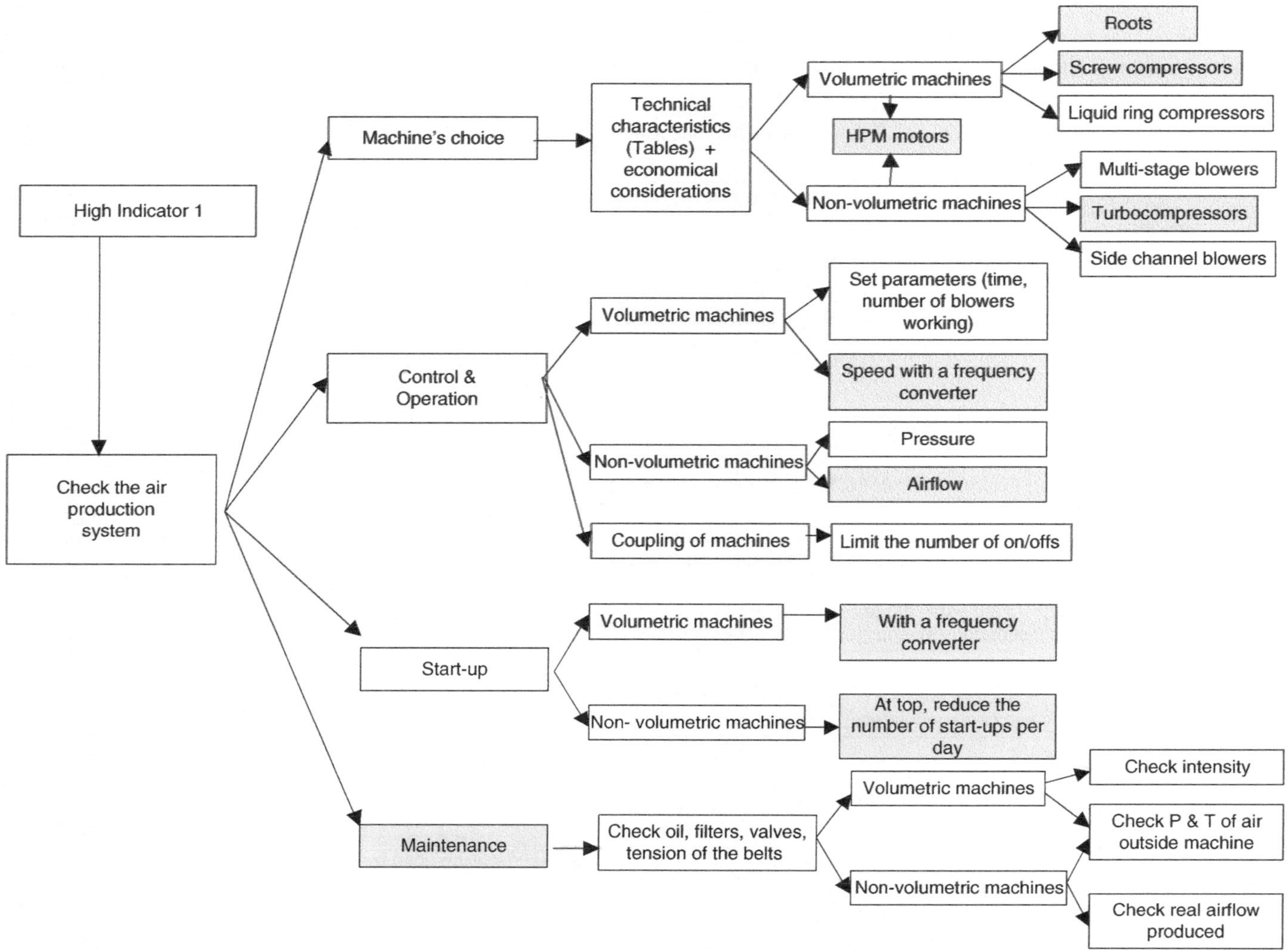

Figure 6.8 Summary diagram of the best practices for improvement of Indicator 1, specific energy consumption for air production (dark boxes should be the preferred choice)

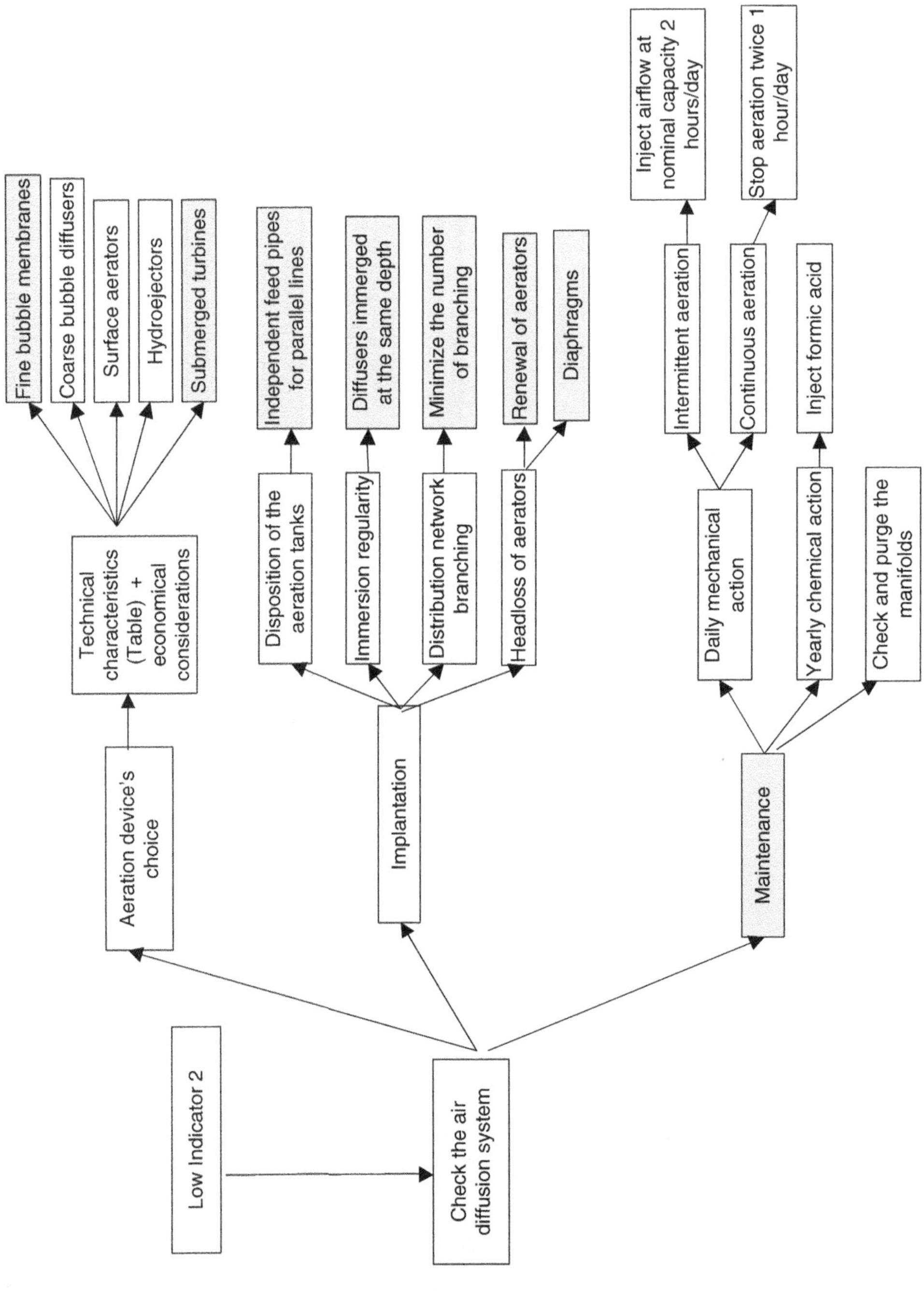

Figure 6.9 Summary diagram of the best practices for improvement of Indicator 2, specific aeration efficiency (dark boxes should be the preferred choice)

6.2.2.2 Optimisation of air diffusion

The second indicator for energy efficiency is the specific energy consumption for the aerator-tank pair, also referred to as specific aeration efficiency (SAE) in kg O_2/kWh. It allows a comparison of all types of aerators for a specific aeration tank or pollution to be treated. When the specific aeration efficiency is low, this indicates that the air diffusion system can be optimised. The choice of the aeration device is crucial. Based on technical characteristics and economical considerations, fine bubble aeration membranes are often the most efficient system.

Nevertheless in some cases, submerged turbines are also a suitable alternative with good energy efficiency, easy maintenance and less frequent replacement. Appropriate design and implantation of the aeration devices can allow additional energy savings: for example independent feed pipes for parallel lines, immersing the diffusers at the same depth, periodical control of headloss. As for the air production system, maintenance is also necessary. A daily mechanical action and yearly chemical cleaning can reduce the inevitable loss of energy efficiency of fine bubble membranes, and as a consequence, a more efficient air diffusion system with time. The best practices for the improvement of specific aeration energy are summarised in Figure 6.9.

6.2.2.3 Optimisation of pollution removal

The third and last indicator of energy efficiency is the specific energy consumption for pollution removal in kWh/kg BOD_5. This ratio characterizes more completely the energetic efficiency of the overall aeration system. High specific energy consumption for pollution removal can be improved by corrective actions of aeration control system and sensors, as well as by reducing the endogenous respiration (Figure 6.10). New regulation systems such as those based on nutriment measurements can allow energy savings of up to 15%. The sensors maintenance is, once again, a key task needed to be completed routinely in order to reduce energy consumption.

Endogenous respiration can be reduced by improvement of the solids handling facilities.

6.2.2.4 Aeration control strategy

When the specific energy consumption for pollution removal is elevated, the main components of the system to be optimised are the sensors and the aeration control system. These two components are related since a good sensor's choice and a correct maintenance are necessary to have a regulation system functioning normally. The aeration control can be enhanced with a good choice of the regulation system and a good automation. A high endogenous respiration is also a cause for having high specific energy consumption for pollution removal.

The most common sensors for aeration control are currently redox (oxidation-reduction potential) and dissolved oxygen sensors. Dissolved oxygen amperometric sensors need frequent calibration and redox sensors are not always reliable. This is why new sensors have been developed to optimise the aeration control. The optical sensor for measuring the dissolved oxygen allows a less frequent calibration and, as a consequence, savings of time and money.

Another promising approach is the use of nutrient sensors, permitting determination of the concentrations of ammonia, nitrate or phosphate in-situ. These new sensors have led to new techniques for aeration control without using redox sensors. For example, the implementation of a new aeration control strategy based on direct measurement of ammonia and nitrate, patented by Degremont (Greenbass™ Biological Aerated Sequential System) ensures a better estimation of the aeration needs and 15% energy reduction, based on operational experience at the WWTP of Dijon, France, 400,000 p.e. (Graveleau *et al.* 2012). This new monitoring and control approach can be used either for fixed air flow or for variable frequency air flow

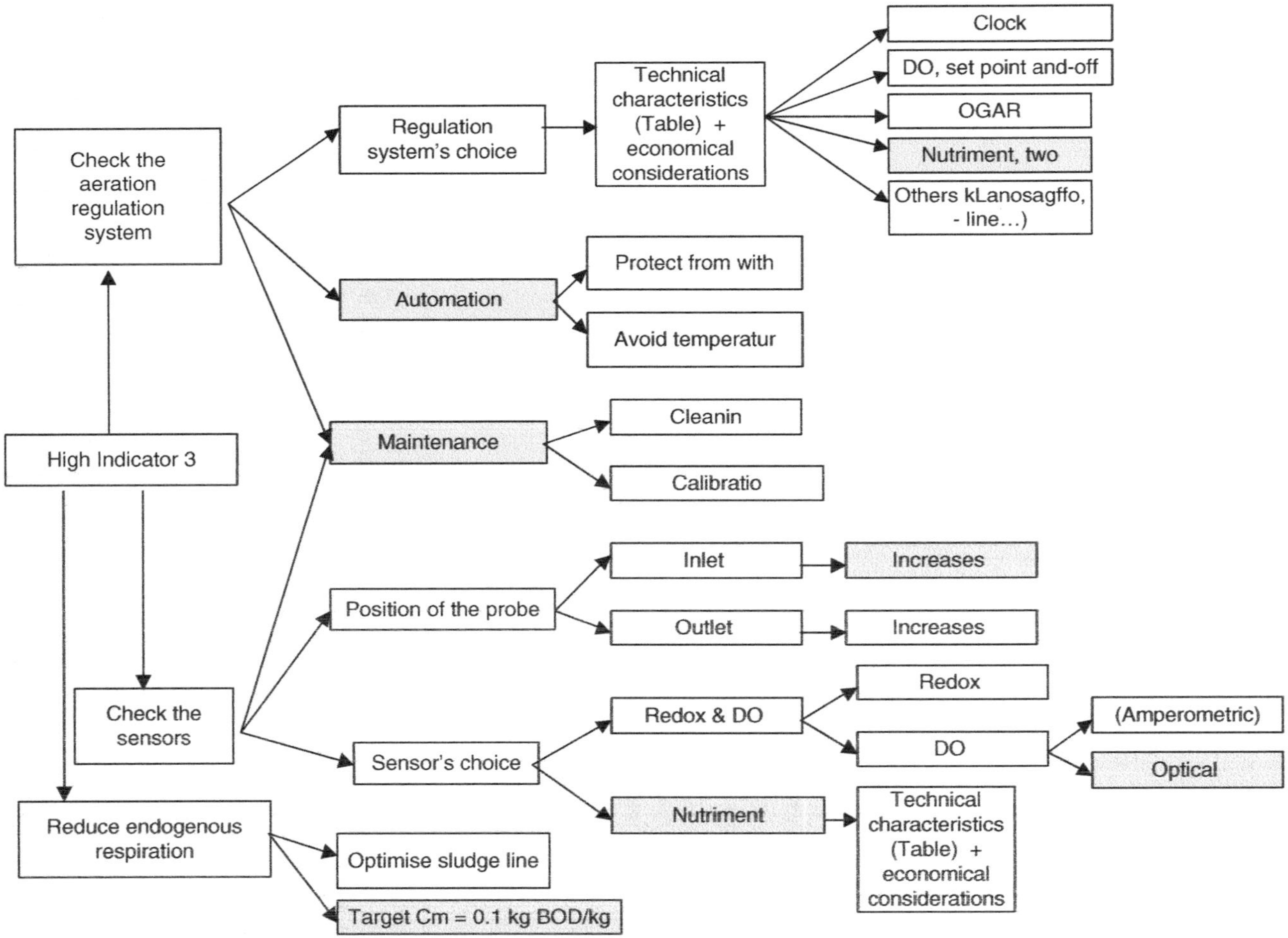

Figure 6.10 Summary diagram of the best practices for improvement of Indicator 3, specific energy consumption for pollution removal (dark boxes should be the preferred choice)

supplying equipment. Energy is saved over time and the applied nutrient load is lower than excepted (night or week end, for example).

6.2.3 Treatment of high ammonium concentration effluents from anaerobic sludge digestion

In the case of WWTPs with anaerobic sludge digestion or of centralized sludge treatment, sidestream treatment of digested sludge or centrate/filtrate with high ammonium concentration can be an attractive alternative for energy reduction, as sludge return flow may account up to 20% of the nitrogen load of the activated sludge system and can increase aeration and energy demand proportionally. In this perspective, one of the most promising biological processes is the nitrification and denitrification of the solids handling waste streams over nitrite.

Several processes have been developed and implemented over the recent years, such as the Sharon® process, high activity ammonia removal over nitrite (Hellinga *et al.* 1998; Mulder *et al.* 2006), Anammox® process, anaerobic ammonium oxidation (Stross *et al.* 1997), OLAND, oxygen limited autotrophic nitrification/denitrification (Kuai & Verstraete, 1998), CANON, completely autotrophic nitrogen removal over nitrite (Sliekers *et al.* 2002), Demon® process, a pH controlled SBR (Wett, 2006) or CLEARgreen® process, a high-load SBR for nitrate shunt (Kaldate *et al.* 2009; Li *et al.* 2010). Compared to conventional nitrification/denitrification reactions, this process has the advantage that up to 25% of the oxygen demand and up to 40% of the organic carbon consumption can be reduced.

The main nitrate shunt process that has been successfully applied at full-scale for treatment of digester supernatant is the SHARON® process in one or two stage reactors (Mulder *et al.* 2006), which is a continuous flow reactor without sludge retention that uses high pH (7.5–8.5) and high temperature (35°C) as main stress factors (Hellinga *et al.* 1998). Good ammonium removal efficiencies of up to 90–95% have been achieved, but the operating costs remain relatively high, mainly for energy and COD source.

The deammonification has been proved to be the most promising process for sidestream treatment, in particular Demon® process, which implementation helped the Strass WWTP in Austria to become energy producer (Wett, 2006). A demonstration study of the CLEARgreen® process at the Glarnerland WWTP in Switzerland (capacity of 80,000 p.e., 22,000 m^3/d) has demonstrated that energy saving of 27.5 MW/yr and up to 95% removal of ammonia can been achieved by sidestream treatment of sludge digestate (Martin Ruel *et al.* 2009). In addition, further decrease of operation costs was from methanol addition and sludge treatment. On-site measurements revealed negligible concentrations of N_2O produced by the SBR reactor.

6.3 ENERGY RECOVERY FROM SEWAGE FLOWS

This type of energy improvement aims at using waste heat and the potential energy that may be contained in large sewage flows in wastewater treatment facilities and sewer systems. Three of the more probable alternatives are micro hydro electricity, heat exchangers and low temperature cycles.

6.3.1 Hydro electricity

There are thousands of sites where wastewater assets offer sources of falling sewage flows. These sources can be used to generate electricity by using turbines which allow converting the potential energy of the down-flowing wastewater into electricity. The power and type of hydraulic turbines depends on two main parameters: the net head (m) and flow (m^3/s) (Pakenas, 1995).

There are several different types of water turbines, for example Pelton, Francis, Axial, and so on. Each type operates over a limited range of site conditions in terms of head and flow. Given a typical efficiency, a 1 kW scheme could acquire its energy from 200 m head with a flow rate of 1 L/s or 2 m head with a flow rate of 100 L/s (ESHA, 2004).

There are two main economic aspects in the installation of turbines on wastewater streams (Fievez, 2009). First, the electromechanical equipment is more expensive than equipment operating with clean water such as river water. Secondly, the overall infrastructure of the plant has to be adapted to the use of hydro electricity. As reported by Fievez (2009), the first use of raw sewage with a drop of 447 m in Switzerland in 1992 was a technical challenge. For the energy recovery from wastewater flow from the famous Verbier resort, a 700 kW Pelton turbine was installed. A major concern was the wear and abrasion of the wheel and injectors, due to the solids contained in the raw sewage. Degrémont has overcome these difficulties and this installation has been in service for 15 years before a major refurbishing of the plant in 2007.

For isolated facilities, which generate wastewater with significant potential (kinetic) energy content (such as in WWTPs located in the mountains), hydroelectricity is the lowest-cost technology for generating electricity, and in particular micro hydroelectricity. Micro hydro is a term used for hydroelectric power installations that typically produce up to 100 kW of electricity. In a micro hydro power plant (Figure 6.11). Sewage flow is captured via a filter/intake structure and passes downhill in a pipeline (1) to the turbine (2). The water exits the pipe under pressure through a nozzle and strikes a water wheel (runner). The force of water fall spins the runner which is directly coupled to electricity generating permanent magnet alternator or enhanced performance induction generator. The electricity is then electronically regulated and transmitted via cable for use (3).

Figure 6.11 Micro hydro electricity layout and example of implementation (courtesy of Suez Environnement)

6.3.2 Heat exchange from sewage flows

Hot water is used for a variety of cleaning applications (showers, dishwashers, laundry) resulting in a wastewater stream that retains a significant proportion of this thermal energy with an average temperature that typically ranges from 11 to 20°C. Accordingly, the inherent thermal energy in wastewater may be recovered and valorised on-site at the wastewater treatment facility or in nearby applications.

In France, wastewater treatment operations are at the heart of an innovative technological solution that recovers the heat from wastewater for use in a new type of district heating system. The process incorporates a novel heat exchanger placed directly in the sewer pipes (Figure 6.12). Using heat pumps and special pipes network, the recovered thermal energy is transferred to a building's HVAC system and used to heat buildings in the winter and cool them in the summer. The process can be adapted to various types of buildings including residential and office buildings, schools, hospitals and even municipal pools.

Figure 6.12 Schematic of thermal energy recovery from wastewater with views of heat exchangers in existing sewers (top) and integrated in new sewer pipes (bottom)

Although capital costs are high and operation is closely dependent on the conditions of the sewer line, heat exchanger pipes systems operating with sewage are already commercialized. Over 30 heat recovery projects are in operation since 1999, mainly in Switzerland and Germany as well as in the US and Canada. In France, the city of Levallois near Paris was the first community to implement this technology in 2009 for heating of a swimming pool recovering 800 MWh/yr energy and avoiding 150 t/yr CO_2 emissions. To date, several other cities have launched projects of heating and cooling of public buildings with a capacity from 412 to 3,626 MWh/yr.

6.4 ENERGY FROM SLUDGE

Sewage sludge is by far the most important source of energy that can be directly generated in a WWTP (Figure 6.13). Although chemical energy contained in the dry solids of the sludge is theoretically enough to cover the WWTP energy requirements (Figure 6.14), this seldom happen in reality. In fact, turning sewage sludge into energy is associated to a number of technical, institutional and environmental challenges that need to be resolved.

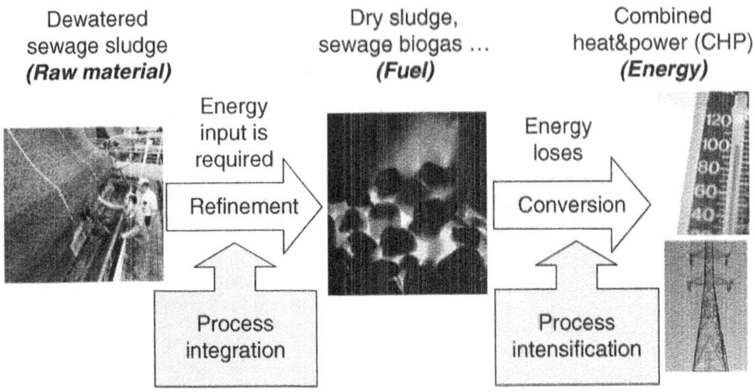

Figure 6.13 Schematics of turning sewage sludge in useful energy

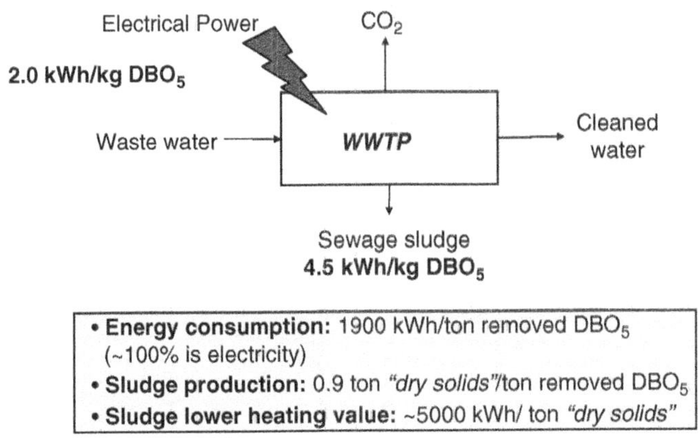

Figure 6.14 Energy balance of WWTP: energy requirements can be covered by the energy potential contained in the dry solids of sewage sludge

Historically, the term used for sludge has evolved through unfinished debates on the status of waste, "biosolids" or compost (Gendebien *et al.* 2010). This vagueness increases with the new applications of sludge as a substitute for fossil fuels (Werther & Ogada, 1999). Consequently, it appears that processing of sludge is very closely linked to the particular context where it is produced. Thus, sludge can be processed for the production of fertilizers (recycling), fuels (energy recovery) or stabilized ultimate waste (disposal). This would be decided depending on the specific sustainable use or final disposal of the product.

If moisture is removed by physical means (i.e. gravitational force, pressure driven or evaporation) sludge moves from liquid to solid as long as dry solids and calorific content increase (Degrémont, 2004). Physical processes are not the only method to obtain suitable fuels from sewage sludge. As mentioned earlier, sludge energy potential can also be extracted from wet sludge by biological means such as anaerobic digestion (Lettinga & van Haandel, 1993; Nayono, 2010), as well as to obtain sewage biogas fuel or by other thermo-chemical techniques, such as pyrolysis and gasification to produce synthetic-gas, also called "syngas" (Higman & Van der Burgt, 2003; Bourgel *et al.* 2011).

One of the most difficult aspects of sewage sludge as feedstock for thermal processes is its heterogeneous nature. The interconnected properties of heating value and moisture content play an important role for both solid and pasty wastes. For solids, particle size, density and specific surface are some of the key parameters to take into account. Additionally, the presence of a number of components in the sludge such as sulphur, chlorides or metals can vary considerably as well. Moreover, variations can be seasonal or even daily. Finally, the long-term supply of the feedstock is another aspect to be considered. All these specificities require a deep knowledge on sewage sludge characterization. Hence, each project for thermal processing of sludge is a unique technical challenge.

6.4.1 Energy from solid fuels derived from sewage sludge

Processing sewage sludge to produce a solid fuel does not intend to generate energy but to avoid its consumption. This can be done if the sludge auto thermal point is reached.[1] However, significant amounts of moisture need to be removed from sewage sludge (Table 6.3). Dewatering is therefore, a very important stage in order to both reduce the volume of sludge and to increase its calorific value. However, energy consumption and final sludge product are quite different among the different available techniques.

Table 6.3 Performances of thickening, mechanical and thermal dewatering of sewage sludge.

Dewatering technique	Specific technique	Specific thermal consumption (kWh_{th}/kg DS)	Specific electrical consumption (kWh_{elec}/kg DS)	Achievable final moisture content (%)	Final shape of sludge
Thickening	Gravitational thickening	N/A	5–10	2–8%	Liquid
Mechanical	Centrifugal thickening[1]	N/A	150–200	2–3	Liquid
	Dynamic thickening[1]	N/A	100–130	4	Liquid
	Mechanical dropping[1]	N/A	50	5–8	Liquid
	Belt filter	N/A	30–40	14–27%	Pasty
	Centrifugal	N/A	25–80	17–32	Pasty/solid
	Press filter	N/A	30–40	28–46	Pasty/solid
Thermal drying[2]	Partial drying	1,140	70	35%	Solid
	Total drying	2,950	180	95%	Solid

DS: dry solids
[1]Techniques used for the thickening of biological sludge
[2]Energy consumption for the thermal drying of sludge having 25% dry solids content. It is assumed specific thermal and electrical consumption of 1000 kWh_{th}/ton of evaporated water and 60 kWh_{elec}/ton of evaporated water respectively

Although several types of mechanical dewatering and thermal dryers are currently installed and operated worldwide, it is still difficult to determine a series of best practices for these processes. It is important to note that process reliability and integration can be even more important than energy recovery implementation.

[1]Which means that the heat generated by the oxidation of the sludge can equal the heat needed to run further sludge oxidation without needing an external fuel

Most positive aspects of WWTP upgrade are related to the use of energy contained in sludge for thermal drying. Such an example is the WWTP of Dijon, France (400,000 p.e.), which was upgraded in 2006 for nutrient removal and thermal sludge drying combined with the upgrade of the existing incineration unit (Peregrina *et al.* 2008). The new sludge treatment configuration enabled to optimise energy consumption efficiency, decreasing the contribution of external sources of energy (mainly natural gas) from 70% to 10–20% of the potential energy contained in the produced sewage sludge.

Waste to energy can be a solution when wastewater treatment plant sludge cannot be recovered for agriculture, whether with or without prior digestion.

6.4.3 Energy from biogas derived from sewage sludge

Processing gas fuel from sewage sludge aims at enhancing the amount of useful energy in order to recover it into the wastewater process as electricity. Gas fuels from sludge can be obtained in two ways, namely, biological or wet route (anaerobic digestion) and thermo-chemical or dry route (pyrolysis and gasification).

6.4.3.1 Biogas production by anaerobic digestion

Thickened sludge can be anaerobically digested to produce a gas fuel (60% CH_4 and 40% CO_2), namely, sewage biogas. However, this operation has been historically applied for sludge mass reduction and stabilization rather than for energy applications (Cillie *et al.* 1969).

The energy balance is quite good since liquid sludge needs little conditioning and above all avoids thermal drying. However, only a partial reduction of the volatile matter is possible. Thus, the disposal of the resulting digested sludge remains a major issue to solve within the sludge management framework.

Depending on the type of sewage sludge, there are different yields of volatile matter digestion (Table 6.4). Consequently, biogas production and volatile matter removal depend on the type of sludge, with the assumption that mineral matter does not participate into the reaction.

Table 6.4 Performance of current anaerobic digesters.

Performance Parameter	Primary sludge	Mixed sludge
Volatile matter removal (%)	50–55	45–50
Sewage biogas production (Nm^3/kg of volatile matter)	0.85–1.2	0.75–1.0

6.4.3.2 Improvement of biogas production from anaerobic digestion

The biogas production from anaerobic digestion of sewage has relatively low efficiency and is commonly in the range of 0.7 to 1.2 Nm^3 par kg removed volatile solids, as illustrated in Table 6.4. Moreover, many of the existing anaerobic digesters are oversized, which is considered as an inefficient use of assets considering the increasing cost of energy. All these challenges are fostering the investigation of new routes for green energy production in WWTPs focused on improvement of anaerobic digestion.

Three major trends are identified for the improvement of sewage biogas use for energy production:

- Maximizing the volatile matter degradation yield by means of pre-treatment for sludge hydrolysis prior to anaerobic digestion (Gavala *et al.* 2003; WERF, 2008).
- Co-digestion of sewage sludge with other bio-wastes (La Cour Jansen *et al.* 2004; Bolzonella *et al.* 2006; Bouchy *et al.* 2012).
- Sewage biogas upgrading and end-uses (Kolish *et al.* 2011).

Further technological development is needed in order to establish the maximum yield of volatile matter into biogas and also the overall energy balance, that is energy input due to pre-treatment and digestion itself compared to the energy output with the sewage sludge produced.

Pre-treatment and co-digestion may increase the energy that can be produced in a WWTP. However, this brings new constrains of operation regarding the mixture of different types of organic solids, as well as the anaerobic digestion of these mixtures. Additional questions would arise regarding the responsibility of the treatment, the resulting additional pollution load and the long-term availability of the external resources.

For example, any improvement in terms of degradability and biogas production rate directly implies higher ammonia pollution to be returned to the head of works. This drawback can be overcome by removing ammonia from the digestate directly using sidestream treatment, usually more efficient than in the main stream when adequate technologies are installed (see §6.2.3).

Sludge pre-treatment

One approach often considered is accelerating the slow hydrolysis conversion of particulate organic compounds into organic fatty acids, commonly considered as the main rate-limiting step of biogas production by anaerobic digestion (Crujer & Zehnder, 1983; Batstone *et al.* 2002). This is usually carried out by including a sludge disintegration phase prior to anaerobic digestion, which solubilizes part of the COD, making it available for digestion under typical sewage digestion operating conditions (often mesophilic, 20–25 days HRT).

During the last twenty years, studies of different types of technologies (physical, chemical, biological) have been carried out to find the best compromise between digestibility and dewaterability (Camacho *et al.* 2009a). A growing number of emerging technologies has been developed based on hydrolysis using mechanical treatment, acid alkaline, heat treatment and combinations (Table 6.5). The most promising strategies have been found to be thermal treatment, ultrasound, lysing centrifuge and mechanical and chemical treatment, in terms of improvement of treatment efficiency (removal of volatile solids, biogas production).

Table 6.5 Technical data for sludge treatment process for the improvement of biogas production in comparison to conventional anaerobic digestion (Camacho *et al.* 2009a).

Type of pre-treatment	VS removal yield[1]	Biogas production[1]	Dewatering[2]	Return to head	Foaming
Thermal	+10–30%	+10–30%	+50–100%	1.9–2.4 g/L $N\text{-}NH_4$	–
Mechanical	+0–22%	+0–22%	+15–30%	N/A	–
Lysing centrifuge	+12–17%	+15–30%	+2%	+8% $N\text{-}NH_4$+11% BOD_5	N/A
Ultra sound	+12–17%	+25–45%	+0–50%	High $N\text{-}NH_4$	–
Ozonation	+0–11%	+0–11%	−3%	+4–17% $N\text{-}NH_4$	+
Biological	+0–44%	+ 0–44%	+0–20%	+60% $N\text{-}NH_4$	+
Chemical & mechanical	+10–16%	+90–100%	N/A	+10–20% $N\text{-}NH_4$	–

[1]compared to conventional AD performance, percentage expressed in relative value
[2]compared to dewatering process without treatment, percentage expressed in relative value
N/A: data non available; VS: volatile solids

Similar results have been reported by Pérez-Elvira *et al.* (2008): the most promising technology was found to be thermal hydrolysis allowing an increase in biogas production by 20–30%, decrease of digester volume of 60% due to the possibility to operate at higher concentrations and low residence time. Moreover, the dewaterability increased by 6–8% allowing the reduction of electrical demand and polymer consumption.

These results explain why the most successful and developed technology so far is thermal pre-treatment which has reached wide commercialisation, mainly with the Cambi© Technology and BioThelys (Gavala *et al.* 2003; Camacho *et al.* 2008; GWRC, 2008; WERF 2008; Water 21, 2011). Many demonstration and full-scale projects generally fail because of the high capital and O&M costs of some of the pre-treatment technologies.

Figure 6.15 illustrates the effect of sludge pre-treatment on energy production obtained by co-generation from biogas (Camacho *et al.* 2009a). The typical range of VS removal for conventional anaerobic digestion of mixed sludge was found to be between 40–55% from internal audits of WWTPs with anaerobic digestion within Suez Environment. The effect of pre-treatment on energy production is estimated from literature data and feedback from operation. The maximal improvement of efficiency of the electrical and thermal production from biogas valorisation is 20%. In order to evaluate the benefit and impact of pre-treatment on the conventional anaerobic digestion line, these technologies were compared with a conventional digestion for two cases: upgrading of an existing anaerobic digester or a new installation (Table 6.6).

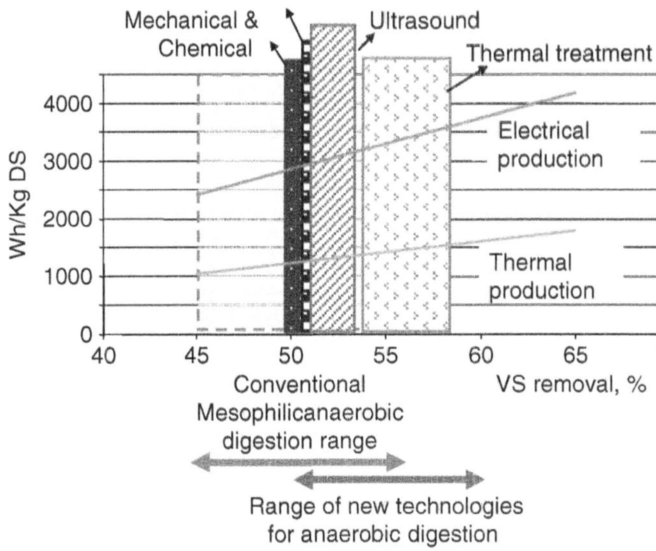

Figure 6.15 Profile of energy production and VS removal for different technologies (Adapted from Camacho *et al.* 2009a)

The selection of the best option for the improvement of energy production by anaerobic digestion of sludge must take into consideration the potential impact on water and sludge treatment lines, in addition to economic and environmental considerations for the given local conditions.

Table 6.6 Relative gain or loss of main performance indicators for new strategies of anaerobic digestion compared to conventional anaerobic digestion in new plants or for upgrading of existing plant (Camacho *et al.* 2009a).

Pre-treatment	Digester volume	Inlet charge	Methane production	Excess energy	Digested Sludge quantity	Dewatering Sludge quantity	Polymer utilisation for dewatering
Thermal:							
new	−70%	+0%	+20%	+44%	−20%	−38%	−10%
upgrading	+0%	+233%	+300%	+379%	+167%	+106%	+199%
Lysing centrifuge:							
new	−33%	+0%	+25%	+48%	−25%	−15%	−10%
upgrading	+0%	+50%	+87%	+123%	+12%	+28%	+35%
Ultra sound:							
new	−27%	+0%	+25%	+46%	−25%	−28%	−27%
upgrading	+0%	+37%	+71%	+100%	+3%	−1%	0%
Chemical & mechanical:							
new	−44~50%	+0%	+26~47%	+37~73%	−26~47%	−13~51%	−13~51%
upgrading	+0%	+79−100%	+126~194%	+145~245%	+6~31%	−2%~+55%	−2%~+54%

Co-digestion

Apart from sludge disintegration, anaerobic co-digestion has been greatly developed over the past few years. Increased sewage biogas production can be obtained from the digestion of other bio-wastes generated during wastewater treatment such as fats, grid and sieved sewage waste. Additional sources for co-digestion can also be external sources of organic solids such as manure, wood residues, food industry waste and agricultural residues as a whole (Camacho *et al.* 2010; Bouchy *et al.* 2010, 2012). In this sense, and considering that the optimum carbon-to-nitrogen ratio (C/N) is between 20 and 30 (Parkin and Owen, 1986), co-digestion of sewage sludge with other organic wastes could improve the nutrient balance, improving at the same time the amount of degradable carbon and consequently, the biogas yield. The typical ratio between external co-substrate and sludge is between 6% and 16% (Silvestre *et al.* 2011).

Previous studies have shown that there are different wastes with different origins that can be co-treated with sewage sludge obtaining promising results (Corti & Lombardi, 2007; Bouchy *et al.* 2009; Mata *et al.* 2011). In some cases, the additional biogas was generated simply due to increased organic matter (higher volumetric load), showing linear production rates based on the ratio of sludge and other substrates. In this sense, a better use of the volume of digesters is obtained. Anaerobic co-digestion is reported to offer several benefits over conventional digestion, such as increased cost efficiency and increased degradation of the treated substrates due to possible positive synergistic effects (Luostarinen *et al.* 2009). However, it is important to note that in some cases negative impacts were observed, showing the importance of both the selection of co-substrate and the mixture ratio (Bouchy *et al.* 2010).

Amongst the wide variety of substrates, various authors reported an increase in the methane yield during the co-digestion of sewage sludge (SS) with different types of fats (Davidsson *et al.* 2007; Luostarinen *et al.* 2009; Kabouris *et al.* 2009). A good option for WWTPs is to use the grease wastes generated inside the plant, such as the grease waste (GW) of the dissolved air flotation (DAF) unit, whose management costs will be reduced (Silvestre *et al.* 2009). Silvestre *et al.* (2011) reported that despite variable characteristics between grease samples from the DAF units of WWTPs, their high organic matter and fatty content resulted in high methane potential (479–710 L CH_4/kg VS), compared with municipal sewage sludge (322 L CH_4/kg VS). Continuous experiments showed that the co-digestion of the two materials together was feasible, with 138% increase of the methane yield when grease waste addition was 23% of inlet VS (HRT 20 days, organic load 3.0 kg $COD/m^3 \cdot d$).

However, the addition of solid wastes could decrease the methane production as hydrolysis becomes the rate limiting step at a given hydraulic retention time. Some of the pre-treatments explained previously can be applied to co-digestion processes in order to overcome this limitation. This strategy was tested by Bouchy *et al.* (2012), applying different disintegration technologies (thermal hydrolysis, ultrasounds and addition of enzymes) to a mixture of grease from DAF unit and municipal sewage sludge. Results showed great improvements in terms of methane productivity (Table 6.7). Ultrasound showed higher biogas productivity than thermal hydrolysis and enzymatic treatment, reaching an increment of 89.5% compared with sewage sludge methane productivity. However, for full-scale operation, the overall most promising option would be the thermal hydrolysis pre-treatment, as, a part from a considerable biogas production increment, it also showed the best improvement of hydrodynamic characteristics, as well as appealing energy integration possibilities (Bouchy *et al.* 2012).

6.4.3.3 Biogas quality enhancement and end uses

To increase the useful energy recovery from sewage sludge, the biogas quality needs to be enhanced to match the requirements of more demanding and intensive end-uses to convert chemical energy to electricity (Figure 6.16). Therefore, testing different systems for biogas cleaning and upgrading is a very relevant topic for research and pilot plant studies.

Table 6.7 Methane productivities and COD solubilisation after pre-treatments (adapted from Bouchy *et al.* 2012).

Pre-treatment	Solubilization (COD_{sol}/COD_{tot})		Methane yield (L CH_4/kgVS)	
	value	% increase	value	% increase
Mixture (grease + sludge) without pre-treatment	6,7%	–	379	–
Thermal Hydrolysis applied to the mixture	10,4%	+55%	575	+51,8%
Ultrasounds applied to the mixture	12,2%	+82%	718	+89,5%
Enzymes applied to the mixture	12,9%	+92%	597	+57,6%

Level of upgrading	Component	Content in raw biogas	End-use
Raw biogas	H_2O	Saturation	Flare
	Particles	Trace	
Low grade biogas	H_2S, NH_3, HC/HC-X & siloxanes	Trace	Heating
Medium grade biogas	CO_2	20-50% vol.	Heating & CHP engines
	N_2	<10% vol.	
	O_2	<2% vol.	
	H_2,CO	Trace	
High grade biogas	CH_4	30-60% vol.	Injection to pipelines, chemical production, vehicles & fuel cells

Figure 6.16 Requirements of upgrading to remove gaseous components depending on the biogas utilization (adapted from Simader, 2004)

For a long time, chemical energy contained in biogas was transformed on-site into electricity in Internal Combustion Engines (ICE) and more recently in micro-turbines (Deublein & Steinhauser, 2008). As shown in Table 6.8, both technologies have a limited electrical efficiency (26–35%) as the majority of the potential energy is converted into heat. In the case of recovery of biogas from WWTPs, this heat is usually in excess of the plant needs. In addition, because of the location of these facilities, the transport of this heat to other sites is economically compromised. The result is a very large loss of heat which causes poor yield of total energy production.

Table 6.8 Comparison of different technologies for on-site Combined Heat and Power in WWTPs.

Technology	Capacity (kW_e)	Net electric efficiency (%)	Overall efficiency (%)	Cost (€/kW_e)
Internal combustion Engines (ICE)	250–2,000	30–35	66–85	1,000–2,000
Micro-turbines	30–200	26–35	60–85	1,500–2,500
Low temperature fuel cells	25–250	40–50	55	5,000–10,000
High-temperature fuel cells	250–2,000	40–50	60–80	5,000–10,000

In order to overcome this limit, other biogas end-use technologies have attracted the interest of WWTP operators in recent years. Fuel cells are becoming one of the most promising alternatives for on-site energy recovery as they have greater electrical efficiencies and lower adverse exhaust emissions than conventional technologies (see also Logan, Chapter 5). In addition, they can cover the whole range of biogas power sites with the same efficiency and can be adapted to all WWTP sizes (Van Hereltet *et al.* 2004). Several fuel cell demonstration or industrial applications in the range of 25 kWe up to 2 MWe have been reported (de Arespacochaga *et al*, 2010a). However, the investment cost of these units, which is 5 to 10 times greater than conventional technologies, remains a limitation for its massive industrial implementation.

Other innovative technologies which are currently being studied are Stirling Engines and Organic Rankine Cycle coupled to ICE (Thomas *et al.* 2009), but few industrial examples are available. Thus, the experience is limited and more experimentation is necessary in order to define their exact application field in wastewater treatment, possibly as supporting technologies to cope with biogas production peaks. Further applications of ORC are described in Section 6.4.4.

On the other hand, off-site energy recovery from biogas has been particularly explored in some European countries such as Sweden, Germany and Austria. By upgrading its lower heating value, biogas can be used elsewhere as a substitute of natural gas (the upgraded fuel is called biomethane). In fact, natural gas-fired appliances (boilers, gas-heat pumps, Natural Gas Vehicles, and so on) can run with equal efficiencies either with natural gas, biomethane or any mixture of the two fuels.

Biomethane production consists of the removal not only of biogas pollutants (H_2S, siloxanes, VOCs) but also of CO_2. It allows for centralized energy recovery, supporting efficient Combined Heat and Power (CHP) systems because it can be brought to areas where heat can be efficiently used alongside with the electricity generated. Furthermore, its ability to be injected into the existing natural gas grids allows for energy-efficient and cost-effective transport (Holm-Nielsen *et al.* 2009). Several biomethane industrial facilities have been running since the late 1990s (de Arespacochaga *et al.* 2010b) and its economics are clearly competitive with on-site energy generation alternatives.

Finally, biogas can be also turned into high-quality chemicals like methanol (Kovac & Kralj, 2010), an important precursor of many synthetic organic compounds produced by the chemical industry and a reagent required for biodiesel production. However, this application is not yet commercially available.

6.4.3.3 Gasification and pyrolysis

Thermo-chemical transformation of sewage sludge is another alternative for producing fuel gas in order to produce high quality energy (Peregrina *et al.* 2011). An important issue with thermo-chemical processes is that the entire organic content of sludge can be transformed into fuel gas (Figure 6.17). Thus, the resulting by-product is carbon free mineral matter that can be landfilled. Another important issue is process integration, which deals with the efficient use of waste heat coming from one operation and that can be used in other process units.

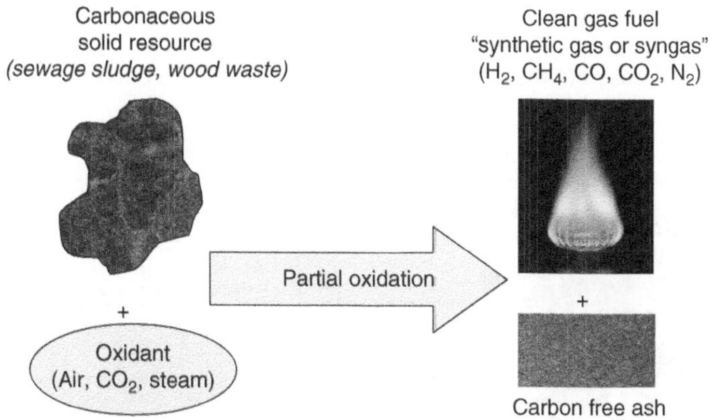

Figure 6.17 Schematics of gasification to getting useful energy from sewage sludge

Process reliability and gas quality are the main issues for process design and operation. Biomass gasification technology needs to be tested and adapted for sewage sludge implementations (Figure 6.18). Last but not least, additional research is needed on alternative recycling routes (agriculture or building uses) for carbon free ashes resulting from gasification.

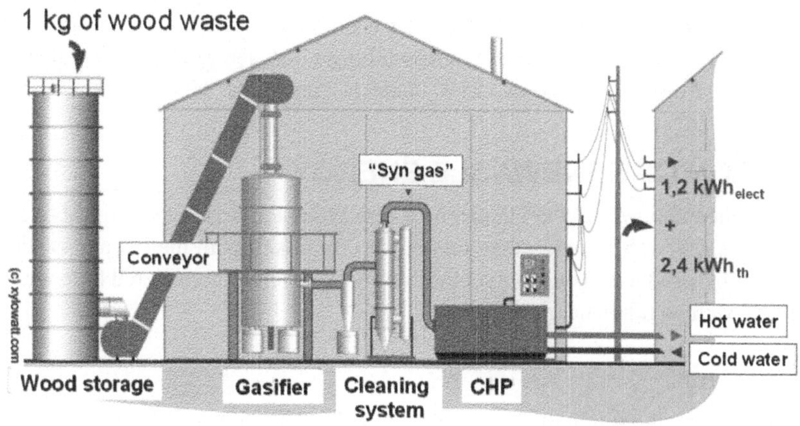

Figure 6.18 Biomass gasifiers need to be adapted and tested for sewage sludge application (*Source*: www.xylowatt.com)

6.4.4 Low temperature heat to electricity

Conventional combined heat and power (CHP) engines can only convert high temperature (>500°C) heat to power. For warm sewage flow or other waste heat generated in WWTP facilities (*i.e.* digestion and drying warm water flows) some heat is available but it is at rather low temperature (<100°C). However, several projects for geo-thermal applications are bringing Organic Rankine Cycle application to the front of the technological stage for low temperature heat to electricity conversion.

The Organic Rankine Cycle (ORC) is a binary system taking heat from a source and transferring it to a working liquid in a closed system to be vaporized and pressurized, typically turn a turbine, and then condense back to liquid form before flowing through the heat exchanger again to repeat the cycle. The diagram on Figure 6.19 breaks out the typical components of an ORC system.

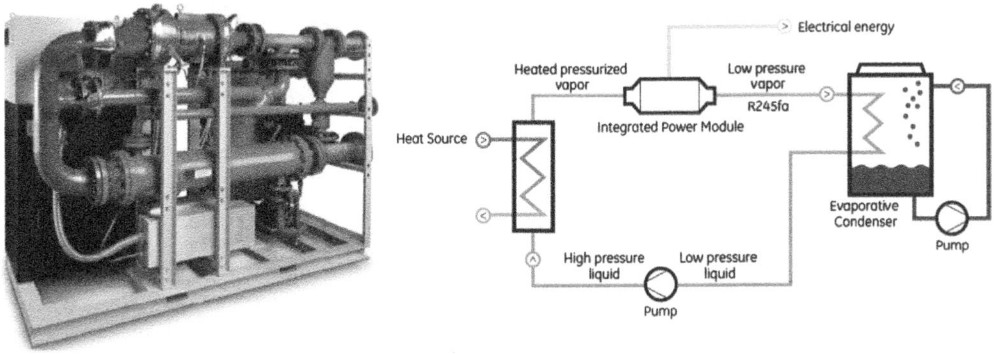

Figure 6.19 Organic rankine layout and clean cycle™ 125 KW by general electric (*Source*: http://www.geheatrecovery.com)

ORC is similar to the cycle of a conventional steam turbine, except for the fluid that drives the turbine, which is a high molecular mass organic fluid, namely, freon, butane, propane, ammonia, and the new "environmentally-friendly" refrigerants. Manufacturers claim that selected working fluids allow exploiting efficiently low temperature (70–100°C) heat sources (Hung *et al.* 1997) to produce electricity in a wide range of power outputs, from few kW up to 3 MW of electric power per unit.

6.5 PRODUCTION OF RENEWABLE ENERGY

Renewable energy, when applied to WWTP, can be a very broad term. However, the above definition only includes those types of renewable energy that are not directly linked with the sewage or with the resulting wastewater treatment by-products. Thus, the renewable energy sources considered here are: solar and wind in a major extent, as well as tidal and geothermal.

6.5.1 Solar energy

Solar energy is produced by the sun in the form of heat or light. Although 3,850 zettajoules (ZJ $= 10^{21}$ J) are absorbed yearly by oceans and land on earth, solar energy accounts less than 1% of the worldwide primary energy consumption. Losses due to sunlight reflection and absorption are not the only reasons for such a low percentage. Technology high capital costs and solar availability in regions are the main reasons for the slow implementation for this source of energy. There is also the variability along the daily and yearly periods.

Solar energy has the highest potential amongst all renewable energy sources. It could be used as a thermal energy or could be transformed into electricity. Solar thermal modules, typically installed on the roofs, are widely used to produce hot water for sanitary purposes and/or hot water/air for heating of industrial building and individual houses. According to the EurObserv'ER survey (2011) the heat-producing applications provided 638.4 MW (megawatts), with a decrease of the newly-assigned surface area in 2010.

Various solar technologies could be used for electricity production such as concentrated solar power (CSP) systems which produce both electricity and heat, and photovoltaic (PV) systems that transform solar radiation directly into electricity. In the European Union, photovoltaic power generated 22.5 TWh (terawatts) at the end of 2010 and was characterised by high growth of +120% for the last year (EurObserv'ER, 2011).

Historically, WWTP are using the thermal component of solar energy in very specific applications namely, sludge drying (Bennamoun, 2012) and water desalination (Blanco *et al.* 2008). Although photovoltaic cells are reliable way of transforming solar energy directly into electricity, capital and operating costs remain high, that is about 3,300-6,000 €/kW installed and 300–550 €/MWh, respectively. For this reason, its implementation as a decentralised or distributed system is and would remain limited to few applications in WWTP facilities.

The most popular use of solar energy is small reverse osmosis desalination plants, mostly photovoltaic (PV) panels (Subramani *et al.* 2011). Because of the need for a large area for the installation of solar panels, the contribution of this green energy to the power supply of large WWTPs is relatively limited. For example since 2009, the WWTP of the city of Chino, California is satisfying 14% of its electrical demand by the green energy produced by 3,047 solar panels installed on a 6475 m^2 area (WERF, 2010b). This plant is an activated sludge with tertiary filtration and disinfection with capacity of 43,200 m^3/d. The 0.7 MW photovoltaic system capital cost was US$3,500/kW.

In addition to the requirement of high surface areas, another disadvantage of PV technologies is the relatively low energy net efficiency, typically 12–15% for crystalline silicon modules and below 10% for flat-module thin film technologies. The large portion of solar power (~70%) is absorbed in module and transformed into thermal energy which increases module temperature to up to 80°C. An ambitious project is under development in France at the new WWTP of Cannes aiming to satisfy 22% of electrical demand by an innovative photovoltaic-thermal system (PVT) with improved energy efficiency. As illustrated in Figure 6.20, this hybrid device, names Solar Active System (SAS), is using PV as a thermal absorber, producing in addition of electricity (14% efficiency), also thermal energy (56% efficiency), enabling thus to improve the recovery of solar power up to 70%, depending on the heat usages.

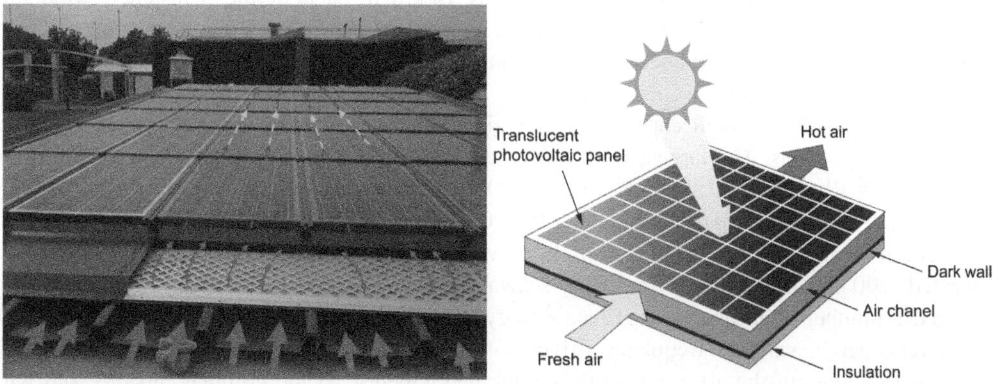

Figure 6.20 View and schematics of the Solar Active System designed for the WWTP of Cannes (France)

The SAS panels can be installed on vertical, inclined or horizontal surfaces of the wastewater treatment plant buildings exposed to the south, in order to collect solar energy and simultaneously transform it into photovoltaic electricity and recuperative heat in the form of air or of warm water. A pilot unit of 100 m^2 will be tested in order to optimise the SAS design. The final objective is the installation of 3000 m^2 of PVT panels with an expected performance of 1500 kWh/m^2.

The new WWTP of Cannes is the first French facility with zero carbon footprint. The plant, with a design capacity of 300,000 p.e., average daily flow 52,000 m^3/d, is a MBR facility in a covered green building with odour treatment and solids handling.

6.5.2 Wind energy

Wind movement, produced in a certain manner by solar energy, can be used by wind turbines to produce electricity. This renewable energy is getting more and more popular with worldwide use approaching 195 GW for a power generation near to 36 GW (EurObserv'ER, 2011). In 2010, 43% of the total wind power was generated in the European Union. Along with environmental concerns, this is mainly due to reduction of capital costs from 600 to 1,000 €/kW per installation. On the other hand, wind electricity prices, which are highly subsidized in Europe, remain high at 80 to 130 €/MWh (EurObserv'ER, 2011).

Wind generators with vertical or horizontal axes are considered as a mature technology. However there is still demand for further improvement of turbine performance and power storage systems. Onshore wind generators produce electricity at relatively low cost and are already largely deployed within the zones with high wind potential. Off-shore wind parks are considered as the booming markets of the years ahead because their higher energy yield and higher installation potential, but they still present numerous challenges.

It is important to underline however, that wind energy is highly dependent on geographical conditions, as it is the case for solar energy. Hence, the implementation of wind turbines is possible only in selected regions where both wind speed and public subsidies meet the conditions of this type of green energy.

Nowadays, the impact of wind power for the improvement of energy efficiency of wastewater treatment plants still remains relatively low, despite the growing interest as a sustainable energy resource. One of the largest WWTPs, which meets about 67% of its average annual electrical demand with wind power, is located near the Atlantic Ocean in Atlantic City, New Jersey in the US (WERF, 2010b). The capacity of this activated sludge plant with disinfection is 1.75 m^3/s with a daily power demand of 2.5 MW. The capital cost for the wind farm was relatively high, US$2000/kW, and was funded by means of a public-private partnership. When the prevailing wind speed exceeds 19 km/h, each of the 5 installed wind turbine produces 1.5 MW of electricity. Several challenges had to be overcome such as turbine design in a hurricane zone, community acceptance and environmental impacts with a periodical monitoring of the impact of wind turbines on birds.

Wind power is used for small capacity desalination plants (Subramani *et al.* 2011) with increased application for large plants. The largest wind powered desalination plant is operated since 2006 in Perth, Australia (Crisp & Rhodes, 2007; Sanz & Stover, 2007). The capacity of the Kwinana Desalination Plant is 143,000 m^3/d with designed expansion to 250,000 m^3/d. To minimize GHG emissions, the plant electrical requirement of 180 GWh/yr is entirely powered with green energy generated by 83 MW wind farm located 200 km north of Perth. This seawater reverse osmosis (SWRO) desalination plant implemented a number of measures to reduce energy consumption: energy recovery devices (ERI® PX Pressure Exchanger®), variable frequency drivers on supply pumps, booster pumps and second pass pumps, as well as state-of-the art low-energy membrane elements. At the nominal capacity and with an overall water recovery rate of 42%, the plant consumes less than 4.2 kWh/m^3 including intake, pre-treatment, both RO passes, post-treatment, potable water pumping and all electrical losses. This is a low energy level compared to the best performance for seawater salinity of 35,000–37,000 mg/L.

6.6 EXAMPLES OF ENERGY SELF-SUFFICIENT WASTEWATER TREATMENT PLANTS

Analysis of current practices shows that energy self-sufficiency of wastewater treatment is a feasible goal, which however requires long optimisation process for existing plants and relatively high investment and innovative technologies for new more energy efficient facilities (Wett *et al.* 2008; GWRC, 2010, 2011; Svardal & Kroiss, 2011). As mentioned previously, the first challenge for the existing WWTPs is the optimisation of treatment performance, reliable benchmarking and the implementation of selected adequate best practices and innovative products.

Robust benchmarking programs and guidelines for energy optimisation have been implemented in Europe since early 90s (Austria, France, Germany, Switzerland, Sweden), demonstrating the high potential of existing plants for energy optimisation from 20 to 50% (Lindtner *et al.* 2004, 2008; Jonasson, 2007; Wett *et al.* 2007). Similar energy reduction programs were launched in Australia, United States and Canada (Keller, 2008; WERF, 2010a). As a result, several wastewater treatment plants in Europe have achieved the goal of high energy efficiency and energy self-sufficiency.

Nowadays, two municipal WWTPs in Austria are energy self-sufficient, Strass WWTP, 220,000 p.e. (Figure 6.21a) and Wolfgangsee-Ischl WWTP, 50,000 p.e. (Wett *et al.* 2007; Novak *et al.* 2011) with a number of other on-going projects, such for example the WWTP of Vienna. Both of the Austrian energy-self sufficient WWTPs are activated sludge with nutrient removal (<1 mgPtot/L, >80% N removal) and energy optimisation lasted for about 20 years. The key energy optimisation measures for these two plants include optimal aeration control, higher recovery of particulate carbon from primary settlers for the anaerobic digestion, optimised mesophilic anaerobic sludge digesters, energy efficient CHP units, DEMON®-process for deammonification of the filtrate of sludge dewatering (Novak *et al.* 2011). The average electric energy produced from sludge digestion is 21.4 kWh/cap · yr for the Strass WWTP (from 2005 to 2007) and 20.6 kWh/cap · yr in the Wolfgangsee-Ischl WWTP for one year period (Sept 2009–Aug 2010). The surplus electric energy of the plants is fed to the grids, but for the peak energy demand, electricity and natural gas are taken from the grids. Since 2008, pre-conditioned organic substrate is directly fed into the digested of the Strass WWTP.

(a) (b)

Figure 6.21 Aerial views of the (a) Strass WWTP (Austria) and (b) As Samra WWTP (Jordan)

Very little information is reported in the literature on the design and operation of new energy sufficient WWTPs. Such a good example is the As Samra WWTP in Jordan (Figure 6.21b), designed and operated by Degremont, which is achieving the target of more than 90% of self sufficiency in energy since its commissioning in 2008 (Fievez, 2009). This new facility treats the wastewater of 2.2 million equivalent inhabitants of Amman and surrounding areas, with a capacity of 267,000 m^3/d, to produce a quality effluent for reuse in agriculture. The plant is an activated sludge with nitrogen removal and disinfection using chlorine, odour treatment and anaerobic digestion of mixed sludge (Figure 6.22). The energy demand of this plant (85 to 95%) is covered by next generation technologies such as hydraulic turbines and biogas driven cogeneration.

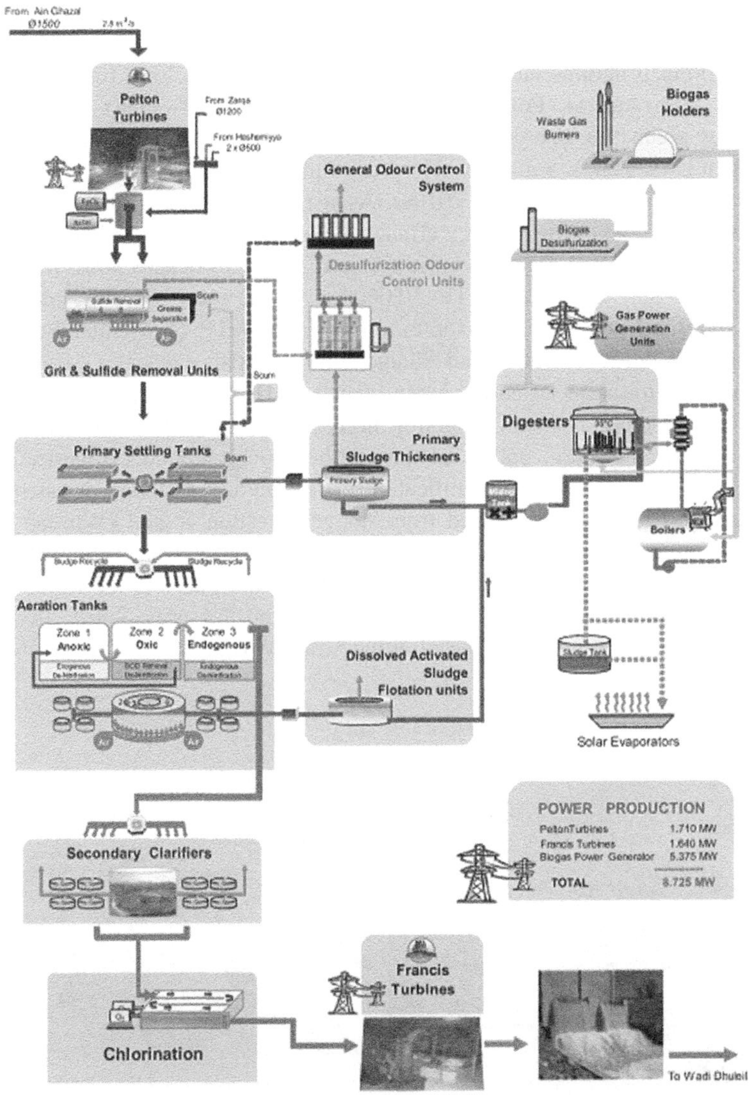

Figure 6.22 Flow sheet of As Samra WWTP in Jordan, 267,000 m^3/d, 2.2 million PE

Inlet turbines

As the altitude of Amman City is about 100 m higher than the plant altitude, the opportunity to install hydraulic turbines at plant inlet was studied at project stage. The available head and flows would theoretically allow the use of Pelton or Francis turbines. Two 875 kW 5-nozzle Pelton turbines (nominal discharge of 1.25 m^3/s, design net head of 79.7 m at 2.5 m^3/s) were selected because of several advantages for operation, e.g. less sensitivity to blocking or clogging by fibrous materials present in raw sewage, easier cleaning, tolerance of high flow variations between 10% and 100% of design discharge. In addition, these turbines were designed to resist to corrosion because of the high concentration of H_2S in sewers. More details on turbine characteristics can be found in Fievez (2009).

Outlet turbines

The height difference between the plant and the natural outlets of 42 m, also allowed for installing hydraulic turbines at plant outlet. Treated wastewater enters a 3000 m long pipe to feed the second plant where two Francis turbines of 838 kW capacity are installed (nominal discharge of 2.3 m^3/s, design net head of 41.64 m at 4.6 m^3/s). The engineering of the Francis turbines is more conventional as they are operating with treated secondary disinfected effluents. However, special anti-corrosion measures have been taken because of the chlorine concentration is higher than that of river water. Finally, the treated water is available in the desert for irrigation after generating about 3 MW of electrical energy.

Cogeneration from biogas

Biogas with an average production rate of 3000 Nm^3/h is generated in 4 digesters with a unit volume of 15,900 m^3. The biogas is desulphurized using two biological treatment lines, reducing the H_2S content to less than 500 ppm. The biogas produced in the digesters is stored in two gas holders each of 5,000 m^3 capacity. The biogas power is transformed into electricity and heat through six biogas cogeneration units of 1000 kWe each. The cogenerated heat is used to maintain the sludge temperature at 35–37°C in the digesters.

The key point to ensure reliable operation of cogeneration from biogas is the grade of biogas. For this purpose, a biogas monitoring system is installed, controlling the H_2S and CH_4 concentrations in the biogas at the inlet of the cogeneration units. Routine and periodic in-depth monitoring is performed of the biogas and motor oil to check whether the presence of these pollutants is limited to acceptable target values. The measured content of total siloxanes of the biogas is less than 1 mg/Nm^3.

Energy management system

Figure 6.23 illustrates the typical distribution of energy production of the As Samra WWTP for 2011. The electrical power provided by the turbines and the biogas generators is introduced to the 11 kV plant electrical network, reducing the electricity abstracted from the Jordanian network (JERCO grid). The energy produced is not sold externally but used toward reducing the plant consumption, and consequently, the electricity bill. An efficient energy management system allows the optimisation of energy production by means of Supervisory Control and Data Acquisition system (SCADA), energy meters and set points. For example, starting an air blower which constitutes the most important individual energy load, will lead to increase in power consumption from the JERCO grid, which is immediately detected. Consequently, the SCADA increases the biogas generators power set point and the biogas production, reducing the grid power consumption.

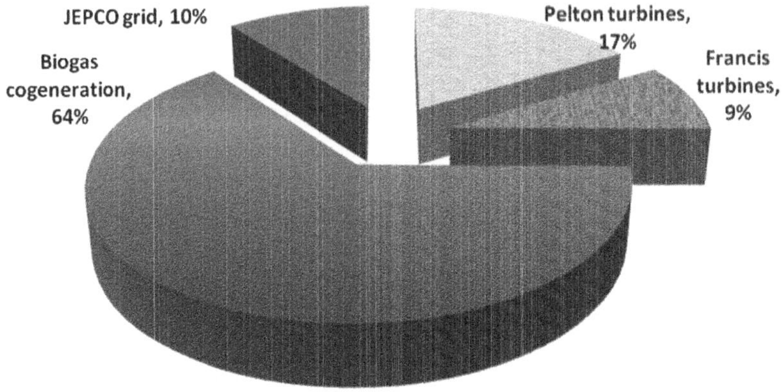

Figure 6.23 Typical distribution of the energy production in the As Samra WWTP in Jordan

The typical specific energy consumption for 2011 was 1.0 $kWh/kgBOD_5$ and 0.78 kWh/m^3 for average daily organic loads of 130–135% (COD and BOD_5) compared to the design values, which is equivalent to approximately 2.9 million equivalent habitants. Per population equivalent, the mean energy consumption is 21.3 $kWh/pe \cdot yr$ (population equivalent related to 110 gCOD/pe.d with a mean COD/BOD_5 ratio of 1.87). This value is slightly higher than the energy consumption of the Strass WWTP of 19.9 kWh/pe.yr (Novak *et al.* 2011), but the As Samra WWTP has also disinfection and odour treatment.

6.7 CONCLUSIONS

Energy self-sufficiency is becoming an important consideration in wastewater treatment. Climate change concerns and energy costs are the most important factors that drive this issue. The major challenge would be to significantly improve energy efficiency of wastewater treatment, and at same time favouring plant performance, improving water quality and minimising GHG emissions.

It is important to underline that energy self-sufficiency of wastewater treatment facilities can not be considered as an objective by itself, but as a component of the global water management strategy, taking into consideration the regional and local specificities, as well as environmental, social and economic aspects. Improvement of wastewater treatment performance is and should be the primary objective, followed by the selection of appropriate best practices and technologies for improvement of energy efficiency, maximization of the use of sludge for energy production and energy recovery from internal or external sources.

A number of mature and innovative technologies, tools and best practices ("toolboxes") are available for the recovery and reuse of energy embodied in wastewater. These "toolboxes" would enable to select various solutions that combine energy efficiency and environmental performance in an optimal balance, depending on the given energy and sustainable development objectives, and in accordance with the specific local needs and the energy management opportunities offered by the existing facilities.

In this chapter, four main components for design and operation of a "positive energy plant" (i.e., WWTP which produces more energy than uses) were identified, including three internal and one external source of energy reduction and generation:

(1) Energy savings from implementation of technologies and best practices for low energy consumption (up to 20% energy recovery):

- Use of best available energy management practices such as optimisation of aeration system design; efficient aeration control; use of premium efficiency motors and variable frequency drives for large pumps and aeration devices (blowers, mechanical aerators.);
- Innovative WWTP waste stream handling technologies such as sidestream treatment of concentrated return flow from anaerobic digestion by deammonification to reduce energy use for biological wastewater treatment;
- Next generation of treatment technologies for domestic wastewater, for instance anaerobic treatment or pre-treatment of the main wastewater flow, or decentralised wastewater treatment and reuse systems.

(2) Energy recovery from sludge (up to 60–80% energy recovery):

- Best practices to improve digester design and operation; enhanced primary settling efficiency;
- Innovative technologies for sludge pre-treatment before anaerobic digestion; co-digestion; combined heat and power generation from digester gas as electricity (cogeneration, fuel cells, microturbines) or mechanical energy (direct drive or stirling engines), gasification;
- Direct reuse of biogas after purification to gas grid;
- Valorisation of biosolids in agriculture or as a solid fuel.

(3) Energy recovery from sewage flows – thermal, hydraulic, potential (up to 10% energy recovery):

- Use of specific local conditions (topography, location, infrastructure) to recover energy from the WWTP influent and/or effluent;
- Innovative waste stream energy harvesting technologies such as hydro turbines and heat exchangers.

(4) Production of renewable energy from external sources such as solar, wind or geothermal energy (up to 10% energy recovery):

- Use of alternative renewable energy technologies which are feasible for the site specific climate conditions and financial incentives, and which require relatively low initial capital costs and site footprint and have short return-on-investment period;
- Innovative technologies such as photovoltaic-thermal system, micro wind turbines; combined waste-to-energy technologies.

Maximising the energy gain from each of the four potential sources of energy savings and generation listed above would allow wastewater treatment plants, especially the largest ones, to recover and eventually to generate all of the energy needed for plant operation, and even yield energy surplus at times. To be attractive, innovative technologies for energy recovery must be cost-effective, reliable, easy to operate, and without adverse impacts on water quality and the environment.

Maximal energy efficiency of wastewater treatment can be achieved only when energy management aims at the optimisation of the entire water cycle, including water reuse and nutrient recovery, as well as semi-centralised and decentralised water loops for the cities of the futures. While in its infancy at present, sustainable wastewater treatment and reuse have come of age and are likely to gain a greater importance in the planning of integrated water and wastewater management systems in the future.

Similar to water reuse, the unlocking and the enhanced reuse of energy contained in wastewater, is a key tool to solving the water-energy nexus which urbanized centres face worldwide.

Acknowledgement

This chapter is the result of the powerful collaboration of a large international and multi-disciplinary team of researchers, technical experts and operators from France, Spain, Germany, Slovakia, Singapore, Chile,

Jordan, UK and the US. The results from several research projects are included, mainly sponsored by R + I Alliance and Corporate R&D programs of Suez Environnement. I would like to especially thank Lynne Bouchy and her colleagues from CETaqua for their reactivity and relevant comments; as well ad Patricia Camacho and Sophie Besnault from CIRSEE – Suez Environnement. Discover the secrets of energy efficiency and practical feasibility of the broad range of innovative solutions were possible thanks to the invaluable assistance of my colleague Gilles Carrand. Special thanks for the fruitful collaboration of the energy teams of Degremont (Eric Fievez and Laure Graveleau) and Lyonnaise des Eaux (Frederic Duong and Irina Jaubert), as well as Hassan Abdallah, the General Manager of As Samra WWTP. We would like also to thank Peter Cornel, Bernhard Wett and Cao Ye Shi for their assistance and fruitful exchanges.

REFERENCES

Asano T., Burton F. L., Leverenz H. L., Tsuchihashi R. and Tchobanoglous G. (2007). Water Reuse – Issues, Technologies, and Applications. Metcalf & Eddy/WECOM, McGraw-Hill, New York.

Balmér P. (2000). Operation costs and consumption of resources at Nordic nutrient removal plants. *Water Science & Technology*, **41**(9), 273–279.

Barillon B., Martin Ruel S. and Lazarova V. (2011). Full scale assessment of energy consumption in MBRs. *6th IWA Specialist Conf. on Membrane Technology for Water & Wastewater Treatment*, Aachen, Germany, 4–7 October 2011.

Batstone D. J., Keller J., Angelidaki I., Kalyuzhnyi S. V., Pavlostathis S. G., Rozzi A., Sanders W. T. M., Siegrist H. and Vavilin V. A. (2002). The IWA Anaerobic Digestion Model No 1 (ADM1). *Water Science and Technology*, **45**(10), 65–73.

Bennamoun L. (2012). Solar drying of wastewater sludge: a review. *Renewable and Sustainable Energy Reviews*, **16**(1), 1061–1073.

Besnault S., Martin Ruel S., Carrand G. and Dauthuille P. (2011). Towards the positive energy WWTP: increase the efficiency of the overall aeration system. *Proc. Singapore International Water Week: Water Convention*, July 1–4, 2011, Singapore.

Bischof F., Höfken M. and Durst F. (1996). Design and construction of aeration systems for optimum operation of large wastewater treatment plants. *Water Science & Technology*, **33**(12), 189–198.

Blanco J., Malato S., Fernández-Ibañez P., Alarcón D., Gernjak W. and Maldonado M. I. (2008). Review of feasible solar energy applications to water processes. *Renewable and Sustainable Energy Reviews*, **13**(6–7), 1437–1445.

Bolzonella D., Pavan P., Battistoni P. and Cecchi F. (2006). Anaerobic co-digestion of sludge with orther organic wastes and phosphorous reclamation in BNR wastewater treatment plants. *Water Science & Technology*, **53**(12), 177–186.

Bouchy L., Pérez A., Camacho P., Rubio P., Silvestre G., Fernández B., Cano R., Polanco M. and Díaz N. (2012). Optimization of municipal sludge and grease co-digestion using disintegration technologies. *Water Science & Technology*, **65**(2), 214–220.

Bouchy L., Rouez M., Pérez A., Frank M., Fernández B., Polanco M. and Camacho P. (2010). Sludge co-digestion: a viable path towards energy self-sufficient WWTP. *Proc. ORBIT 2010, Organic resources in the Carbon Economy*. 29 June–3 July, Heraklion, Greece.

Bourgel C., Véron E., Poirier J., Defoort F., Seiler J. M. and Peregrina C. (2011). Behavior of phosphorus and other inorganics during the gasification of sewage sludge. *Energy and Fuels*, **25**(12), 5707–5717.

Camacho P., Ewert W., Kopp J., Panter K., Perez-Elvira S. I. and Piat E. (2008). Combined Experiences of Thermal Hydrolysis and Anaerobic Digestion Latest Thinking on Hydrolyis of Secondary Sludge Only. WEFTEC, October 2008, Chicago, US.

Camacho P., Gerard R., Garel C., Martin S., Monnot L. and Dauthuille P. (2010). Anaerobic co-digestion of lipid-rich wastes and excess sludge on wastewater treatment plant. *Water Convention, Singapore Water Week*, June 22–July 2, 2010, Singapore.

Camacho P., Li T., Martin S. and Dauthuille P. (2009a). Evaluation of new strategies for the improvement of anaerobic digestion. *Water Convention, Singapore Water Week*, 22–26 June 2009, Singapore.

Camacho P., Martin S. and Dauthuille P. (2009b). Technical expertise on the new strategies for the improvement of biogas production by anaerobic digestion. *IWA Conf. Water and Energy*, Copenhagen, Denmark, 29–31 October 2009.

Cao Y. S. (2011). Mass Flow and Energy Efficiency of Municipal Wastewater Treatment Plants. IWA Publishing, London.

Carrand G. and Martin Ruel S. (2006). Best practices of fine bubble aeration. Report R + I Alliance EN 0502, Suez Environnement.

Cillie G. G., Henzen M. R., Stander G. J. and Baillie R. D. (1969). Anaerobic digestion—IV. The application of the process in waste purification. *Water Research*, **3**(9), 623–643.

Crisp G. and Rhodes M. (2007). Perth Seawater Desalination Plant – Blazing a Sustainability Trail. IDA World Congress, Maspalomas, Gran Canaria, Spain, October 21–26.

Crujer W. and Zehnder A. J. B. (1983). Conversion processes in anaerobic digestion. *Water Sciences and Technologies*, **15**(8–9), 127–167.

De Arespacochaga N., Cortina J. L., Larrotcha E., Sánchez M., Peregrina C. and Bouchy L. (2010b). Sewage biogas to biomethane for grid injection: challenges and opportunities in Europe. *Proc. 7th Int. Conf. Organic Resources for the Carbon Economy ORBIT*, June 29–July 3, 2010, Heraklion, Crete, Greece.

De Arespacochaga N., Cortina J. L., Larrotcha E., Sánchez M., Peregrina C., Bouchy L. and Audic J. M. (2010a). Are High Temperature Fuel Cells Ready To Run On Sewage Biogas? *IWA Water and Energy Conf.*, Amsterdam, 10–12 November 2010.

De Arespacochaga N., Mesa C., Hornero A., Larrotcha E., Sánchez M., Peregrina C. and Bouchy L. (2011). Lessons Learned From the Operation of a Pilot Sewage Biogas-powered SOFC: Possibilities and Limitations. International Water Week, Aquaterra, 1–3 November, 2011, Amsterdam, The Netherlands.

De Baere L. (2000). Anaerobic digestion of solid waste: state-of-the-art. *Water Science & Technology*, **41**(3), 283–290.

Degrémont. (2004). Mémento technique de l'eau (Water Treatment Handbook). 10th édn, Tec & Doc Lavoisier. Cachan, France, ISBN: 2743007176, pp. 1874.

Deublein D. and Steinhauser A. (2008). Biogas from Waste and Renewable Resources – An Introduction. Wiley–VCH Verlag GmbH, Weinheim, pp. 389–400.

EIA, US Energy Information Administration (2012) Petroleum & Other Liquids: Daily Europe Brent Spot Price FOB. http://www.eia.gov/dnav/pet/hist/LeafHandler.ashx?n=PET&s=RBRTE&f=D (accessed 18 January 2012)

ESHA, European Small Hydropower Association. (2004). Guide on How to Develop a Small Hydropower Plant. Part 1 and 2. pp. 196 .

ESSEA (2010). Carbon Sequestration and the Copenhagen Accord. Earth System Education Alliance Courses, http://essea.strategies.org/module.php?module_id=160 (accessed 18 January 2012)

EuroObserv'ER (2010). Biogas Barometer, Le journal des énergies renouvelables N°200, Nov2010, http://eurobserv-er.org/pdf/baro200b.pdf (accessed December, 2011).

EuroObserv'ER (2011). Wind Power Barometer le journal de l'éolien N° 8–2011, http://www.eurobserv-er.org/pdf/baro201.pdf (accessed January 2012).

European Commission (2008). Energy efficiency for the 2020 goal. http://europa.eu/legislation_summaries/energy/energy_efficiency/en0002_en.htm (accessed 5 January 2012)

Fievez E. (2009). AS SAMRA WWTP: towards the energy self-sufficient wastewater treatment plant. *IWA Conf. Water and Energy*, Copenhagen, Denmark, 29–31 October 2009.

Fdz-Polanco F., Velazquez R., Pérez-Elvira S. I., Casas C., del Barrio D., Cantero F. J., Fdz-Polanco M., Rodriguez P., Panizo L., Serrat J. and Rouge P. (2008). Continuous thermal hydrolysis and energy integration in sludge anaerobic digestion plants. *Water Science & Technology*, **57**(8), 1221–1226.

Gavala H. N., Yenal U., Skiadas I. V., Westermann P. and Ahring B. K. (2003). Mesophilic and thermophilic anaerobic digestion of primary and secondarysludge. Effect of pre-treatment at elevated temperature. *Water Research*, **37**(19), 4561–4572.

Gendebien A., Davis B., Hobson J., Palfrey R., Pitchers R., Rumsby P., Carlton-Smith C. and Middleton J. (2010). Milieu Ltd for European Commission, DG Environment Report: DG ENV.G.4/ETU/2008/0076r, pp. 116.

Graveleau L., Descamps P., Fievez E., Gérard R., Hazard J. P. and Secondat M. (2012). New aeration control strategy for energy saving in biological wastewater treatment. World Congress on Water, Climate and Energy, 13–18 May 2012. Dublin, Ireland.

GWRC (2008). State of Science Report: Energy and Resource Recovery from Sludge. In: Y. Kalogo and H. Monteith (eds), Report for the Global Water Research Coalition (GWRC), WERF and STOWA, Alexandria, VA, USA.

GWRC (2010). Energy Efficiency in the European Water Industry: A Compendium of Best Practices and Case Studies. Report prepared for GWRC by Frijns J. (KWR) and Uijterlinde C. (STOWA), KWR 2010.011.

GWRC (2011). Water & Energy in the Urban Water Cycle: Improving Energy Efficiency in Municipal Wastewater Treatment. Report prepared by Nanyang Environment & Water Research Institute (NEWRI), PUB Singapore and Global Water Research Coalition (GWRC), Singapore.

Hellinga C., Schellen A., Mulder J. W., van Loosdrecht M. C. M. and Heijnen J. J. (1998). The SHARON process: an innovative method for nitrogen removal from ammonium-rich waste water. *Wat. Sci. Tech*, **37**(9), 135–142.

Higman C. and Van der Burgt M. (2003). Gasification. Gulf Professional Publishing, Elsevier Science, Burlington, US. ISBN: 0750677074. pp. 391.

Holm-Nielsen J. B., Al Seadi T. and Oleskowicz-Popiel P. (2009). The future of anaerobic digestion and biogas utilization. *Bioresource Technology*, **100**, 5478–5484.

Hung T. C., Shai T. Y. and Wang S. K. (1997). A review of organic rankine cycles (ORCs) for the recovery of low-grade waste heat. *Energy*, **22**(7), 661–667.

International Energy Agency (2007). Oil market report, October 2007.

Jonasson M. (2007). Energy Benchmark for Wastewater Treatment Processes – a Comparison between Sweden and Austria. Dept. of Industrial Electrical Engineering and Automation Lund University, Lund, Sweden. http://www.iea.lth.se/publications/MS-Theses/Full%20document/5247_full_document.pdf (acceded 15 January 2012).

Kaldate A., Vlaeminck S., Gonzalez Ospina A., Dyson J., Verstraete W. and Graveleau L. (2009). Fast start-up or a Deammonification SBR treating sludge digestate. WEFTEC 2009, 10–14 October 2009, Orlando, FL, USA.

Keller J. (2008). Paradigm Shift from Wastewater Treatment to Resource Recovery Systems. Australian-German workshop on sustainable urban water management, Melbourne, Australia.

Kolisch G., Schirmer G. and Salomon D. (2011). Options for an optimised utilisation of digester gas at sewage plants. *Water Practice & Technology*, **6**(2).

Kovac A. and Kralj D. (2010). Methanol production from biogas. *International Journal of Mathematics and Computers in Simulation*, **2**(4), 34–41.

Kuai L. and Verstraete W. (1998). Ammonium removal by the oxygen-limited autotrophic nitrification-denitrification system. *Applied and Environmental Microbiology*, **64**(11), 4500–4506.

La Cour Jansen J., Gruvberger C., Hanner N., Aspegren H. and Svärd Å. (2004). Digestion of sludge and organic waste in the sustainability concept for Malmö, Sweden. *Water Science & Technology*, **49**(10), 163–169.

Lazarova V. and Kamisoulis G. (2005). Financial aspects of the operation of sewage treatment plants. Programme for the Assessment and Control of Pollution in the Mediterranean Region, WHO, MED POL Phase III.

Lazarova V., Carrand G., Martin S., Bonroy J. L. and Dauthuille P. (2009). Enhancement of Energy Efficiency of Membrane Bioreactors. *IWA Conf. Water and Energy*, Copenhagen, Denmark, 29–31 October 2009.

Lazarova V., Li T., Martin S. and Dauthuille P. (2010). Toward energy self-sufficiency of wastewater treatment. *Water Convention, Singapore Water Week*, Singapore, June 22–July 2, 2010.

Lettinga G. and van Haandel A. C. (1993). Anaerobic digestion for energy production and environmental protection. In: Renewable Energy Sources for Fuels and Electricity, T. B. Johansson, A. K. N. Reddy and R. H. Williams (eds), Island Press, Washington, DC, pp. 817–839.

Li T., Martin S., Graveleau L., Saldate A. and Dauthuille P. (2010). Sidestream management in wastewater treatment plant towards a global energy efficiency. *Water Convention, Singapore Water Week*, June 22–July 2, 2010.

Lindtner S., Kroiss H. and Nowak O. (2004). Benchmarking of municipal waste water treatment plants (an Austrian project). *Water Science & Technology*, **50**(7), 265–271.

Lindtner S., Schaar H. and Kroiss H. (2008). Benchmarking of large municipal wastewater treatment plants treating over 100,000 PE in Austria. *Water Science & Technology*, **57**(10), 1487–1493.

Martin M. and Aguilera G. (2011). Energy audits of WWTPs. Internal reports Suez Environnement.

Martin Ruel S., Gonzalez Ospina A., Brauchli H. and Audic J. M. (2009). Technical, Economical and Environmental Study of Nitrate Shunt in Full-Scale SBR Treating Digester Centrate. *IWA 6th Leading-Edge Conference on Water and Wastewater Technologies*, Singapore, 23–25 June 2009.

Meda A. and Cornel P. (2010). Energy and Water: Relationships and Recovery Potential. *IWA Water and Energy Conf.*, Amsterdam, 10–12 November 2010.

Mulder J. W., Duin J. O. J., Goverde J., Poiesz W. G., van Veldhuizen H. M., van Kempen R. and Roeleveld P. (2006). Full-scale Experience with the Sharon Process trough the Eyes of the Operators. WEFTEC 2006, October 21–25, 2006, Dallas, TX, USA.

Nayono S. E. (2010). Anaerobic Digestion of Organic Solid Waste for Energy Production. KIT Scientific Publishing. Karlsruhe, Germany. ISBN 3866444645. 131 pp.

Nowak O., Keil S. and Fimml C. (2011). Examples of energy self-sufficient municipal nutrient removal plants. *Water Science & Technology*, **64**(1), 1–6.

Peregrina C. (2008). Energy in wastewater treatment plants. Internal Report, CIRSEE – Suez Environnement, pp. 33.

Peregrina C., Audic M. and Dauthuille P. (2011). Gasification as an optimum alternative for the sludge management. *Water Practice & Technology*, **6**(4), doi:10.2166/wpt.2011.080.

Peregrina C., Large M., Audic J. M., Monnot L., Pezzoni J. M. and Lesole M. (2008). WWTP upgrading towards positive energy balance: the case study of Dijon (France). *Singapore Water Week*, 24–25 June 2008.

Pérez-Elvira S. I., Fdz-Polanco M., Rodriguez P., Rouge P. and Fdz-Polanco F. (2008). Hydrothermal multivariable approach: Full-scale feasibility study. *IX Taller y Simposio Latinoamericano de Digestión Anaerobia* (IX Latin American Workshop and Symposium on Anaerobic Digestion), 19–23 October 2008, Isla de Pascua, Chile.

Rydh J. and Åkesson A. (2007). Energy Conservation in Wastewater Treatment Operation – A Case Study at Sjölunda and Klagshamn WWTPs. Lund University, Lund, Sweden, CODEN:LUTEDX/(TEIE-5248)/1–102.

Sanz M. A. and Stover R. L. (2007). Low Energy Consumption in the Perth Seawater Desalination Plant. IDA World Congress, Maspalomas, Gran Canaria, Spain October 21–26.

Seghezzo L., Zeeman G., van Lier J. B., Hamelers H. V. M. and Lettinga G. (1998). A review: the anaerobic treatment of sewage in UASB and EGSB reactors. *Bioresource Technology*, **65**(3), 175–190.

Silvestre G., Rodriguez-Abalde A., Fernandez B., Flotats X. and Bonmatí A. (2011). Biomass adaptation over anaerobic co-digestion of sewage sludge and trapped grease waste. *Bioresource Technology*, **102**, 6830–6836.

Simader G. (2004). Avanced Prediction, Monitoring and Corolling of Anaerobic Digestion Processes Behavior Towards Biogas Usage in Fuel Cells (AMONCO). Report of the first business interest group meeting, ENK6-CT-2001-00518. pp. 68.

Sliekers O., Derwort N., Campos-Gomez J. L., Strous M., Kuenen J. G. and Jetten M. S. M. (2002). Completely autitrophic nitrogen removal over nitrite in a single reactor. *Water Research*, **36**(10), 2475–2482.

Strouss M., Van Gerven E., Zheng P., Kuenen J. G. and Jetten M. S. M. (1997). Ammonium removal from concentrated waste streams with anaerobic ammonium oxidation (anammox) process in different reactor configurations. *Water Research*, **31**(8), 1955–1962.

Svardal K. and Kroiss H. (2011). Energy requirements for waste water treatment. *Water Science & Technology*, **64**(6), 1355–1361.

Tilche A. and Galatola M. (2007). The Potential of Anaerobic Digestion for Reducing GHG Emissions: A Qualitative Assessment for Europe in a Life Cycle Thinking Perspective. *11th IWA World Congress on Anaerobic Digestion*, Brisbane, Australia, 23–27 September 2007.

Thomas B., Bekker M., Kelm T., Oechsner H. and Wyndorps A. (2009). *Gekoppelte Produktion von Kraft und Wärme aus Bio-, Klär- und Deponiegas in kleinen, dezentralen Stirling-Motor-Blockheizkraftwerken* (Coupled heat and power production from biogas, sewage gas and landfill gas in small, decentralized Stirling engine cogeneration), Final report, BWPLUS Projekt No. 25008–25010.

Van Herelt J., Membrez Y. and Bucheli O. (2004). Biogas as a fuel source for SOFC co-generators. *Journal of Power Sources*, **127**, 300–312.

Verstraete W., Van de Caveye P. and Diamantis V. (2009) Maximum use of resources present in domestic 'used water'. *Bioresource Technolology*, **100**(23), 5537–5545.

Water 21 (2011). Anaerobic digestion. Wastewater Treatment: Technology Review 2011, Water 21.

WERF (2008). State-of-the-Science Energy and Resource Recovery from Sludge. In: Y. Kalogo and H. Monteith (eds) WERF Report OWSO3R07, WERF and STOWA, pp. 236.

WERF (2010a). Overview of State Energy Reduction Programs and Guidelines for the Wastewater Sector. In: R. Cantwell (eds), WERF Report OWSO6R07b, Water Environmental Research Foundation and IWA Publishing, pp. 120.

WERF (2010b). Energy Efficiency in Wastewater Treatment in North America: A Compendium of Best Practices and Case Studies of Novel Approaches. In: R. Cantwell (ed.), WERF Report OWSO4R07e, Water Environmental Research Foundation and IWA Publishing, pp. 80.

Werther J. and Ogada T. (1999). Sewage sludge combustion. *Progress in Energy and Combustion Science*, **25**(1), 55–116.

Wett B. (2006). Solved up-scaling problems for implementing deammonification of rejection water. *Water Science & Technology*, **53**(12), 121–128.

Wett B., Buchauer K. and Fimml C. (2007). Energy self-sufficiency as a feasible concept for wastewater treatment systems. *Proc. IWA Leading Edge Technology Conference*, Singapore, Sept. 2007, 21–24.

Chapter 7

Performance improvement of full scale membrane bioreactors

Christoph Brepols

7.1 INTRODUCTION

In recent years an increasing number of membrane bioreactors (MBR) for municipal and industrial waste water treatment have been built in Asia, Europe and the Americas. As confidence in the technology increased, so have the number and the capacity of full scale applications, which can now be found all over the world. This chapter illustrates strategies for improving nutrient removal rates and reducing energy consumption using the example of three full scale MBRs which have been operated for more than ten years under a common authority in the Erft river region of Germany.

The Erft river catchment area is a model territory for the introduction of MBR technology, as the following brief geographic profile reveals. The hydrology of the Erft region is shaped by mining of lignite and its conversion into energy. Low groundwater levels, low natural flow in many rivers and stringent discharge criteria for local wastewater treatment plants (WWTPs) are the consequence. In recent decades, advanced nutrient removal, often in combination with tertiary effluent treatment, has been established at all WWTPs operated by the Erftverband, a special statutory water association. Initial interest in MBR technology was spurred by technical limitations of conventional activated sludge (CAS) processes, such as bulking sludge.

The Erftverbands first MBR for municipal wastewater treatment, which went into operation in 1999, was also the first full-scale installation of its kind in Germany. A few years later, in 2004, the Erftverband launched the Nordkanal plant, which at the time was the world's largest municipal MBR. The third Erftverband MBR, at Glessen, was completed in 2008. These three MBRs are different in size and layout. Their designs reflect increasing knowledge of MBR technology for municipal applications. Over the years, process layouts have improved and efficiencies have increased. Still, enhancing treatment reliability and performance while reducing energy consumption and thus operational cost is a work in progress for the operators.

7.2 METHODOLOGY

The three MBRs which provide much of the source information for this chapter are operated under common technical supervision. In an ongoing effort to improve operations, overcome existing difficulties and better deal with unexpected complications, numerous studies and trials at full-scale have been conducted. In the course of such studies, the plants always had to remain fully operative. For this reason, in some cases pilot or lab-scale studies or computer models have been used to test and better understand critical operating conditions. In this discussion, selected data and findings are presented, which derive from full-scale trials conducted to help enhance biological nutrient removal and energy efficiency. The main technical specifications of all three MBRs are displayed in Table 7.1 followed by brief descriptions of their layout.

Table 7.1 Main technical specifications of the MBRs.

	Rödingen, 3,000 PE	Glessen, 9,000 PE	Nordkanal, 80,000 PE
Commissioned	1999	2008	2004
Dry-/rain weather inflow	300/3,200 m^3/d	2,000/6,500 m^3/d	12,000/45,000 m^3/d
Bioreactor	400 m^3 + 80 m^3 (filtration tanks)	1,600 m^3 + 320 m^3 (filtration tanks)	9,300 m^3
Membrane	5,280 m^2 hollow fibre modules, separate filtration units	12,100 m^2 hollow fibre modules, separate filtration units	84,480 m^2 hollow fibre modules, integrated filtration
Effluent requirements (peak concentrations)	COD < 30 mg/L NH_4-N < 4 mg/L at 5°C P_{tot} < 1.5 mg/L	COD < 30 mg/L NH_4-N < 1.5 mg/L at 5°C P_{tot} < 0.6 mg/L	COD < 90 mg/L NH_4-N < 10 mg/L at 10°C N_{tot} < 18 mg/l P_{tot} < 1.5 mg/L

PE: people equivalent

The Rödingen plant was designed to pioneer MBRs at full-scale. It employs two equally equipped bioreactor tanks and two separate filtration lines and features several options to combine the biological process and membrane filtration units (Figure 7.1). From today's perspective portions of the equipment are out-dated. The equipment is not variable enough to deal with flow variations optimally, and therefore, operations are often laborious.

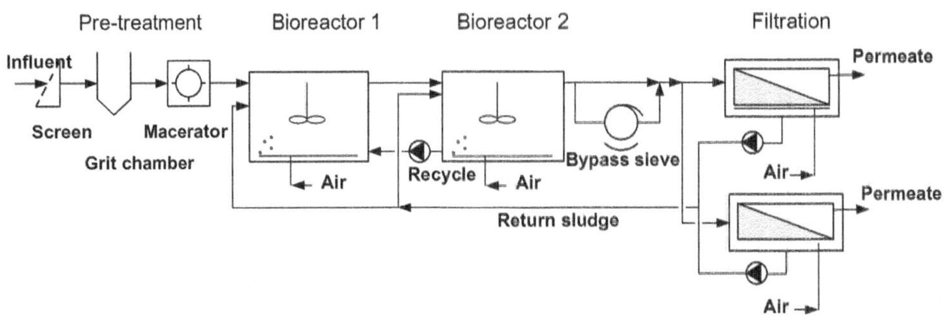

Figure 7.1 Process flow scheme of the Rödingen MBR

At Nordkanal MBR, the membrane filters are directly integrated into the bioreactors. Figure 7.2 gives an impression of the process layouts of the Nordkanal MBR. Two filtration lines, each consisting of 24 hollow fibre membrane cassettes, are immersed into each of the four parallel nitrification basins.

Figure 7.3 gives an idea of the size of the facility with the aeration-, permeate- and air-scouring-pipes installed atop the bioreactors.

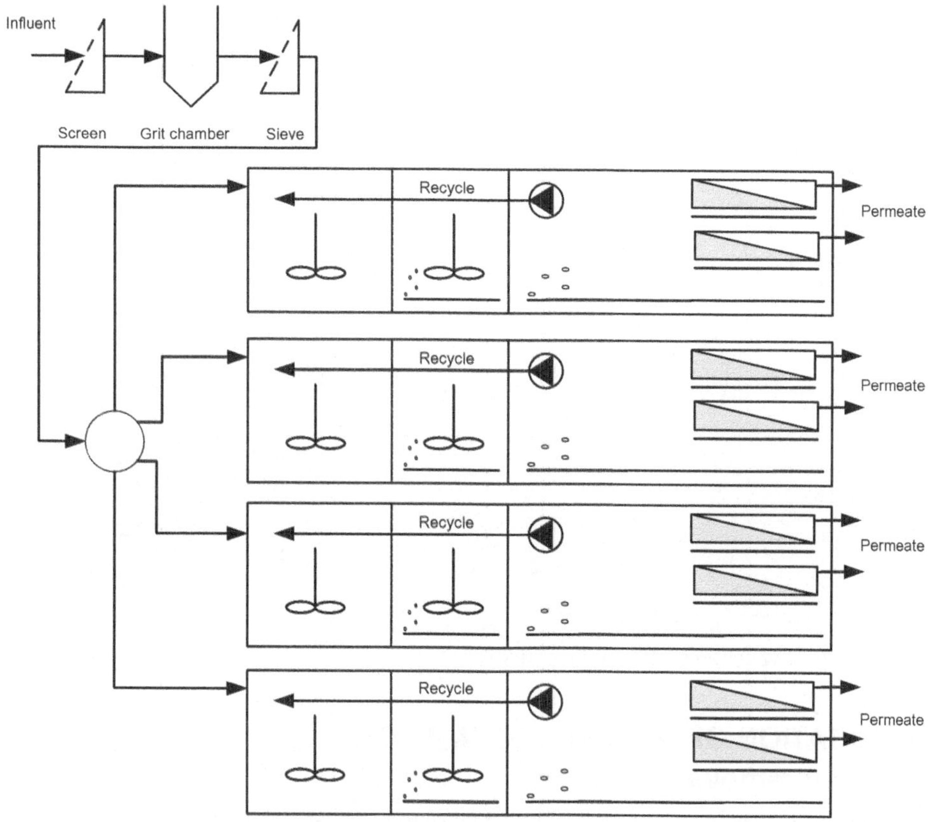

Figure 7.2 Process flow scheme of the Nordkanal MBR

Figure 7.3 Nordkanal MBR, view of the nitrification/filtration basins

The Glessen MBR was designed to retrofit an existing CAS plant. The existing secondary clarifier was replaced by separate filtration units while the old bioreactor with surface aerators is still in use. The membrane filtration has been installed in four separate basins (Figure 7.4). This multi-lane design facilitates the operational flexibility and reliability that is required to deal with considerable variations of flow, load and temperature while respecting the strict discharge limits and while the inevitable regular membrane cleaning is performed.

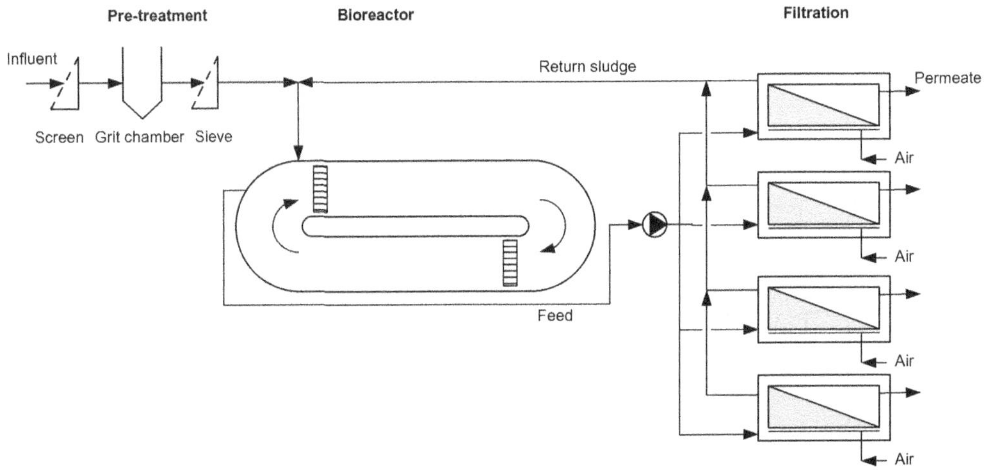

Figure 7.4 Process flow scheme of the Glessen MBR

7.3 OPTIMISATION OF MBR PERFORMANCE

7.3.1 Nitrogen removal

The three MBRs have been designed for biological nutrient removal and a low food to biomass ratio F:M resulting in sludge retention times (SRT) of 25 days or more. The biomass loading is similar to CAS plants with simultaneous aerobic stabilisation, and so are the average removal rates. However, the manifold biomass concentration leads to reduced bioreactor volumes; the hydraulic residence time (HRT) is significantly shorter (see Table 7.2).

Table 7.2 Quantities of key design parameters, MBRs and CAS plants with simultaneous stabilisation.

Parameter	Rödingen	Glessen	Nordkanal	CAS
Sludge concentration, kg MLSS/m^3	12–18	8–12	12	3–5
F:M ratio, kgBOD/(kgMLSS d)	0.04	0.03	$<$0.05	$<$0.05
Hydraulic retention time (HRT), h	3.6	6.2	4.6	$>$14
Sludge retention time (SRT), d	25	$>$25	25	$>$25

In a CAS the large bioreactors and additional volume of the secondary clarifiers provide a hydraulic buffer which compensates high variations in effluent concentrations. MBRs lack this buffer. As a result, variations of effluent concentrations are more pronounced. There is less time to react to sudden changes in influent conditions, especially when extremely low peak effluent concentrations must be respected. Apart from that, the excess oxygen supply induced by the membrane filtrations' air scouring also affects bioreactor aeration and involuntary oxygen carryover via the sludge recycle may impede denitrification. As a result, process control for bioreactor aeration, nitrogen removal and phosphorus precipitation at MBRs, can be more challenging than at CAS plants.

At both the Rödingen and Glessen facilities, difficulties involving nitrogen removal became evident soon after initiation of service, while at Nordkanal MBR, the denitrification rate was found to be higher than expected. This section describes the distinct situations at each plant and the adjustments that were made in their operation.

The Rödingen MBR was designed to test different strategies for nitrogen removal. The plant is equipped to operate either with pre-denitrification or intermittent denitrification. Although only ammonia effluent concentrations and thus full nitrification is mandatory, denitrification was meant to reduce the overall aeration demand in the bioreactors. Like many smaller sized WWTPs in the region, the Rödingen MBR is exposed to low winter temperatures and pronounced daily variations of flow and load. At Rödingen, the size of the bioreactors invariably fixed the anoxic to aerobic volume ratio of pre-denitrification at 100%. The treatment process often lacked aerobic volume, and thus, nitrification capacity to meet the discharge limits.

Intermittent denitrification finally turned out to be the superior strategy: It allows the operator to adjust the nitrification capacity by prolonging the aerobic phases. Denitrification is omitted entirely when wastewater temperatures decline below approximately 10°C.

At the Nordkanal MBR denitrification is mandatory. With respect to its size and the different filtration layout, the plant was equipped with pre-denitrification. It has been found, that nitrogen removal rates are slightly higher than expected, because the on/off operation of the filter lines within the nitrification basins seems to create unintended temporary anoxic niches for nitrate removal (Brepols, 2010).

At Glessen MBR, the ammonia effluent limit values are even more stringent than those at Rödingen. Like Rödingen, Glessen MBR uses intermittent denitrification. Although the process control strategy for nitrogen removal was designed with the assistance of an activated sludge model computer simulation, it was unsatisfactory from the beginning. The original idea was to determine the dissolved oxygen concentration in the bioreactor and the duration of the nitrification and denitrification phase with respect to the measured effluent concentrations of ammonia and nitrate. An ammonia process photometer in the final effluent was used as a feed-back signal for this control loop. During normal dry weather, the nitrification in the bioreactor is further enhanced by nitrification in the aerated filtration basins such that ammonia concentrations in the permeate are extremely low and with negligible variations (see Figure 7.5). As a consequence, there was barely a signal to trigger the automatic control actions. Unfortunately, recurring peak load exceeded the nitrification capacity. The hydraulic retention time in the filtration tanks and sample preparation time created a delay of more than 45 minutes before a critical increase of the effluent concentration could be detected. Under these conditions, the surface aerators increased the oxygen supply too slowly. The whole control action was too delayed to react adequately to load variations, resulting in frequent ammonia effluent peaks.

In order to ensure the compulsory value of <1.5 mg/L NH_4-N, the dissolved oxygen concentration in the bioreactor was increased and aerobic conditions in the bioreactor were maintained over long periods, both at the cost of a low denitrification rate. In May 2009, the old ammonia analyser was replaced by an ion-selective ammonia probe inside the bioreactor. This drastically reduced the response time of the system, furthermore the set-points of the control loop were adjusted in favour of extended denitrification phases, while ammonia effluent peaks could still be diminished successfully. Figure 7.6 shows that the total nitrogen effluent was reduced from approximately 7 to 3 mg/L without a significant increase in the ammonia average (Brepols, 2010). With reference to the following section, it may be noted, that these operating conditions also reduced the energy consumption for bioreactor aeration.

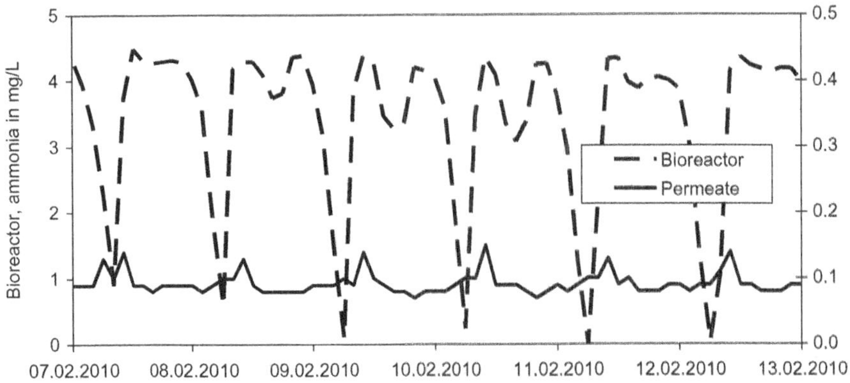

Figure 7.5 Daily variations of the ammonia concentration in the bioreactor and the permeate effluent at the Glessen MBR

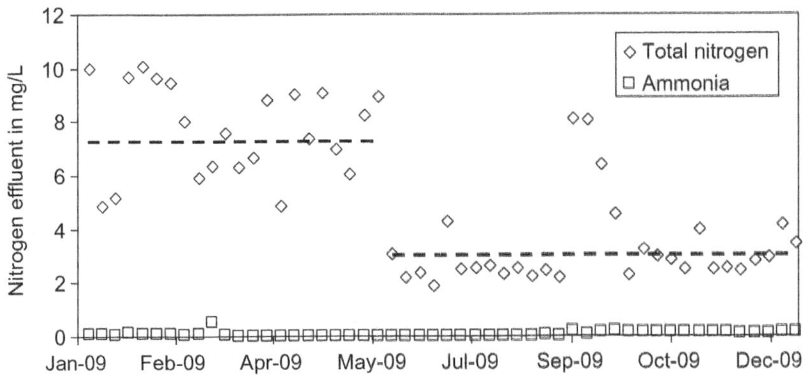

Figure 7.6 Nitrogen effluent concentrations before and after optimizing the process control loop at Glessen MBR

7.3.2 Energy consumption

It is widely recognised, that energy consumption for aeration of activated sludge at municipal WWTPs clearly exceeds that of any other treatment step (Müller *et al.* 1994). At MBRs, the overall aeration demand is further increased by the air scouring that is required for operation of immersed membrane filtration.

When MBRs operate on mixed sewer systems, they have to deal with significant flow variations between average dry-weather and maximum storm weather conditions. Membrane filtration has to be designed for extreme conditions with maximum inflow at minimum wastewater temperatures. Plant equipment and operational set-points are thus, in many cases, tailored to handle the most unfavourable conditions without regard for energy saving.

As a rule, operators are also expected to extend membrane service life, which is the single most important factor in plant economics (Brepols *et al.* 2010; Judd, 2011). This consideration also contributes to conservative design and operational practice that stand in the way of increase energy efficiency. The actual influence of these aspects is illustrated by the example of the Nordkanal MBR.

A detailed energy audit at Nordkanal MBR (Pyöry, 2009) has revealed that the main driver of energy consumption is the air scouring of membranes, which takes an average share of 49.2% of the overall energy consumption. Process aeration in the bioreactor (12.1%), biomass circulation (1.3%), bioreactor mixing (11.5%) and permeate suction (2.7%) take significantly lower shares. The remainder (23.2%) is consumed by pumping stations, pre-treatment, sludge dewatering and miscellaneous process units.

Energy consumption and operational cost are to a large extent pre-defined by the MBR design and the intended operational regime. De Wever *et al.* (2009) and Verrecht *et al.* (2010) demonstrated the influence of various design options. Verrecht *et al.* (2008) concluded from a review of design and operational practice at several MBRs that there is a trade-off between (design) membrane flux and aeration energy demand. The average energy consumption is affected by the design hydraulic peak load, as well as the actual variations in flow. In the following comparison, the hydraulic utilisation ratio (HUR) is used to describe the relationship between average daily dry weather flow Q_d and maximum design flow Q_m as the quotient of the two values. Table 7.3 shows key operational figures for the three Erftverband MBRs and their average HUR in comparison to average values from 14 CAS plants with additional tertiary treatment.

Table 7.3 Operational key figures of three MBRs and weighted average values for CAS plants with tertiary treatment.

Parameter	Rödingen	Glessen	Nordkanal	CAS
Capacity, P.E.	3,000	9,000	80,000	40,500
Maximum design flow, m^3/d	3,200	6,400	45,000	51,000
Actual average hydraulic utilisation ratio Q_d/Q_m	0.15	0.32	0.34	0.25
Actual average specific energy consumption, kWh/m^3	2.2	1.0	0.9	0.5

The HUR at Glessen and Nordkanal MBR are slightly above the CAS average. The Rödingen MBR receives a considerably low load in dry weather, resulting in a low HUR. Its equipment is aged and in some cases sub-optimal. Despite the differences in age, equipment, capacity and load, the actual specific energy demand decreases to about 0.5 kWh/m^3 at all three MBRs as soon as the membrane filtration is operated near its maximum design flux. At present, this seems to be the technically attainable minimum value.

Other authors have reported similar findings. For several MBRs in Germany, Veltmann *et al.* (2011) have recently reported energy consumption between 0.74 and 1.83 kWh/m^3 of wastewater treated. Air scouring contributed from 34% to 71.5 % of total energy demand. It has also been reported that energy consumption is strongly related to plant hydraulic load. Another detailed energy audit of a full scale MBR system in Belgium (Fenu *et al.* 2010) gives an average energy consumption of 0.64 kWh/m^3. The plant is operated side by side with a CAS. The MBR mainly treats the dry weather inflow of the entire facility. In this case, the average hydraulic utilisation ratio of 0.65 clearly exceeds that of the MBRs presented in Table 7.1. The inflow variations at the plant and the HUR strongly influence energy demand. There are several points which contribute to this observation.

Membrane module hydraulics and filtration performance are largely influenced by the air scouring. The appropriate aeration rate is often tied in with warranty conditions by the membrane supplier and thus cannot be deliberately changed by operators, although more recent recommendations describe altered air scouring regimes that would at least temporarily reduce the required aeration rates. At Nordkanal MBR, blowers for air scouring are operated at a constant speed and thus at a constant SADm (specific aeration demand per m^2 membrane surface). In practical operations, the applied membrane net flux can be variable in a certain range, for example between 17 and 24 $Lm^{-2}h^{-1}$ (and only sporadically up to 35 $Lm^{-2}h^{-1}$) as at Nordkanal. Figure 7.7 shows how the effective aeration demand per unit permeate SADp decreases with increasing operational flux. As a consequence, the rate of energy consumption also decreases with an increase in hydraulic load.

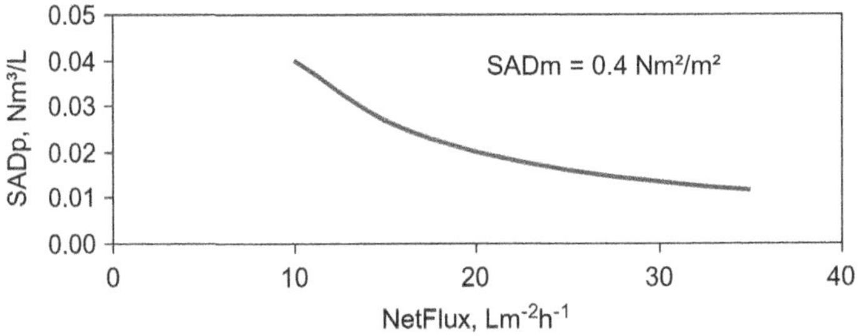

Figure 7.7 Variation of the SADp at variable membrane net flux and constant SADm

In general, hydraulic capacity of membrane filtration is automatically adjusted by reducing the number of active filtration lines, but other factors increasing energy use also reside in the automatic process control. Figure 7.8 compares average blower runtime of all eight filtration lines at Nordkanal MBR to hydraulic loading. Ideally, aerator runtime would be equal to permeate pump runtime and thus close to the dashed line indicated in the graph. In practice, several operational measures lead to increased aerator runtime: (1) the automatic sequence chain starts blower operation before permeate production begins; (2) when an increase in inflow is measured at the pumping station, several filtration lines switch to a so-called stand-by mode where blowers are already running and permeate pumps are (still) off; (3) at very low inflow, permeate production is interrupted temporarily to adjust to the inflow while blower operation continues; (4) inflow is evenly distributed to the four bioreactor lines. This further reduces inflow to lines that are actually in operation.

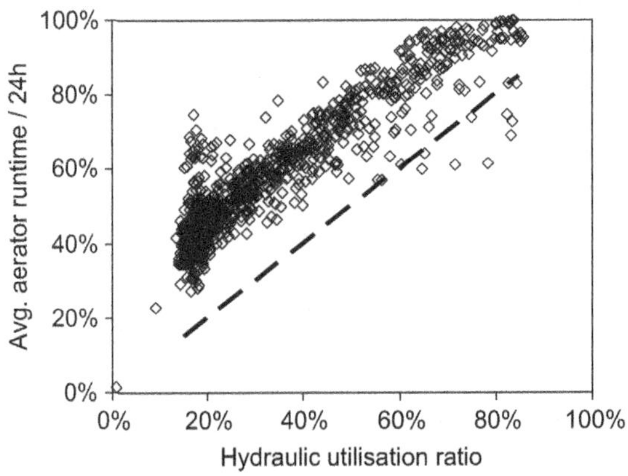

Figure 7.8 Average daily aerator runtime compared to actual hydraulic utilisation ratio at Nordkanal MBR

All these measures are established to ensure sufficient filtration performance under adverse conditions. During storms, a steep inflow increase may occur, and the full filtration performance has to be maintained when wastewater temperatures drops below 10°C. However, dry weather conditions with less than 30% of the maximum design flow prevail more than 60% of the time (Figure 7.9). Consequently, such dry weather conditions, with excess air scouring of up to 50%, lead to a specific energy demand of 1.7 kWh/m^3 and have a dominating influence on overall energy consumption. It can thus be summarised, that there appears to be huge potential for reducing energy consumption by carefully managing process control and operating set-points in a multi-lane membrane filtration as at Nordkanal MBR.

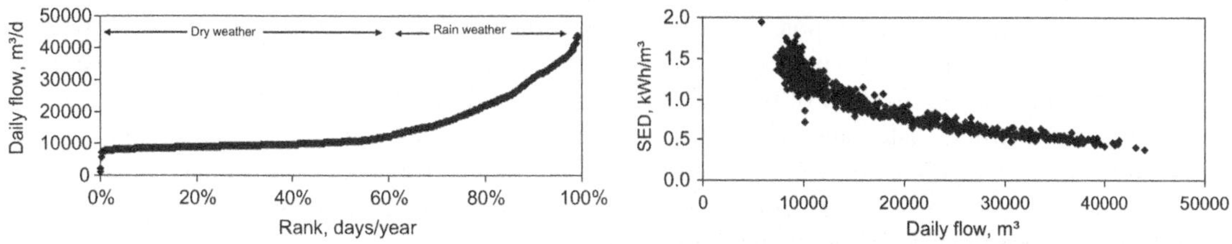

Figure 7.9 Distribution of dry and wet weather flows (left) and the specific energy demand SED (right) at Nordkanal MBR

7.4 IMPROVEMENT OF MBR ENERGY EFFICIENCY

7.4.1 A new operating paradigm

Over time, the benefits of MBR have become more self-evident while energy requirement of membrane filtration remains a major argument against the use of MBRs. Reducing energy use is not only an economic but also an environmental necessity, and generally, increasing energy efficiency of all kinds of WWTPs has become highly relevant. As has been seen, energy consumption is tied in with operational reliability in various ways, but there are also other factors playing an important role in plant reliability.

The effort to improve membrane cleaning and maintenance described by Brepols *et al.* (2008), improvements of mechanical wastewater pre-treatment and ancillary filtration equipment resulted in a stable and enduring performance of the membrane filters. Years of operational experience with membrane filtration at Nordkanal raised the confidence that the filters are capable of treating the maximum inflow even under extremely unfavourable conditions. This changed the equation between reliable membrane filter operation and desirable energy reduction. As a result, the operational strategy of the Nordkanal MBR was revised and five key measures to practically reduce energy consumption were identified (Janot *et al.* 2009):

(1) Decreasing and adjusting the actual specific aeration demand per membrane unit SADm by equipping the blowers with variable-frequency instead of constant-speed drive.
(2) Decreasing actual aeration demand per unit permeate produced SADp by increasing the lower set-point of the membrane flux.
(3) Improving flow distribution to the four bioreactor lines to reduce the number of filtration lines that are in stand-by mode when inflow is low.
(4) Increasing the permeate yield by: (a) maximising the duration of the filtration cycles and (b) reducing frequency, flux and duration of the backwash.
(5) Temporarily increasing the maximum operational net flux even further when wastewater temperatures are high and filter performance is good and finding the optimal MLSS concentration in the bioreactor/filtration compartment.

Although these measures have been adopted by Veltmann *et al.* (2011) as general recommendations for achieving energy efficient operation of MBRs, they bear a risk of irreversibly damaging membrane performance through increased fouling and clogging of the filtration units if applied uncritically and incautiously. Long-term success thus relies on careful, continuous and realistic estimation of actual filtration performance. Unfortunately, many of the causal relationships between plant equipment and operation, biological process properties and hydraulic membrane performance remain unclear (Drews, 2010).

For the moment, the key to success lies mainly within the experience and responsibility of the plant operators. For the future, there is potential for further automation. Melin *et al.* (2007) and Huyskens *et al.* (2011) have described how the use of sophisticated control algorithms could further increase the reliability, variability and efficiency of the filtration process in MBRs.

The five key measures listed above aim at the automation of the altered air scouring regime and the increased net flux. In combination with the other measures listed in Table 7.4, it is estimated that a 20 to 30% cumulative reduction in energy consumption might be achieved at Nordkanal MBR, reducing the average specific consumption to less than 0.7 kWh/m^3.

Table 7.4 Energy savings potential, Nordkanal MBR.

Measure	Estimated average energy savings, kWh/m^3
Altered air scouring regime	0.10
IE3 premium efficiency blowers for air-scouring	0.06
Increased net flux	<0.05
Variable speed controllers for bioreactor mixing	0.04
Improved bioreactor process aeration	<0.01
Improved sludge management and dewatering	<0.01
Replacement of other minor equipment	<0.01

Several authors have reported an even further decrease of the specific energy consumption at other installations. Pyöry (2009) estimates that ideally, an average energy consumption of 0.45 kWh/m^3 could be achieved. Garces *et al.* (2007) and Tao *et al.* (2009) report that a 35% to 40% reduction in energy consumption for air scouring can be attained by applying a different aeration regime, while Tao *et al.* (2009) gives a value of 0.4 kWh/m^3 for an energetically optimised MBR with a capacity of 23,000 m^3/d.

7.4.2 Practical application

In the year of 2011 the average operating flux at Nordkanal was increased to peak values of 40 $Lm^{-2}h^{-1}$ without seeing a negative influence on membrane permeability. Although permeate pump runtime was reduced accordingly, this did not produce a measurable effect on the entire plant's energy consumption because the blower runtimes remained relatively unaffected. It must be acknowledged, that simply tuning the set-points of the filtration process control is not sufficient to yield substantial energy reductions. Initial efforts in modifying the plants supervisory control and data acquisition (SCADA) system and the programmable logic controller (PLC) units revealed that the system was not capable of handling the intended modifications without putting operational reliability at risk. A reasonable energy reduction thus requires more profound changes in the plant's equipment.

A complete overhaul of the MBRs SCADA and PLCs is now in the making. It requires extreme caution, as the plant will remain under full operation at any time. It is expected that the savings listed in Table 7.4 will only be achieved gradually. Some of them, like the improved bioreactor process aeration and sludge management and dewatering, have only a minor influence on the average consumption, but are important in reducing peak consumption. The pay-off period for these changes is estimated to be shorter than three years.

Further energy savings may be more expensive and have to wait for inevitable module replacements and the retrofit of their peripheral equipment, which will pave the way to more energy efficient generations of membrane filtration modules. In recent years much development of commercial membrane filter systems has already focused on reducing aeration whilst maintaining membrane permeability (Judd, 2011). It can be expected that this development will continue.

7.5 CONCLUSIONS

As with any WWTP, the operation of the MBRs requires a constant effort of analysing, maintaining and improving current treatment performance. Conventional biological processes require as much attention as does the membrane. Optimised biological process control improves effluent quality and helps reduce energy demand. High operational fluxes can be sustained under favourable conditions. Reduced energy consumption of the filtration units can thus be facilitated by improving multi-lane operation at large MBRs and adjusting the operational set-points. Unfortunately, the attainable effects of process optimisations are often limited by the original plant design and equipment. Assessing current plant performance, improving membrane filter maintenance and plant equipment, rethinking flow management and the automation concept may open new ways to reduce the energy consumption without compromising reliability. It could be seen that, even within an existing MBR, there is still potential to economically reduce energy consumption.

REFERENCES

Brepols C. (2010). Operating Large Scale Membrane Bioreactors for Municipal Wastewater Treatment. IWA Publishing 2010, London, UK.

Brepols C., Drensla K., Janot A., Trimborn M. and Engelhardt N. (2008). Strategies for chemical cleaning in large scale membrane bioreactors. *Water Science & Technology*, **57**(3), 457–463.

Brepols C., Schäfer H. and Engelhardt N. (2010). Considerations on the design and financial feasibility of full-scale membrane bioreactors for municipal applications. *Water Science & Technology*, **61**(10), 2461–2468.

De Wever H., Brepols C. and Lesjean B. (2009). Decision tree for full-scale submerged MBR configurations. Final MBR-Network Workshop 'Salient outcomes of the European R&D projects on MBR technology' 31 March – 1 April, Berlin, Germany.

Drews A. (2010). Membrane fouling in membrane bioreactors – characterisation, contradictions, cause and cures. *Journal of Membrane Science*, **363**, 1–28.

Fenu A., Roels J., Wambecq T., De Gussem K., Thoeye C., De Gueldre G. and Van De Steene B. (2010). Energy audit of a full scale MBR system. *Desalination*, **262**(1–3), 121–128.

Garces A., De Wilde W., Thoeye C. and De Gueldre G. (2007). Operational cost optimisation of MBR Schilde. 4th International Water Association Conference on Membranes for Water and Wastewater Treatment, 15–17 May 2007, Harrogate.

Huyskens C., Brauns E., Van Hoof E., Diels L. and De Wever H. (2011). Validation of a supervisory control system for energy savings in membrane bioreactors. *Water Research*, **45**(3), 1443–1453.

Janot A., Drensla K. and Engelhardt N. (2009). *Energieeinsparung bei Membranbelebungsanlagen durch ein wirtschaftliches Steuerungssystem* (Energy savings in membrane bioreactor plants through an economic control system). In: Begeleitbuch zur 8. Aachener Tagung Wasser und Membrane, 27–28 Octobre 2009, Aachen.

Judd S. (2011). The MBR Book. 2nd edn, Elsevier, Oxford, UK.

Melin T., Lyko S., Batsch A., Wintgens T., Dott W., Al-Halbouni D., Hollender J., Pinnekamp J., Grinwis S., Schürmann B., Tacke D., Drensla K. and Janot A. (2007). *Strategien zur Foulingkontrolle bei Membranbelebungsanlagen in der kommunalen Abwasserbehandlung* (Strategies on fouling control in membrane bioreactor for municipal wastewater treatment), *Abschlussberichtes für das Ministerium für Umwelt, und Naturschutz, Landwirtschaft und Verbraucherschutz des Landes Nordrhein-Westfalen (MUNLV)*, (final report to the ministry of environment and nature protection, agriculture and consumer protection of north rhine - westpahlia, MUNLV), http://www.lanuv.nrw.de/wasser/abwasser/forschung/pdf/abschlussberichtstrategien.pdf, Aachen 2007, (accessed 28 February 2012).

Müller E., Thommen R. and Stähli P. (1994). *Energie in ARA* (Energy in WWTP), *Bundesamt für Konjunkturfragen* (Federal Office for Economic Issues), 3003 Bern (CH).

Pöyry (2009). *Abschlussbericht zum Forschungsvorhaben Ermittlung energetischer Beurteilungskriterien für Membrankläranlagen mit Hohlfasermodulen und deren Verifizierung an einer Großanlage* (Final report on the research project to investigate energetically evaluation criteria for membrane bioreactors with hollow fibre modules and their verification at a large scale plant), Auftraggeber, Erftverband Bergheim, Verfasser: Pöyry GKW GmbH, Essen, den 27.02.2009.

Tao G., Kekre K., Oo M-H., Viswanath B., Lew C-H., Kan L-M. and Seah H. (2009). Large scale membrane bioreactor plant design (retrofit) and optimisation, *Proceedings of the 4th IWA Membrane Technology Conference*, Beijing, China, Sept 1–3.

Veltmann K., Palmowski L. and Pinnekamp J. (2011). *Reduzierung des Energieverbrauchs großtechnischer Membranbelebungsanlagen* (Reduction of the energy consumption of full scale membrane bioreactor plants). *Korrespondenz Abwasser* (Correspondence Wastewater), **58**(6), 566–571.

Verrecht B., Judd S., Guglielmi G., Brepols C. and Mulder J. W. (2008). An aeration energy model for an immersed membrane bioreactor. *Water Research*, **42**(19), 4761–4770.

Verrecht B., Maere T., Nopens I., Brepols C. and Judd S. (2010). The cost of a large-scale hollow fibre MBR. *Water Research*, **44**(18), 5274–5283.

Chapter 8

Energy optimization of large scale membrane bioreactors – importance of the design flux

Laurence Palmowski, Katrin Veltmann and Johannes Pinnekamp

8.1 INTRODUCTION

The advantages of membrane bioreactors (MBR) for the treatment of municipal wastewater include a small footprint, a full retention of suspended solids as well as a low microbial contamination of the effluent (Stephenson *et al.* 2000; Pinnekamp & Friedrich, 2006; Judd & Judd, 2006) and have led to an increase in the number of MBR plants worldwide (Atkinson, 2006).

The energy demand of MBR plants is, however, their main drawback. Average values for the specific energy consumption of MBR are reported in the literature in the range of 0.64–2.3 kWh/m^3 of treated effluent, depending on the size and operating conditions of the plant (Krause, 2005; Hübner and Rolfs, 2007; Novotny, 2010; Brepols, 2011). This energy demand is two to four times higher than that of conventional plants with an average specific consumption of 0.28–0.55 kWh/m^3 (Judd & Judd, 2006; Novotny, 2010).

The main reason for the higher energy demand of MBR is the need for an energy intensive cross-flow (CF) aeration of the membrane modules in order to reduce membrane fouling and maintain satisfactory hydraulic conditions. To be able to treat rain weather flows, sufficient equipment capacity, including CF-aeration has to be available. This imperative usually leads to a sub-optimal use of the equipment at time of low influent flows.

Approaches to increase energy efficiency of MBR, with a particular focus on the operation at dry weather flow and the importance played by the design flux are discussed in the following based on long-term data from large scale MBR plants.

8.2 METHODOLOGY AND STUDY SITES

8.2.1 Study sites

7 full-scale municipal MBR plants located in Germany, with a total membrane surface area of 158,000 m^2 were considered. Details of the plants are presented in Table 8.1.

Table 8.1 Details of investigated MBR plants.

Plant	Population Equivalent (PE)	Dry weather inflow (m^3/d)	Membrane Area (m^2)	Type of Module (–)
MBR Plant A	80,000	25,000	84,480	Hollow fibre
MBR Plant B	11,625	1,040	10,240	Flat sheet
MBR Plant C	1,995	285	2,100	Flat sheet
MBR Plant D	9,700	4,100	23,040	Flat sheet
MBR Plant E	6,200	850	13,440	Flat sheet
MBR Plant F	11,500	2,030	12,480	Flat sheet
MBR Plant G	6,500[(1)]	3,121	12,096	Hollow fibre

[(1)]This plant is designed to treat wastewater from 9,000 PE.

For each plant, data was collected over a period of 12 to 48 months and the plant operation assessed according to following criteria (Veltmann *et al.* 2011):

- *Utilisation of hydraulic capacity* (in %), which corresponds to the ratio between the average influent flow and the maximum influent flow used for design. In terms of energy optimization, a strong focus is required on plants with a low utilisation of their hydraulic capacity in order to avoid a sub-optimal use of the energy intensive CF-aeration.
- *Specific energy consumption* (in kWh/m^3), which is defined as the energy consumed per volume of effluent treated. If not otherwise mentioned, the specific energy consumption refers in the following to the specific energy consumption of the overall MBR plant, including pre-treatments, pumping, aeration, filtration and fouling control.
- *Net flux* (in $L/m^2 \cdot h$), defined as the volume of permeate produced per hour (considering back-washing and relaxation time) per membrane surface area.

8.2.2 Model WWTP

Because of the large scale nature of the plants investigated, modifications of the MBR process or operation could not be undertaken easily. Therefore, a model wastewater treatment plant (WWTP) was developed to estimate the impact of some modifications on the energy consumption of MBR. This model was based on recommendations published in the manual "Energy in WWTP" by MURL (1999) for conventional activated sludge systems and extended to cover all aspects of membrane bioreactors (see also Pöyry, 2009). The model WWTP comprises the following units, for which the energy consumption was calculated (see Figure 8.1):

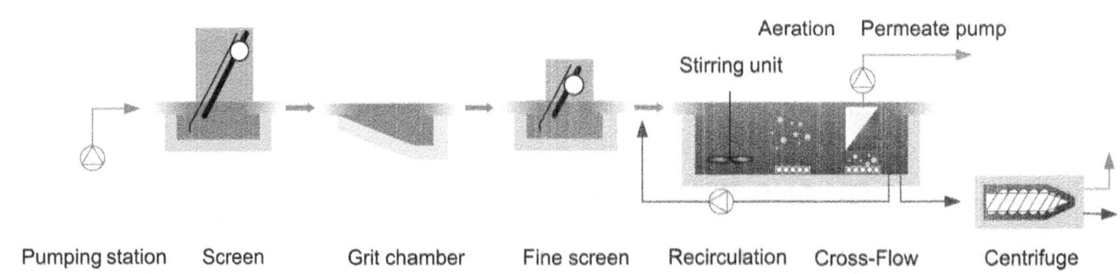

Figure 8.1 Schematic of the model wastewater treatment plant (Pinnekamp *et al.* 2011)

- *Mechanical pre-treatments*: pumping station, rack or coarse screen, grit chamber and fine screen
- *Biological stage*: stirrers, oxygen supply (nitrification and carbon oxidation), cross-flow aerators for the membrane modules, permeate pumps, recirculation pumps, excess sludge pumps, chemical cleaning operation of the membrane
- *Infrastructure*: lab, workshop, lighting, heating, aeration, exhaust air treatment
- *Sludge dewatering*: centrifuge
- *Options*: additional primary clarifier (including primary sludge pumps and scrapers) and sludge digestion (power and heat consumption, power and heat generation, additional centrifuge)

The design of the biological treatment stage was carried out with design2treat (ISA, 2000) under consideration of the recommendations given by Böhnke (1989) and Dohmann *et al.* (2003). The criteria set by MUNLV (2003) for MBR were incorporated. Specific loads per capita were set according to the specifications from ATV-DVWK-A 131 (2000). Flat sheet and hollow fibre membranes were considered separately. The net design flux was assumed to be $29\ L/m^2 \cdot h$ for flat sheet membranes and $20\ L/m^2 \cdot h$ for hollow fibre membranes. The required membrane area was determined accordingly. The model outcomes were calculated for several plant sizes (different PE).

The results represent ideal energy consumption as they do not consider energy losses. They are however very useful in order to undertake a preliminary screening of the many possible options to optimize MBR design and operation.

8.3 MAIN RESULTS

8.3.1 Hydraulic capacity utilisation

The average utilisation of the hydraulic capacity for the 7 MBR plants lies by 30%, with an individual average for each plant ranging from 14% to 45% (see Table 8.2 and Veltmann *et al.* 2011). This low capacity utilisation is justified by the need to deal with strong rain events and cater for future flow increases. However, this situation may lead to disadvantageous operational conditions of the membrane unit if the energy intensive components are not adjusted according to the incoming flow.

Table 8.2 Energy consumption figures of various MBR plants[(1)].

Parameter	Unit	Plant A	Plant B	Plant C	Plant D	Plant E	Plant F	Plant G
Average value[(2)] of the specific energy consumption $E_{spec,average}$	kWh/m^3	0.97	0.80	1.83	0.74	1.59	1.19	0.96
Median value of the daily specific energy consumption $E_{spec,Median}$	kWh/m^3	1.27	0.90	1.98	0.76	1.89[(3)]	1.23	1.09
Average energy consumption per PE and year	kWh/PE/yr	65.0	45.3	111.3	166.8	97.8	65.0	120.0
Average daily hydraulic capacity utilisation	%	31	26	35	45	14[(3)]	24	34

[(1)]Depending on the MBR plant configuration, the energy consumption data correspond to slightly different treatment schemes (e.g. primary clarifiers do not exist in all MBR plants).
[(2)]Average was calculated from the sums over the entire period considered. This period ranged from 12 to 48 months depending on the plant.
[(3)]Average was calculated from weekly values.

8.3.2 Energy consumption

High specific energy consumption is the result of the sub-optimal MBR operation at low influent flows. Figure 8.2 shows the daily specific energy consumption of Plants A and B as a function of the influent flow, represented as hydraulic capacity utilisation. With increasing hydraulic capacity utilisation (i.e. with increasing inflow), the specific energy consumption decreases until it reaches an asymptote. The same trend was observed for all investigated MBR plants and can be explained by the presence of some equipment, which consumes the same amount of energy regardless of the volume treated. This is the case of the energy intensive CF-aeration if no measure is taken to adjust the aeration to the flow or to reduce the overall filtration time.

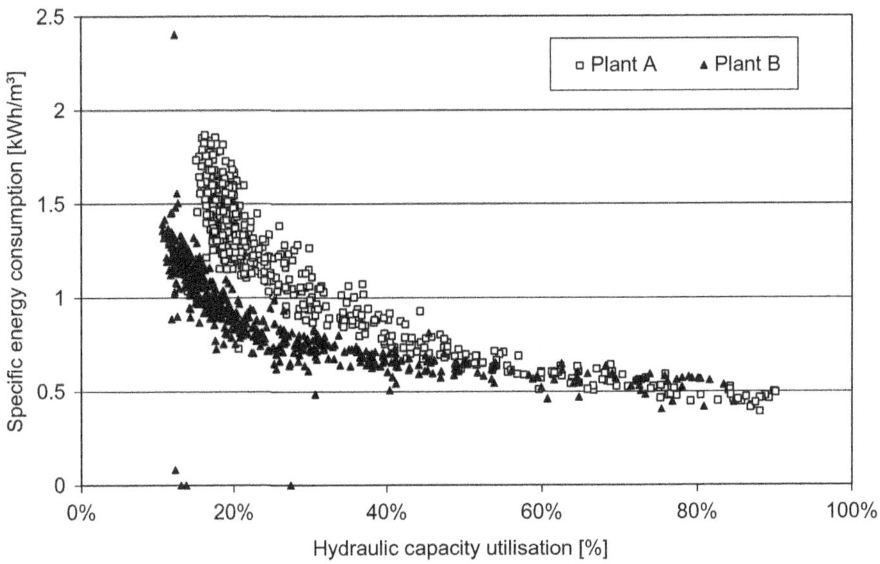

Figure 8.2 Specific energy consumption vs. hydraulic capacity utilisation for two MBR plants (all data are daily average data)

Table 8.2 summarises the energy consumption figures for the 7 MBR plants. Further data collected across the study sites have shown that blowers for CF-aeration are accountable for more than 50% of the plant energy needs (see Figure 8.7 and Veltmann *et al.* 2011), highlighting the need for optimization of the CF-aeration, especially at low influent flows.

8.3.3 MBR operation regimes to respond to influent flow fluctuations

Filtration with variable flux

The first option to respond to low utilisation of the plant hydraulic capacity consists in adjusting the flux to the influent flow. Ideally, the aeration should be adjusted to the influent flow accordingly, in a so-called proportional aeration (Verrecht *et al.* 2008). However, membrane manufacturers have up to now specified that CF-aeration of a module should occur at a constant rate to guarantee sufficient shear forces to limit fouling, no matter how high the flux is. As a result of the current manufacturers' recommendation, this option, which does not cut down the CF-aeration time, leads to high energy consumption even for low influent flows.

Assuming a constant CF-aeration, the impact of flux variations on the specific energy consumption was calculated using the model WWTP. Figure 8.3 gives an example for flat sheet membranes: reducing the flux from 29 to 20 $L/m^2 \cdot h$ causes a 15% increase in the specific energy consumption of the entire plant.

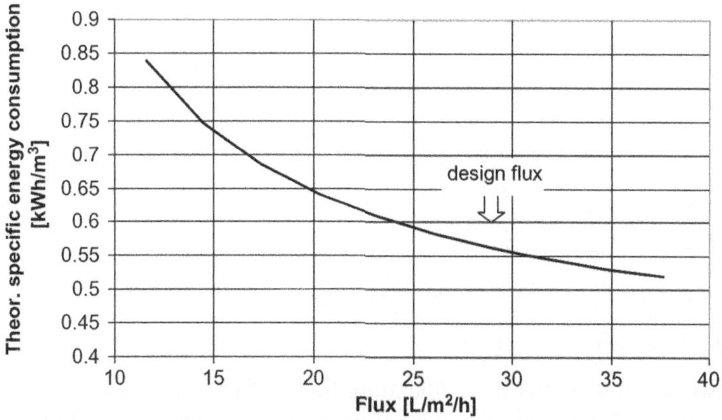

Figure 8.3 Specific energy consumption of an entire MBR plant based on calculations for a model MBR plant with flat sheet membranes

Filtration with constant flux over shorter periods of time

The second option is to carry out filtration over limited periods of time at a constant flux, which should be close to the design flux specified by the membrane manufacturer. In this approach, which has been adopted by some of the investigated MBR (Plants B, D and F), the energy intensive CF-aeration is utilised for a larger amount of permeate. Filtration pauses are introduced, until sufficient water is accumulated to run a full filtration cycle.

Little aeration may be required during these filtration pauses to avoid anoxic conditions and maintain the biomass in suspension. Through an optimized process control and a frequent rotation of the active chambers, the idle period in each chamber was diminished, thus reducing the aeration needs during filtration pauses to almost zero at the MBR plant in Seelscheid, Germany (Palmowski *et al.* 2011). A negative impact on permeability as a result of these measures was not encountered. Because the filtration is not continuously in operation, some buffer capacity is required in the plant for this option. This approach leads to significant energy savings, as is illustrated in the following for Plant B.

8.3.4 Comparison of the two filtration regimes on the basis of large scale operational data

Investigations were undertaken at Plant B to compare the two filtration regimes: filtration with variable flux vs. filtration with constant flux over shorter periods of time.

Net flux

Figure 8.4 illustrates the evolution of the net flux over a 14-month period. The average net flux amounts to 27.0 $L/m^2 \cdot h$ during filtration with high flux, while it was only 20.5 $L/m^2 \cdot h$ during filtration with variable flux.

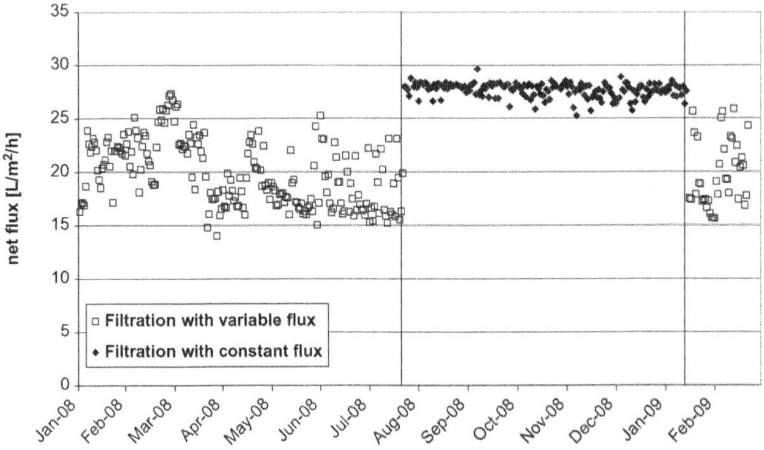

Figure 8.4 Average daily net flux under different filtration regimes at Plant B

CF-aeration and permeate pumps

The impact of the two regimes on the daily operating times of the CF-aeration and permeate pumps are presented in Figures 8.5 and 8.6. The operating times reached values up to 24 h/d during strong rain events for both regimes. Differences are however noticeable for dry weather flows, where lower values were observed during filtration with constant flux for shorter periods. Average values are summarised in Table 8.3.

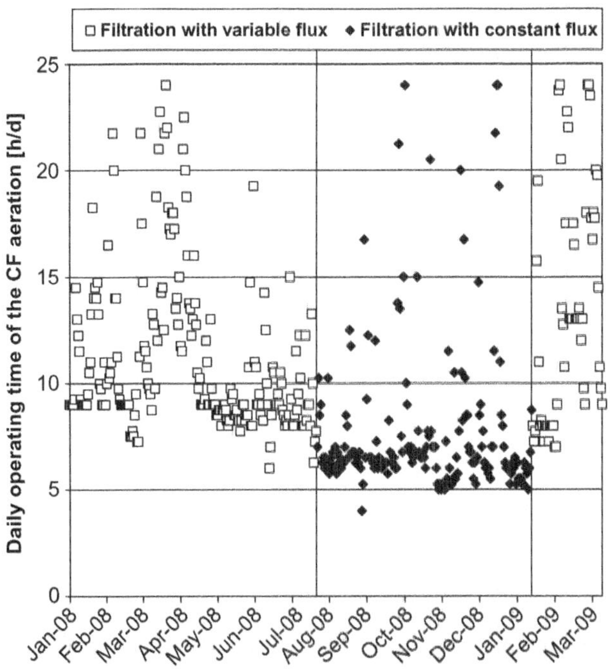

Figure 8.5 Daily operating time of the CF-aeration for the two filtration regimes

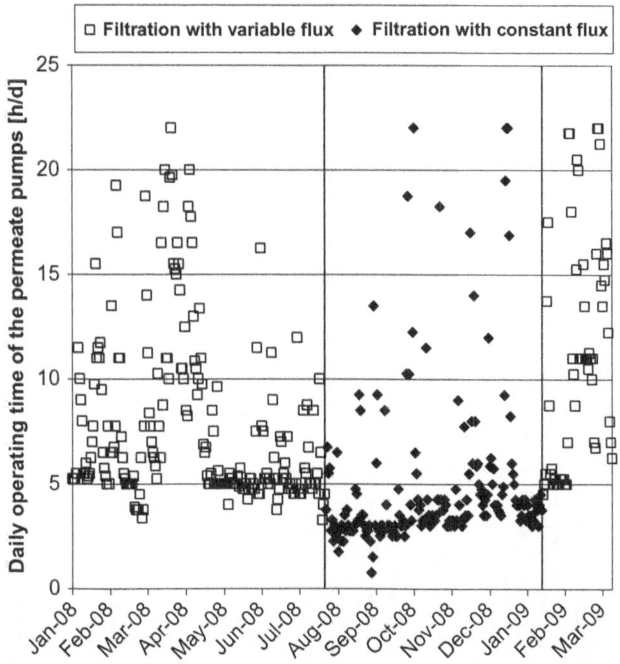

Figure 8.6 Daily operating time of the permeate pumps for the two filtration regimes

Table 8.3 Comparison of the two filtration regimes at Plant B over the 14-month period.

Filtration regime	Daily operating time of the permeate pumps	Daily operating time of the cross-flow aeration	Specific energy consumption of the entire CF-aeration
Filtration with variable flux	8.1 h/d	11.3 h/d	0.637 kWh/m^3
Filtration with constant flux over shorter periods	5.0 h/d	7.9 h/d	0.582 kWh/m^3

Amongst the energy consumed by the entire plant, the percentage required for the CF-aeration is lower during filtration at a constant flux in comparison with the filtration at a variable flux (see Figure 8.7), 66% and 82% respectively.

Energy savings

Permeate pumps are regulated with frequency converters to adjust the flux to the inflow. There is therefore no direct correlation between their operating time and their energy consumption. As neither the frequency levels nor the energy used for the pumps were recorded, the energy consumptions of the permeate pumps under the two filtration regimes could not be compared. This comparison was however possible for the CF-aeration, as the CF-blowers are not equipped with frequency converters. The average savings realised by the filtration regime with high flux over shorter periods add up to 0.055 kWh/m^3, which represent 94.2 kWh/d or 34,400 kWh per year for Plant B (see Table 8.3).

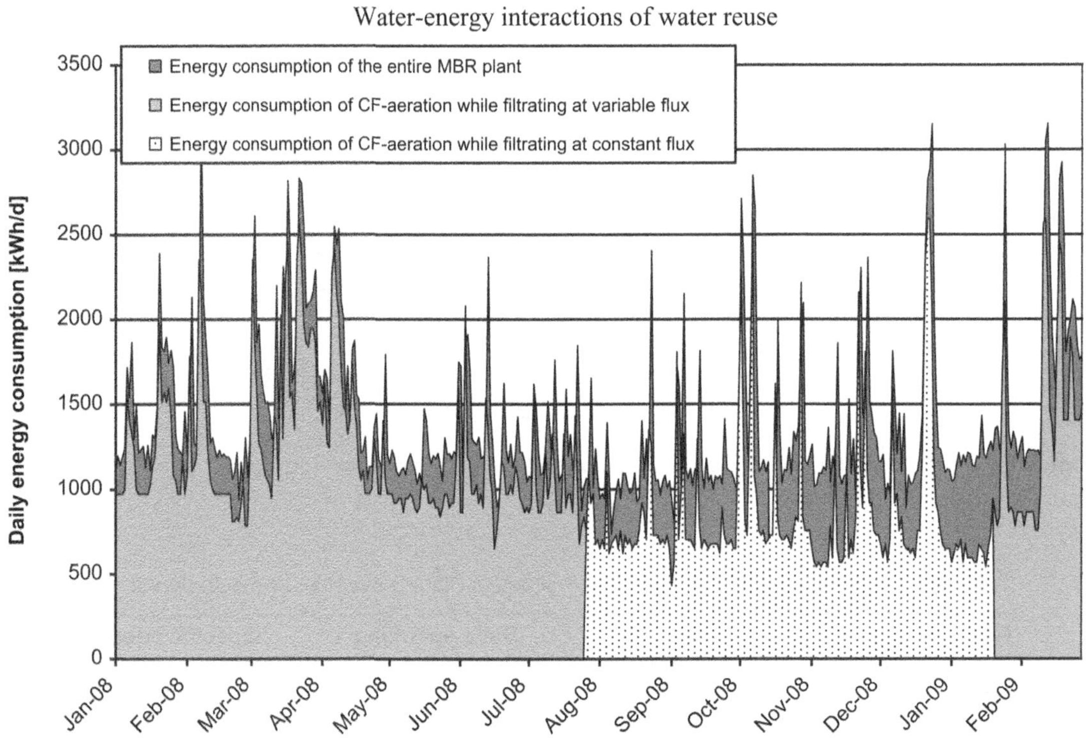

Figure 8.7 Daily energy consumption of the entire MBR plant as well as of the CF-aeration during the two filtration regimes

Using the model WWTP and the measured average flux values of 27.0 L/m$^2 \cdot$ h and 20.5 L/m$^2 \cdot$ h (see Figure 8.4), ideal specific energy consumptions of 0.57 and 0.64 kWh/m^3 were obtained from Figure 8.3 for the filtration regime with high flux and variable flux, respectively. The difference amounts to 0.07 kWh/m^3, which is slightly higher than the measured savings of 0.055 kWh/m^3 (Table 8.3). This discrepancy is probably due to inaccuracies in the model and to the fact that flux fluctuations occur in the real plant and were not considered when using Figure 8.3 to determine the energy consumption.

8.4 CONCLUSIONS

The problem of the high specific energy consumption of MBR at low influent flows highlighted the need to adjust the membrane filtration to the influent conditions. The best approach at the time being is a filtration at high flux over shorter time periods. In comparison to the filtration with variable flux, the filtration at high flux over short periods of time led to energy savings of 0.055 kWh/m^3, which represent 94.2 kWh/d or 34,400 kWh/year saved for Plant B. These measured savings were in good agreement with those of the model WWTP with MBR. The approach to adjust the intensity of cross-flow aeration to the flux, which fails today due to the requirements on the aeration rate imposed by membrane manufacturers, might be an option for the future.

Acknowledgements

The authors would like to thank the Ministry of Environment and Conservation, Agriculture and Consumer Protection of North Rhine Westphalia, Germany, for their financial support.

REFERENCES

Atkinson S. (2006). Research studies predict strong growth for MBR markets. *Membr. Technol.*, **2006**(2), 8–10.

ATV-DVWK-A 131 (2000). Bemessung von einstufigen Belebungsanlagen (Design of single stage activated sludge system). Editor: GFA-Gesellschaft zur Förderung der Abwassertechnik e.V., Germany.

Böhnke B. (1989). Bemessung der Stickstoffelimination in der Abwasserreinigung – Ergebnisse eines Erfahrungsaustausches der Hochschulen (Design of nitrogen elimination in wastewater treatment – Outcomes of an exchange of experiences from universities). *Korrespondenz Abwasser* (Wastewater Correspondence), **9**, 1046–1061.

Brepols C. (2011). Operating Large Scale Membrane Bioreactors for Municipal Wastewater Treatment. IWA Publishing, London, UK.

Dohmann M., Höhn P., Owerdieck C. and Buer T. (2003). Bemessung von Membran-Bioreaktoren in kommunalen Kläranlagen. Abschlussbericht zum Forschungsvorhaben AZ IV-9-042 161 im Auftrag des Ministerium für Umwelt und Naturschutz, Landwirtschaft und Verbraucherschutz des Landes Nordrhein-Westfalen (Design of membrane bioreactors for municipal WWTP. Final report of the research project AZ IV-9-042 161 on behalf of the Ministry of environment and conservation, agriculture and consumer protection of North Rhine Westphalia), Germany.

Hübner M. and Rolfs T. (2007). Betriebserfahrungen mit den Membrankläranlagen Woffelsbach und Konzen (Experiences with the membrane plants Woffelsbach and Konzen). *7. Aachener Membrantagung* (7th Aachen Membrane Conference), Germany, A6-1–A6-12.

ISA (2000). Design2treat software, http://www.design2treat.de/ (accessed in July 2010).

Judd S. and Judd C. (2006). The MBR Book: Principles and Applications of Membrane Bioreactors in Water and Wastewater Treatment. Elsevier, Oxford, UK.

Krause S. (2005). Untersuchungen zum Energiebedarf von Membranbelebungsanlagen (Research into energy consumption of Membrane Bioreactors). Dissertation am Fachbereich Bauingenieurwesen und, Geodäsie der Technischen Universität. Darmstadt.

MUNLV (2003). Abwasserreinigung mit Membrantechnik – Membraneinsatz im kommunalen und industriellen Bereich (Wastewater Treatment with Membrane Technology – Membrane Applications in Municipal and Industrial Areas). Editor: Ministry of environment and conservation, agriculture and consumer protection of North Rhine Westphalia, FiW–Publishing, Aachen, Germany.

Novotny V. (2010). Water and Energy in the Cities of the Future – Achieving Net Zero Carbon and Pollution Emissions Footprint. IWA World Water Congress and Exhibition, Montréal, Canada, 19–24. Sept. 2010.

Palmowski L., Veltmann K., Pinnekamp J. and Weber M. (2011). Energieoptimierung großtechnischer Membranbelebungsanlagen am Beispiel der Kläranlage Seelscheid (Energy optimization of large scale membrane bioreactors using the example of the WWTP Seelscheid). 44. Essener Tagung für Wasser- und Abfallwirtschaft, 23.–25. März 2011, Aachen, 10/1–10/15.

Pinnekamp J., Veltmann K. and Palmowski L. (2011). Prozessoptimierung von Membranbelebungsanlagen (Process optimisation of membrane bioreactors). Report currently under preparation, Research project funded by MUNLV, NRW, Germany.

Pinnekamp J. and Friedrich H. (2006). Membrane Technology for Waste Water Treatment. FiW Publishing, Aachen, Germany.

Stephenson T., Simon J., Jefferson B. and Brindle K. (2000). Membrane Bioreactors for Wastewater Treatment, IWA Publishing, London, UK.

Veltmann K., Palmowski L. M. and Pinnekamp J. (2011). Modular operation of membrane bioreactors for higher hydraulic capacity utilisation. *Water Sci. Technol.*, **63**(6), 1241–1246.

Verrecht B., Judd S., Guglielmi G., Brepols C. and Mulder J. W. (2008). An aeration energy model for an immersed membrane bioreactor. *Water Res.*, **42**, 4761–4770.

Chapter 9

Designing a membrane bioreactor to minimize energy use while meeting a low nitrogen limit

Sudhanva Paranjape and Rod Reardon

9.1 INTRODUCTION

The use of membrane bioreactors (MBRs) for water reclamation continues to increase due to the good water quality and compact size of this technology. An MBR combines a low-pressure membrane process, such as a microfiltration or ultrafiltration membrane, with an activated sludge process, where the membrane provides the solids separation. MBR facilities, however, are prone to higher energy use than conventional activated sludge plants for the following reasons: (1) a separate aeration system (immersed) or pumping system (external) is required to scour the membranes and control fouling; (2) high return activated sludge (RAS) rates, typically between 300% to 400% of influent flow, are required to maintain the solids balance between the membrane tanks and bioreactors at the high mixed liquor suspended solids (MLSS) concentrations used in MBRs, typically between 6 to 12 g/L; and (3) additional pumping is usually needed to move water through the membranes. Table 9.1 provides the energy consumption reported in literature for wastewater treatment plants employing various unit processes. According to this data, MBR facilities have significantly higher energy consumption than conventional treatment facilities.

Table 9.1 Literature values of energy consumption for wastewater treatment by facility type.

Wastewater treatment facility type	Energy consumption, kWh/m^3 (kWh/kgal)	Typical range, kWh/m^3 (kWh/kgal)
Lagoons	0.213 (0.8)	0.08–0.32 (0.3–1.2)
Trickling filter	0.266 (1.0)	0.186–0.426 (0.7–1.6)
Activated sludge	0.453 (1.7)	0.346–0.64 (1.3–2.4)
Advanced treatment w/nitrification	0.506 (1.9)	Not available
Oxidation ditch/Extended aeration	0.773 (2.9)	0.48–1.04 (1.8–3.9)
Membrane Bioreactors (MBRs)	0.986 (3.7)[a]	1.20–1.49 (4.5–5.6)[b]

(*Source*: EPRI, 1994; EPRI, 1996; ECW, 2002; Pearce, 2008)
[a]Data for large MBR Plants
[b]Date for small MBR Plants

This chapter discusses measures used to reduce overall energy usage for an immersed MBR system being retrofitted into an existing municipal water reclamation facility. The facility was being upgraded to produce effluent water quality not exceeding, on an annual average basis, 5 mg/L of BOD_5, 5 mg/L of total suspended solids (TSS), 3 mg/L of total nitrogen (TN), and 2.5 mg/L of total phosphorus (TP) with high-level disinfection for reclaimed water. Table 9.2 provides the design flows and plant influent pollutant loadings for the municipal water reclamation facility being upgraded.

Table 9.2 Design flows and loads for the municipal water reclamation facility being upgraded.

Parameter	Units	Value
Design Flow (AADF)[a]	m^3 (mgd)	17,034 (4.5)
Design Flow (MMADF)[b]	m^3 (mgd)	17,034 (4.5)
Design Flow (PHF)[c]	m^3 (mgd)	22,712 (6.0)
BOD Mass Loading (MMADF)	kg/d (lb/d)	5,353 (11,800)
TKN Mass Loading (MMADF)	kg/d (lb/d)	758 (1,670)
TSS Mass Loading (MMADF)	kg/d (lb/d)	8,800 (19,400)
TP Mass Loading (MMADF)	kg/d (lb/d)	126 (277)

[a]AADF – Annual Average Day Flow
[b]MMADF – Maximum Month Average Day Flow
[c]PHF – Peak Hour Flow

The following design elements make significant contributions to overall MBR energy consumption:

(1) Preliminary and primary unit processes;
(2) Membrane selection;
(3) Biological Nutrient Removal (BNR) process configuration;
(4) Process aeration design;
(5) Aeration control strategy for both process air and membrane scour.

9.1.1 Preliminary and primary unit processes

For typical wastewater treatment facilities, the preliminary unit processes (named also pre-treatment) typically comprise screening (coarse, fine or both) followed by grit removal. Primary clarification is an optional unit process for additional treatment of raw, screened, and degritted wastewater before biological treatment of the wastewater in an activated sludge process. The decision to use primary clarification for a conventional activated sludge plant is primarily made based on economics and also largely depends on the type of sludge handling facilities to treat/dispose of the sludge.

Properly designed and operated primary clarifiers have been reported to remove 50% to 70% of suspended solids including grit, rags, and other colloidal, inert, and biodegradable suspended solids that will increase the sludge yield in the activated sludge process and may interfere with aeration systems in the activated sludge process. A proportional amount of BOD_5 (25% to 50%) is also removed from the raw wastewater with removal of suspended solids.

Using primary clarification is a matter of choice for an MBR facility, but should be seriously evaluated for its ability to provide reduced capital and operating costs as well as additional assurance to the overall facility performance. From an energy perspective, primary clarifiers provide an economical way of removing BOD_5 and TSS. With 30 to 50% removal of BOD_5 by primary clarification, the required aeration capacity, bioreactor volume and energy consumption are proportionally reduced. This is true for any treatment facility and offers energy savings by reducing the loading on the aeration system.

Table 9.3 presents a comparison of anticipated energy savings with and without primary clarifiers.

Table 9.3 Comparison of predicted aeration energy consumption for plants with and without primary clarifiers.

Parameter	Without primary clarification	With primary clarification[a]
Total actual oxygen demand (AOR), kg/d (lb/d)	7,152 (15,790)	5,560 (12,280)
Total SOR required, kg/d (lb/d)	19,380 (42,780)	14,520 (32,050)
Estimate of required bioreactor Volume, m^3 (ft^3)[b]	4,702 (166,040)	3,580 (126,340)
Required airflow rate, Nm^3/h (scfm)[c]	10,640 (6,190)	8,300 (4,815)
Total blower power, kW (hp)[d]	216 (290)	168 (226)
Annual power consumption, kWh	1,898,000	1,476,000

[a]Based on typical performance of primary clarifiers: 30% BOD_5 and 7.5% TKN removal
[b]MLSS concentration = 8 g/L, Aerobic SRT = 12 days
[c]Alpha = 0.50; spatial dissolved oxygen (D.O.) concentration = 2.0 mg/L; AOR/SOR = 0.38
[d]Blower efficiency = 70%, blower suction pressure = 99 kPa (14.4 psia), blower discharge pressure = 157 kPa (22.8 psia)

9.1.2 Membrane selection

Selecting the right technology and equipment for an MBR application is one of the key elements. In a municipal bid environment, selection of the most energy efficient membrane system can be done using an evaluated bid process in which the equipment vendors must provide not only a competitive capital cost, but also guaranteed operating costs including power, chemicals usage, and typical labor costs for maintenance.

For immersed MBRs, energy demand for membrane air scouring varies considerably depending on aeration intensity and duration, and membrane system configuration. Table 9.4 illustrates the variability in air scour systems offered by three of the leading membrane manufacturers and its impact on energy consumption. From an energy perspective, Membrane Manufacturer A has the lowest power consumption as compared to the other two membrane systems.

Table 9.4 Summary of membrane air scour systems offered by three of the leading membrane manufacturers.

Parameter	Membrane Manufacturer A	Membrane Manufacturer B	Membrane Manufacturer C
Total membrane surface area, m^2 (ft^2)	31,334 (348,160)	32,637 (362,632)	30,989 (344,320)
Total system average scour air, Nm^3/h (scfm)	11,966 (6,960)	10,685 (6,215)	10,806 (6,285)
Scour air, $Nm^3/h/m^2$ (scfm/ft^2) of membrane surface area	0.38 (0.020)	0.33 (0.017)	0.35 (0.018)
Scour air daily effective operating hours	12	24	19
Average operating power for scour air, kW	117.9	89.4	156.7
Annual power demand for scour air, kWh	516,271	783,144	1,086,715
Average specific energy, kWh/m^3 (kWh/kgal)	0.083 (0.314)	0.127 (0.476)	0.176 (0.661)

9.1.3 BNR process selection

Achieving consistently low effluent TN levels ($\leq$3 mg/L) requires a robust process capable of consistent performance. The four or five stage Bardenpho™ process coupled with an immersed membrane system would offer a robust process for meeting Florida's advanced wastewater treatment (AWT) limits. A five stage Bardenpho™ process consists of a sequence of anaerobic, pre-anoxic, aerobic, post-anoxic, and re-aeration zones.

The anaerobic zone primarily provides reaction time for the adsorption of volatile fatty acids (VFAs) by the polyphosphate accumulating organisms (PAOs) with the concurrent release of phosphorus. The anaerobic zone also allows for additional fermentation of the influent BOD_5 for generating easily bio-degradable short-chain VFAs, primarily acetic and propionic acids.

The mixed liquor then enters the pre-anoxic zone where BOD_5 is used by heterotrophic bacteria to reduce the nitrates recycled from the aerobic zone to nitrogen gas. The denitrification in the pre-anoxic zone is typically not limited by the availability of carbon.

The remaining BOD_5 and ammonia is oxidized in the aerobic zone concurrently with the uptake of orthophosphate by the PAOs. Ammonia is converted to nitrate, and a portion of this nitrate is recycled back to the pre-anoxic zone for denitrification.

The post-anoxic zone removes the nitrates leaving the aerobic zone that were not recycled back to the pre-anoxic zone. Denitrification in the post-anoxic zones is generally limited by the availability of carbon. While endogenous denitrification can be achieved relatively easily in warmer temperatures (wastewater temperatures >20°C); it poses a challenge in colder temperatures. In such circumstances, addition of supplemental carbon to the second anoxic zones may be necessary to increase the denitrification rates to desired levels.

The fifth stage is the re-aeration zone where any bubbles of nitrogen gas are stripped from the mixed liquor suspended solids, and any ammonia released within the post-anoxic zone is converted to nitrate. In case of a Bardenpho/MBR process, the MBR tank replaces the reaeration tank. The size and thus the hydraulic retention time (HRT) for the MBR tank is governed by the size of the membrane equipment. Longer HRT in the MBR tank can result in undesirable endogenous decay and nitrification. In addition, higher RAS recycle rates (~4Q) are necessary to return sufficient MLSS to the process tanks.

The high recycle impacts the HRT of the various process reactors. Careful consideration should be given to sizing the second anoxic tank to achieve almost full denitrification to compensate for any additional nitrates that may be formed in the MBR tank and also to reduce the creation of any additional ammonia with endogenous decay that can be then converted to nitrates in the MBR tank.

A process configuration such as the one shown in Figure 9.1 offers balance between energy required for pumping versus the impact of recycle streams on process performance (air demands, nitrogen removal, and supplemental carbon). In this instance, RAS pumping would use a "pump to" configuration.

While RAS recycle directly to the first anoxic tanks, this minimizes recycle pumping, the oxygen load decreases the nitrogen removal potential or increases external carbon supplements by preferentially removing COD from the primary effluent. A double recycle configuration directs the oxygen load in the RAS back to the aeration tank where it offsets process air demands. The RAS from the membrane tanks is saturated with oxygen from the scour air introduced in the membrane tank. Modeling predicted the dissolved oxygen (DO) concentration in the RAS to be around 6 mg/L. At a RAS flow rate of four times the plant influent (4Q) or 68,100 m^3/d (18 mgd) under average day flows and 90,800 m^3/d (24 mgd) under maximum day flows, about 410 kg/d (900 lb/d) and 540 kg/d (1,200 lb/d) of DO would be returned back to the system, respectively.

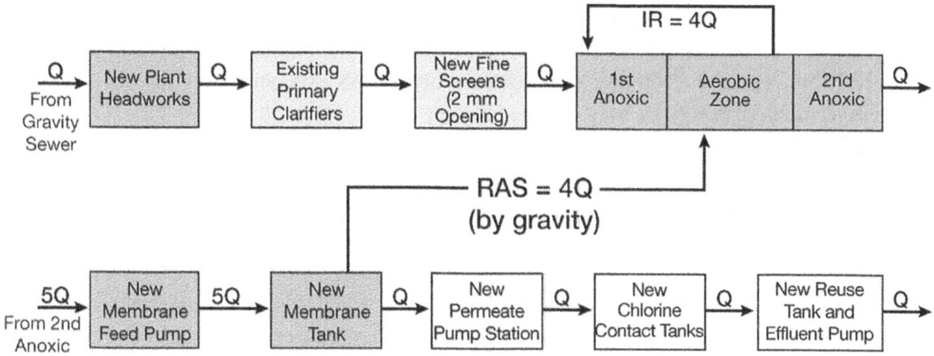

Figure 9.1 A BNR/MBR process schematic providing a balance between energy required versus process performance

9.1.4 Process aeration design

Because aeration efficiency has a significant impact on overall facility power use, careful consideration should be given to the selection of the process air blowers and diffusers during design. This is true for any activated sludge facility with or without membranes. However, in MBR processes, the effects of high MLSS concentrations on oxygen transfer efficiency and reduced aeration tank volume on diffuser density are two additional aspects that need to be carefully evaluated.

Four types of blowers typically used in municipal wastewater treatment for aeration: positive displacement, single-stage centrifugal, multi-stage centrifugal, and turbo type blowers were evaluated. While the turbo type blowers are integrally built with variable frequency drives (VFDs); external VFDs were considered for the other blower types for air control. Fine pore diffusers are more efficient than coarse bubble diffusers. Two types of fine pore diffusers were evaluated, 228.6 mm EPDM membrane discs, and 150-mm polyurethane strip diffusers. A present worth analysis was completed to select the most efficient type of aeration system.

Alpha factor

Field oxygen transfer, used to size aeration equipment (blowers and piping), is often calculated during design using manufacturer suggested standard oxygen transfer rates (SOTR) for the aeration equipment (diffuser system) and applying correction factors such as alpha (α), beta (β), and theta (τ) to account for the effects of wastewater characteristics.

Alpha is the most variable factor of the three and the most difficult to accurately test. Stenstrom & Gilbert (1981) noted that alpha is not constant and varies with several process conditions including MLSS concentrations, diffuser types, location within the aeration tank, and others. Assuming an alpha for design is one of the more challenging aspects in the design of aeration systems. Stenstrom & Gilbert (1981) demonstrated that drastic under or over design of the aeration systems can result with inaccurate assumptions of alpha. However, there is no evidence of variation in alpha among different types of fine bubble diffusers (Rosso *et al.* 1995).

A 10% variation in alpha results in an almost equal amount of variation in the power consumption of the aeration system, or in other words a 10% higher alpha used in the design could predict lower air demand due to 10% higher oxygen transfer efficiency, whereas in reality the actual transfer efficiency could be lower than design and cause an oxygen deficient system that can affect the nitrification process.

Several researchers (Muller *et al.* 1995; Gunder, 2001; Krampe & Krauth, 2003; Germain, 2004) have shown an exponential relationship between alpha and MLSS concentrations. (Judd, 2006). Cornel *et al.* (2003) measured the variation of alpha (α) values with MLSS concentration in two full scale MBR plants. An average alpha of 0.6 ± 0.1 was reported for an MLSS concentration of 12 g/L versus 0.7 ± 0.1 at 8 g/L. Therefore, from an energy perspective, designing and operating MBR plants in the range of 8 g/L will be more efficient over time as opposed to designing at 12 to 15 g/L to reduce tank volumes.

Huibregtse *et al.* (1983) noted that higher oxygen transfer efficiency for fine pore diffused aeration is achieved when the maximum number of diffusers is operated at the lowest airflow rate per diffuser. Standard oxygen transfer efficiency (SOTE) for fine pore diffuser systems in a grid placement decreases significantly with increased airflow rate (Morgan & Bewtra, 1960). Groves *et al.* (1992) noted that total floor coverage diffuser layout had higher oxygen transfer efficiencies regardless of the type of diffusers used. In short, the larger the floor space of the aeration tank that is covered by diffusers, the less variation in the minimum to maximum airflow per diffuser (proportional to the BOD loading). Consequently, the smaller the variation in pressure drop across the diffuser, the higher the oxygen transfer efficiency of the system. Accordingly, a packed grid of panel or strip diffusers should be more efficient than a packed dome or disk diffuser grid system due to the higher air flux in the latter.

Evaluation of blower and diffuser type

Table 9.5 provides a summary of results of an analysis comparing the energy consumption of the two different types of diffuser systems evaluated. As the floor coverage increases, the transfer efficiency increases, and hence, power cost decreases significantly. This type of analysis is necessary for all activated sludge systems, regardless whether it is conventional or uses an MBR.

Table 9.5 Impact of diffuser type on energy consumption.

Parameter	Diffuser type	
	Membrane disc diffuser	**Strip or panel type diffuser**
Clean water SOTE, %/m (%/ft)	7.09 (2.16)	9.7 (2.97)
Diffuser submergence, m (ft)	3.93 (12.90)	4.05 (13.4)
Diffuser SOTE, %	27.9	39.9
Diffuser floor coverage, %	12.0	35.1
Aeration air demand, Nm^3/h (scfm)[a,b]	4,714 (2,742)	3,294 (1,916)
Pressure required @ top of dropleg, kPa (psig)	46.8 (6.8)	48.2 (7.0)
Blower discharge pressure, kPa (psig)	52.9 (7.68)	54.3 (7.88)
Total blower motor power, kW (hp)[c]	95.5 (128)	68.6 (92)
Annual power consumption, kWh	839,100	598,600
Specific energy, kWh/m^3 (kWh/kgal)	0.136 (0.51)	0.096 (0.36)

[a]Average AOR = 4,330 kg/d (11,600 lb COD/d);
RAS to aeration tank; RAS flow = 85,200 m^3/d (22.5 mgd); internal recycle flow = 68,900 m^3/d (18.2 mgd)
[b]alpha = 0.50; beta = 0.95; ave. D.O. = 2.0 mg/L; aerobic SRT = 12 days; water temperature = 22°C
[c]blower efficiency = 70%

To estimate the annual energy consumption, a schedule of operating conditions (five conditions) for process aeration over one year was created from a frequency analysis of the daily peaking factors for oxygen demand from a historical three-year period. The operating schedule was developed to provide a more accurate representation of the operating conditions experienced by the aeration system over the duration of a year. The blower manufacturers were requested to provide the blower break horsepower (bHP) and overall wire to water efficiencies for each of the five conditions listed in the operating schedule. This information was used to calculate the total wire power demand (kW) and annual energy consumption (AEC) for average day condition. Based on this analysis, the single-stage centrifugal blowers (SS) had the least net present worth (NPW) (see Table 9.6).

Table 9.6 Evaluation of blower types.

Operating condition					Total wire power (kW)[a,b]					
N°	**Annual operating hours**	**Airflow, Nm^3/h (scfm)**	**No. of blowers running**	**Airflow per blower, Nm^3/h (scfm)**	**CS PD**	**PD w/VFD**	**Turbo**	**MSC**	**MSC w/VFD**	**SS**
1	864	8,768 (5,100)	3	2,923 (1,700)	175	183	182	189	180	165
2	864	7,822 (4,550)	3	2,608 (1,517)	175	151	140	179	153	134
3	2,208	6,695 (3,894)	3	2,231 (1,298)	175	141	131	163	155	121
4	4,392	5,918 (3,442)	2	2,923 (1,700)	117	116	121	126	120	106
5	432	2,450 (1,425)	1	2,450 (1,425)	58	56	51	58	57	47
						Net Present Value, NPV (Million US$)				
					1.78	1.82	1.89	1.83	1.85	1.70

[a]PD = rotary positive-displacement; MSC = multi-stage centrifugal w/inlet throttling; Turbo = turbine blower w/ integral variable speed drive; SS = single stage centrifugal blower; CS = constant speed; VFD = variable Frequency drive.
[b]For single-stage centrifugal blowers, only two operational blowers are required during the maximum airflow requirement of 8,768 Nm^3/h (5,100 scfm), each with capacity of 4,384 Nm^3/h (2,550 scfm).

9.2 AERATION CONTROL STRATEGY

9.2.1 Process air control strategy

The use of on-line and *in-situ* instruments to continuously monitor various process parameters in an activated sludge system and combining these instruments with a control system can provide overall process optimization and energy efficiency benefits. On-line instruments should be reliable and easy to maintain. Control systems should be simple and easy for plant operators to understand.

A comprehensive survey of 24 full-scale treatment plants was conducted by WERF and presented in the report "03-CTS-8 – Online Nitrogen Monitoring and Control Strategies". The survey identified five basic aeration control strategies for nitrification:

(1) Control based on DO concentration setpoint.
(2) Control based on ammonium concentration setpoint.
(3) Control based on both DO and ammonium concentration setpoint (cascade loop).
(4) Phase length control (intermittent aeration).
(5) Control based on oxidation-reduction potential (ORP) control (Palmer *et al.* 2007).

However, the majority of the plants surveyed based their aeration control strategy on a DO concentration setpoint by either manual or automatic adjustments of blower airflows and only monitored the effluent nitrogen species levels to further adjust blower airflow to increase or decrease the DO setpoint.

To minimize effluent nitrogen while simultaneously minimizing energy demand, two basic aeration control strategies were evaluated:

- simple DO control
- cascade control using DO and ammonia (NH_3) concentration signals

In cascade control there are two proportional-integral-derivative controllers (PIDs) arranged with one PID controlling the setpoint of another. A PID controller acts as an outer loop controller, which controls the primary physical parameter, such as NH_3. The other controller acts as the inner loop controller, which uses the output of the outer loop controller as the setpoint, and typically controls a more rapid changing parameter, like DO concentration. By definition, a PID controller attempts to correct the error between a measured process variable and a desired setpoint by calculating and then signalling a corrective action that can adjust the process accordingly to minimize the error. In either case, *in-situ* or on-line instrumentation is required to monitor and report the values of the control parameter.

The proposed control logic can provide the operator with three control choices for the airflow to each of the tanks as follows:

(1) Aeration Zone DO – Control of the process air is based on Aeration Zone DO. A reverse acting PID controller will control process air using DO concentrations as the process variable and process airflow as the control variable. The operator provides the DO setpoint.
(2) Swing Zone DO – Control of the process air is based on Swing Zone DO Control of the process air which is provided by a reverse acting PID controller using DO concentration as the process variable and process airflow as the control variable. The operator provides the DO setpoint.
(3) Swing Zone NH_3 Control with Swing Zone DO Max – Control of the process air is based on the Swing Zone NH_3 concentration. A direct acting PID controller using the NH_3 concentration as the process variable and process airflow as the control variable provides aeration system control. A maximum DO limits airflow using the Swing Zone DO concentration. The operator provides the NH_3 and maximum DO setpoints.

9.2.2 Membrane scour air control strategy

All commercial MBR systems, both hollow fiber and flat plate, use air scour provided by coarse bubble aeration underneath the membranes to control fouling. Since the intensity of air scour has a significant effect on both membrane performance and energy demand, careful attention should be given to the air scour design and operating protocol. To date, most commercial systems use either a constant or, an intermittent air scour. One manufacturer uses an alternating, intermittent aeration, in which a constant airflow is switched between two or four sets of modules, such that the membranes are aerated for one-quarter or one-half the time. Table 9.4 shows the effect of intermittent air scour on energy demand as compared to constant aeration. Membrane manufacturer B recommends continuous air scour for its membrane whereas, membrane manufacturer A and C recommend intermittent air scour. The intensity of air scour together with the duration determines the amount of power consumed.

Research has shown that development of a "thin" cake layer on the surface of the membrane protects them from soluble microbial products (SMPs), and thereby helps to maintain membrane permeability (Giraldo *et al.* 2007). The formation of this cake layer is a direct function of the intensity of the air scour applied and also depends on the MLSS concentration. Higher concentrations of MLSS and SMPs result in the need for higher air scour rates. With lower air scour intensity, the chances of forming a "thick" cake layer with higher trans-membrane pressure (TMP) are greater. In order to obtain a balance, the research suggests a feed-forward control of the air scour intensity proportional to the plant influent flow rate.

Other research suggests that flat plate MBR systems operate at high permeabilities (>2 L/m^2 · h/kPa or 17 gfd/psi) and have high air scour demands as compared to hollow-fiber MBR systems, which operate at lower permeabilities (<2 L/m^2 · h/kPa or 17 gfd/psi) and have lower air scour demands (Judd, 2006). The volume of permeate generated was found to be almost linear to the air scour demand, with about 0.4 m^3 (105 gal) generated per Nm^3/bar of air scour. Membrane technology is still maturing, and higher energy costs and sustainability concerns will certainly lead to more efficient energy use by this technology in the future.

9.3 CONCLUSIONS

As with any treatment technology, MBRs have both benefits and challenges. Energy consumption is one of the major challenges of this technology. However, as described above, many key unit processes in an MBR facility can be evaluated with the aim of reducing energy consumption. Implementation of the energy reduction measures as discussed above could reduce the overall energy use of the example municipal wastewater treatment facility by 20%, as compared to typical energy consumption values found in the literature shown in Table 9.1, while still meeting the mandated nitrogen limit and providing a high quality effluent for reuse.

These energy reduction measures (Table 9.7) include primary clarification, high efficiency aeration system using full-floor cover fine bubble diffusers, high efficiency turbo blowers, efficient membrane air scour scheme, using a process configuration to minimize pumping and reduce process air demand by implementing appropriate process instrumentation to control the aeration system.

Table 9.7 Summary estimated power consumption for example municipal WWTP.

Unit Process	Estimate of annual power consumption, kWh/yr	Specific energy, kWh/m^3 (kWh/kgal)
Plant headworks[a]	338,267	0.055 (0.206)
Primary clarification[b]	85,826	0.0138 (0.052)
Activated sludge process[c]	1,453,271	0.236 (0.885)
MBR process & equipment[d]	1,830,924	0.296 (1.11)
Effluent disinfection & pumping[e]	1,034,089	0.167 (0.629)
HVAC, lighting and other miscellaneous[f]	180,945	0.029 (0.11)
Total Plant	4,923,332	0.793 (2.99)

[a]Plant headworks includes coarse screens, screening compactor, and master pump station with odor control system
[b]Primary clarification includes primary clarifier sludge scrapper mechanisms, sludge transfer pump and odor control system
[c]Activated sludge process includes fine screens, screenings compactor, anoxic basin mixers, process blowers, internal recycle pumps, scum pumps, and other miscellaneous motors
[d]MBR process & equipment includes membrane feed pumps, permeate pumps, air scour blowers and other miscellaneous motors, and so on
[e]Effluent disinfection & pumping includes effluent transfer pumps, plant water pumps, and disinfectant feed pumps
[f]HVAC, lighting & other miscellaneous includes lighting and heating/ventilation for the operations room, blower building, and other buildings

Figure 9.2 shows a graphical representation of the energy distribution within the major unit processes for the example facility. The energy consumption of the MBR process and equipment is equal to or slightly greater than the energy consumption of the activated sludge process (~0.773 kWh/m^3 as presented in Table 9.1).

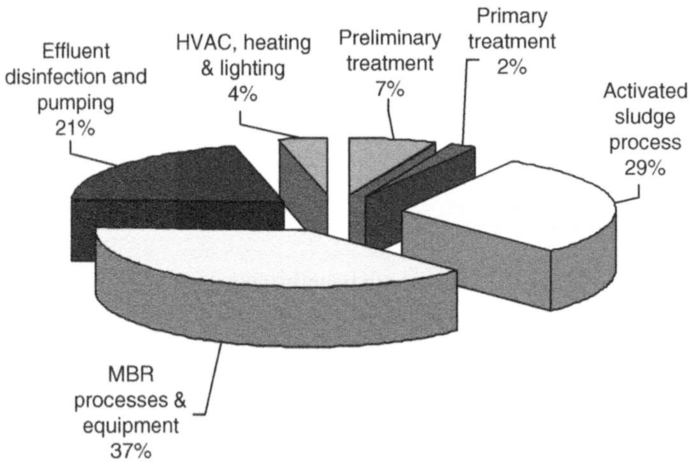

Figure 9.2 Estimate of power distribution for a 4-stage Bardenpho™/MBR process with DO control

Figure 9.3 shows a graphical representation of the energy distribution for the MBR equipment. Almost two-thirds of the total energy is spent on pumping (feed and permeate pumps) with the remaining third of the total energy required for air scour demand. Compared to a conventional activated sludge, the MBR process requires around 30–35% more energy and 70% of energy is used for pumping.

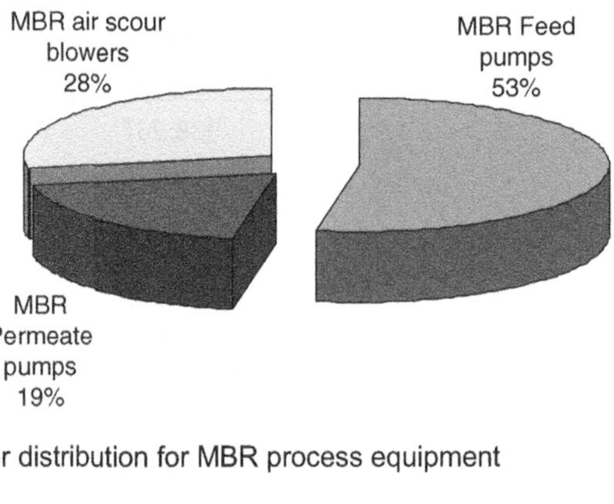

Figure 9.3 Estimate of power distribution for MBR process equipment

REFERENCES

Cornel P., Wagner M. and Krause S. (2003). Investigation of Oxygen transfer rates in full-scale membrane bioreactors. *Wat. Sci. Tech.*, **47**(11), 313–319.

EPRI (1994). Energy Audit Manual for Water/Wastewater Facilities, CEC Report CR-104300. EPRI, Palo Alto, CA.

EPRI (1996). Water and Wastewater Industries: Characteristics and Energy Management Opportunities; Product ID # CR-106491. EPRI, Palo Alto, CA.

Energy Center of Wisconsin (ECW). (2002). Roadmap for the Wisconsin Municipal Water & Wastewater Industry, October 2002. ECW, Wisconsin.

Germain E. (2004). Biomass Effects on Membrane Bioreactor Operations. EngD thesis, Cranfield University. U.K.

Giraldo E. and LeChevallier M. (2007). Let them wear cake. *Water Environment & Technology*, **19**(3), 46–51.

Groves K. P., Daigger G. T. and Simpkin T. J. (1992). Evaluation of oxygen transfer efficiency and alpha-factor on a variety of diffused aeration systems. *Water Environ. Res.*, **64**(5), 691–698.

Gunder B. (2001). The Membrane Coupled-Activated sludge Process in Municipal Wastewater Treatment. Technomic Publishing Company Inc., Lancaster, U.K.

Huibregtse G. L., Rooney T. C. and Rasmussen D. C. (1983). Factors affecting fine bubble diffused aeration. *J. WPCF*, **55**, 1057–1064.

Judd S. (2006). The MBR Book – Principles and Applications of Membrane Bioreactors in Water and Wastewater Treatment. Elsevier Ltd., Oxford.

Krampe J. and Krauth K. (2003). Oxygen transfer into activated sludge with high MLSS concentrations. *Wat. Sci. Tech.*, **47**(11), 297–303.

Metcalf & Eddy (2003). In: Wastewater Engineering, Treatment and Reuse. G. Tchobanoglous, F. L. Burton, H. D. Stensel (eds), 4th edn, McGraw Hill Education, New York, NY, USA.

Morgan P. F. and Bewtra J. K. (1960). Air diffuser efficiencies. *J. WPCF*, **32**, 1047–1059.

Muller E. B., Stouthammer A. H., van Verseveld H. W. and Eikelboom D. H. (1995). Aerobic domestic wastewater treatment in a pilot plant with complete sludge retention by cross flow filtration. *Water Environ. Res.*, **24**(4), 1179–1189.

Palmer T. M., Ross M. C., Nutt S. G. and Kharkar S. (2007). Online Nitrogen Monitoring and Control Strategies, Project No. 03-CTS-8. Water Environment Research Foundation, Alexandria, VA.

Pearce G. (2008). Introduction to membrane bioreactors., *Filtration & Separation*, **45**(1), 32–35.

Rosso D., Iranpour R. and Stenstrom M. K. (1995). Fifteen years of off-gas transfer efficiency measurements on fine pore aerators: key role of sludge age and normalized air flux. *Water Environ. Res.*, **77**(3), 266–273.

Stenstrom M. K. and Gilbert R. G. (1981). Effects of alpha, beta and theta factor upon the design, specification and operation of aeration systems. *Water Research*, **15**(1), 243–254.

Chapter 10

Finding the balance between greenhouse gas emission and energy efficiency of wastewater treatment

Marlies J. Kampschreur and Mark C.M. van Loosdrecht

10.1 INTRODUCTION

Due to increased interest in climate change mitigation measures, the emission of greenhouse gasses is an important topic, also in the field of wastewater treatment. Greenhouse gas (GHG) emission during wastewater treatment is caused by direct emission of nitrous oxide (N_2O) and methane (CH_4) from wastewater treatment plants. Additionally, carbon dioxide (CO_2) emission is caused due to use of energy which is generated by fossil fuels. Currently, wastewater treatment operators focus strongly on the reduction of energy use at WWTPs. This is done by application of more energy-efficient process operation, which leaves the processes generally the same. Simultaneously, there is a trend to convert WWTPs into energy recovery plants. In that case, the potential energy available in the organic carbon in wastewater is harvested.

These initiatives to reduce energy use can impact (positively and negatively) the emission of greenhouse gasses other than CO_2 (thus N_2O and CH_4). This means that there is a possibility that reduction of CO_2 by energy savings can nevertheless lead to an increased GHG emission, due to an increased emission of N_2O or CH_4.

10.1.1 Energy production at wastewater treatment plants

Currently, most activated sludge plants that apply both COD and nutrient removal consume 1 to 2 MJ/kg COD (Chemical Oxygen Demand) · year, which is equivalent to 15 to 30 kWh per population equivalent (p.e.) per year. This corresponds to 0.5 to 1% of the electricity consumption of a household. Theoretically, the energy content of organic carbon in wastewater equals approximately 14 MJ/kg COD · year. This is based on 0.35 Nm^3 CH_4/kg COD and a Higher Heating Value for methane of 39.96 MJ/Nm^3 CH_4. Obviously, not all this energy can be harvested and used for electricity generation. Nevertheless, there is clearly potential energy available in sewage for net generation of energy or methane containing biogas in WWTPs while still reaching the effluent standards.

Hence, energy-consuming WWTPs can potentially be converted to power generating plants. This production of renewable energy will also decrease the greenhouse gas footprint of the wastewater treatment.

10.1.2 Greenhouse gas emission at wastewater treatment plants

In this chapter, the possible effects of increased energy efficiency in WWTPs on the non-CO_2 greenhouse gas emission are discussed. Wastewater treatment has a climate footprint, in which N_2O, CO_2, and CH_4 are relevant gasses. CO_2 emission can be assessed based on energy demand of a wastewater treatment plant (WWTP). The CO_2 that is directly produced by conversion of organic matter is short-cycle CO_2 (not originating from fossil fuels), which is considered not to contribute to the greenhouse effect. CH_4 and N_2O are important greenhouse gasses, having respectively a 21 fold and 310-fold stronger global warming potential than CO_2 over a period of 100 years (IPCC, 2001).

CH_4 emission from full-scale WWTPs in the Netherlands was measured to range between 190 and 530 g CH_4/per year (STOWA, 2010). Up till now only one other study is known that measured the emission of CH_4 from a full scale WWTP in the US (Czepiel *et al.* 1993), in which the emission was 39 g CH_4/per year. The WWTP of that study was small (12.500 p.e.) compared to the ones studied in the STOWA-research.

Field methane measurements in sewers in Australia revealed that sewage methane can represent an additional GHG emission of 12 to 100% on top of the emission at the WWTP itself and should not be overlooked (Guisasola *et al.* 2008). Methane emission is very dependent on local conditions in the sewer system (e.g. temperature, residence time, contribution of industrial discharges) and the process design of the sludge line and presence of sludge digestion.

During the measurement campaign in the Netherlands also CO_2 and N_2O were measured besides CH_4. This showed that CH_4 emission varied between 5% and 36% of the total GHG footprint of the wastewater treatment plants (STOWA, 2010). The factors that are applied by Waterboards in the Netherlands, 0.0070 kg CH_4/kg COD for WWTP without sludge digestion and 0.0085 kg CH_4/kg COD for WWTPs with sludge digestion, correlate well with recent measurements (STOWA, 2010).

The exact emission of N_2O during wastewater treatment was very dependent on local conditions (e.g. design characteristics, operation strategy, season) ranging from 0 to 5% of the nitrogen load in full-scale WWTPs (Kampschreur *et al.* 2009a).

10.2 SOURCES OF CH_4 AND N_2O EMISSIONS

CH_4 and N_2O can be produced and emitted at several locations in WWTPs (see Figure 10.1). CH_4 is expected to be prevalently emitted in the sewerage system and during sludge handling. As a consequence, CH_4 emission is expected to be dependent on the residence time of wastewater in the sewerage system and the wastewater temperature. Sludge digestion also leads to increased CH_4 emission due to carry-over of dissolved CH_4 in reject water to the aerated compartments, where it is stripped to the gas phase. N_2O emissions are associated with several processes in wastewater treatment plants. Previously, denitrification was considered the dominant source of N_2O emission. Recent measurement at full-scale WWTPs clearly indicate that nitrification might be a more dominant pathway for N_2O production (Ahn *et al.* 2010).

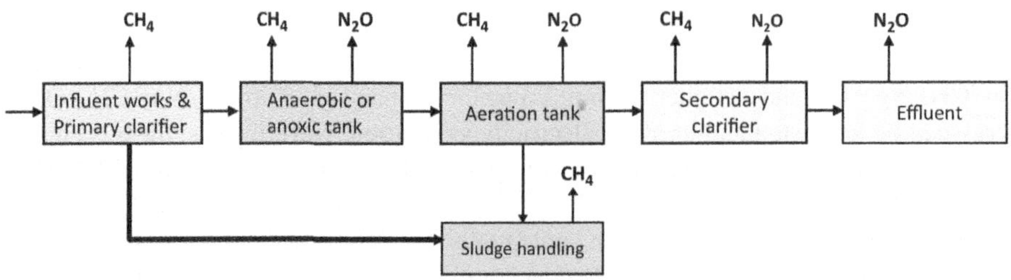

Figure 10.1 Potential sources of CH_4 and N_2O emission at WWTPs

The most important operational parameters that lead to N_2O emission in WWTPs are (i) low dissolved oxygen (DO) concentration in the nitrification stage, (ii) increased nitrite concentrations in both nitrification and denitrification tank, (iii) low COD/N ratio in the denitrification stage and (iv) increased ammonium concentrations in the nitrification tank (see Figure 10.2).

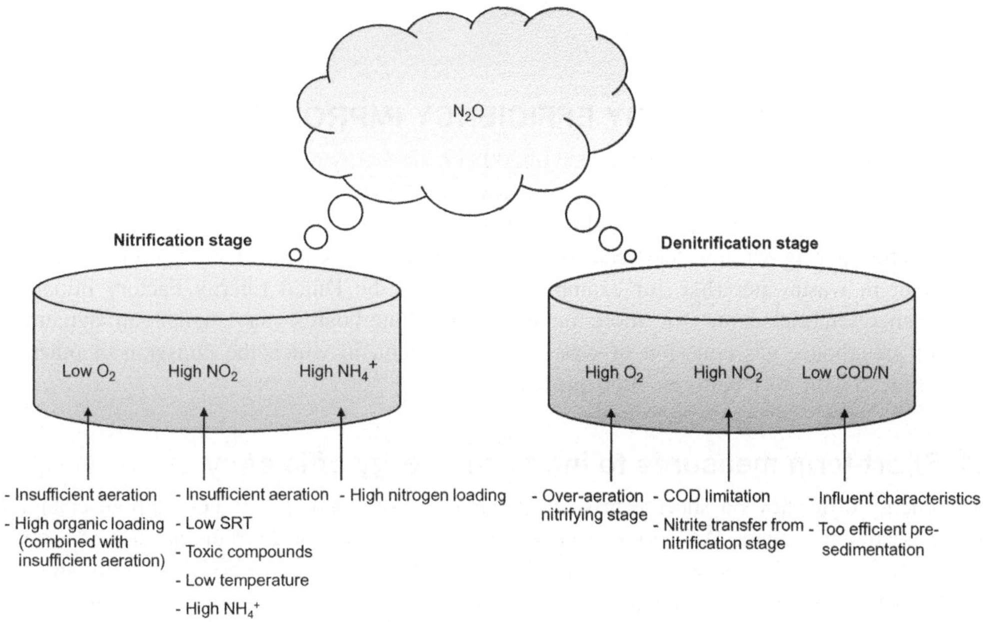

Figure 10.2 Main parameters leading to N_2O emission (updated from Kampschreur *et al.* 2009a)

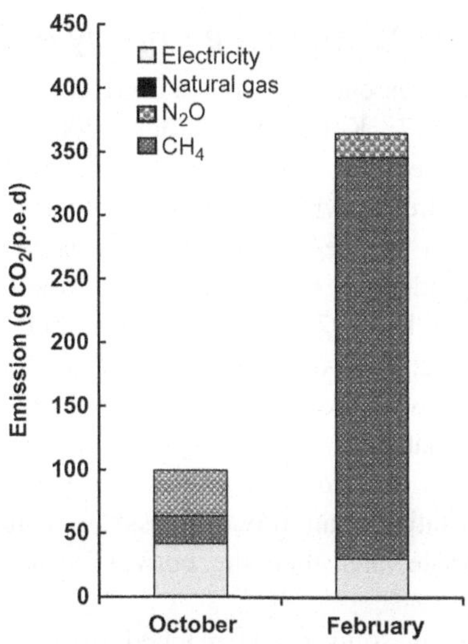

Figure 10.3 GHG emission in October and February 2009 at Kralingseveer WWTP

The emission of N_2O is very dynamic and can vary largely for the same WWTP depending on conditions like organic, nitrogen and hydraulic loading and process control at the plant. The large variation in emission can be seen in the difference in the emission of an activated sludge plant during a measurement campaign in October and February (Figure 10.3). The exact reason for the large difference is not yet clarified, but can be partially explained by the high hydraulic loading and low temperature in February. It is clear that for proper evaluation of emissions, a long term monitoring program would be required. For more details see STOWA (2010).

10.3 INITIATIVES TO ENERGY EFFICIENCY IMPROVEMENT

In the Netherlands the waterboards have agreed in 2008 on an energy reduction of 2% per year, leading to 30% energy usage reduction until 2020. Consequently, many initiatives for reducing the energy demand from wastewater treatment have arisen. Additionally, the Dutch waterboards strive to move towards even energy neutrality during wastewater treatment by harvesting the chemical energy that is present in organic matter in wastewater (like for example proposed in the Dutch Energy Factory initiative, see http://www.energiefabriek.com/ for more details). The Waterboards also signed an agreement on reduction of greenhouse gas emission of wastewater treatment, in which the emission of other GHGs should be reduced with 30% between 1990 and 2020.

10.3.1 Short-term measures to increase energy efficiency

Increased energy efficiency on short-term is mostly realised in existing WWTPs without changing the wastewater treatment concept. As aeration energy is a large part of the total energy use of a treatment plant (60–80%), an easy solution is decreasing the aeration energy by applying appropriate aeration control and energy efficient aeration equipment. Moreover, lowering the oxygen set-point in the aeration tank when nitrogen removal is sufficient to meet effluent standards will decrease energy demand. In addition, implementation of energy efficient equipment like pumps can be applied to reduce energy demand.

10.3.2 Long-term measures to increase energy efficiency

On the long-term, significant decrease in energy usage and GHG emission can be achieved by adequate process selection (Jetten *et al.* 1997; Keller & Hartley, 2003). A WWTP can become overall energy-generating by recovering the energy available in organic carbon in wastewater combined with energy-efficient nitrogen removal. Already with proven technologies, a WWTP can be converted to an energy generating plant. Application of the AB-process combined with nitritation-anammox on the reject water led to net electricity production in a 200,000 p.e. WWTP in Strass (Wett *et al.* 2007). In AB-type wastewater treatment (Böhnke, 1977), the activated sludge plant consists of a high-loaded first stage for COD removal (A-stage, SRT 0.5 day) followed by a low-loaded second stage (B-stage, SRT 7 days) where nitrogen removal takes place. By application of the AB-process, both the dissolved and particulate organic matter is collected in the sludge, while in most activated sludge plants the dissolved organic matter is oxidized. Therefore, subsequent sludge digestion in an AB-plant generates more energy. The nitritation-anammox process is a more energy-efficient method for removing ammonium from the wastewater than the conventional nitrification-denitrification process (van Dongen *et al.* 2001).

Increased use of the energy in organic solids can be reached by more efficient pre-sedimentation (e.g. by application of flocculants) or by application of the AB-process (van Nieuwenhuijzen *et al.* 2004). This leads to energy saving by using less aeration energy for COD removal and energy production from the organic solids.

Electricity production from organic solids in sludge is generally realised by application of anaerobic digestion followed by biogas combustion in an improved gas engine. The methane yield of fermentation can be increased by sludge disintegration technologies like thermal degeneration or sonification.

Innovative technologies might increase the efficiency of energy production from organic solids by application of a fuel cell or supercritical gasification (Matsumura *et al.* 2005). These innovative technologies are not yet applied at full-scale.

The application of the nitritation-anammox process for nitrogen removal from reject water can also lead to significant reduction in energy consumption. Application of the DEMON process (where nitrititation-anammox is applied in a SBR-system) at the 200,000 p.e. Strass WWTP led to 12% reduction of energy consumption (Wett *et al.* 2007).

10.4 GHG EMISSION DUE TO ENERGY SAVING AT WWTPs

10.4.1 Effects of short-term measures on GHG emission

Saving on aeration energy is an effective measure to reduce CO_2 emission and can be realised on a short time-scale. Application of energy efficient aeration equipment will not impact the N_2O emission as long as the dissolved oxygen (DO) set-point remains unchanged.

Changes in the aeration control itself however, will have an impact on the GHG emission of the WWTP. The CH_4 emission will probably remain unchanged. The N_2O emission however, will be different as changes in aeration control directly impacts the liquid concentrations of oxygen, ammonium, nitrite and nitrate in the activated sludge units. The oxygen set-point is an important factor; low DO concentrations can indirectly promote N_2O emission due to nitrite accumulation. Low DO concentrations also directly lead to increased N_2O emissions due to the fact that nitrifier denitrification (which has N_2O as the final product) is promoted under these conditions (Colliver & Stephenson, 2000). Nitrifier denitrification is a biochemical pathway that is applied by nitrifying bacteria, in which nitrite instead of oxygen is used as final electron acceptor. The final product of the nitrite reduction by nitrifying bacteria is N_2O.

Additionally, the overall nitrogen removal can be decreased due to less aeration. N_2O emission, expressed as % of the N-load, seems to be lowest in WWTPs that have a low biomass-specific nitrogen load and low nitrogen concentration in the effluent. This suggests that when energy efficiency is maximized by reducing aeration and thereby nitrogen removal, GHG emission may increase. The reduction in CO_2 emission is in that case easily compensated by increased N_2O emission due to its 310 times stronger greenhouse effect.

On the other hand, prevention of over-aeration can lead to both decreased CO_2 and N_2O emission. Too much aeration leads to unnecessary CO_2 emission due to the electricity that is used. In addition, increased N_2O emission occurs during over-aeration at peak-loading events, possibly due to a stripping effect or due to oxygen carry-over towards the denitrification unit. Over-aeration can occur for example during rainfall events. As a consequence of rainfall events, initially COD and N loading of a WWTP increase. This phenomenon is known as first-flush: organic matter that is sedimented during previous dry weather periods in the sewer system is transported towards the WWTP. Consequently, ammonium concentrations rise in the activated sludge stage.

High ammonium concentrations induce maximum nitrifying activity which is shown to enhance N_2O emission (Yu *et al.* 2010). High ammonium also often leads to increased nitrite levels, which, already at low levels (0.5 mg N-nitrite/L) leads to substantial N_2O formation. Aeration control based on ammonium levels in these conditions might add to the increased emission. Because at increased ammonium levels the oxygen setpoint is set higher, stripping of N_2O is promoted. Sometimes, the oxygen set-point is set unnecessarily high, as the limiting step is not oxygen availability but nitrification capacity. Both in a nitritation reactor and a nitritation-anammox reactor it was observed that a higher aeration flow did not lead to a change in the nitric oxide (NO) concentration, and possibly neither in the N_2O concentration, in the off-gas (Kampschreur *et al.* 2009b). NO was measured in this study as it is an intermediate in the conversion of nitrite to N_2O by both nitrifying and denitrifying micro-organisms.

Therefore, higher aeration linearly increased the NO emission load (as the load is the gas flow times the concentration in the gas), as shown in Figure 10.4. Also measurements at a full-scale activated sludge plant indicated that over-aeration of the nitrification stage led to dramatic increase of N_2O (STOWA, 2010).

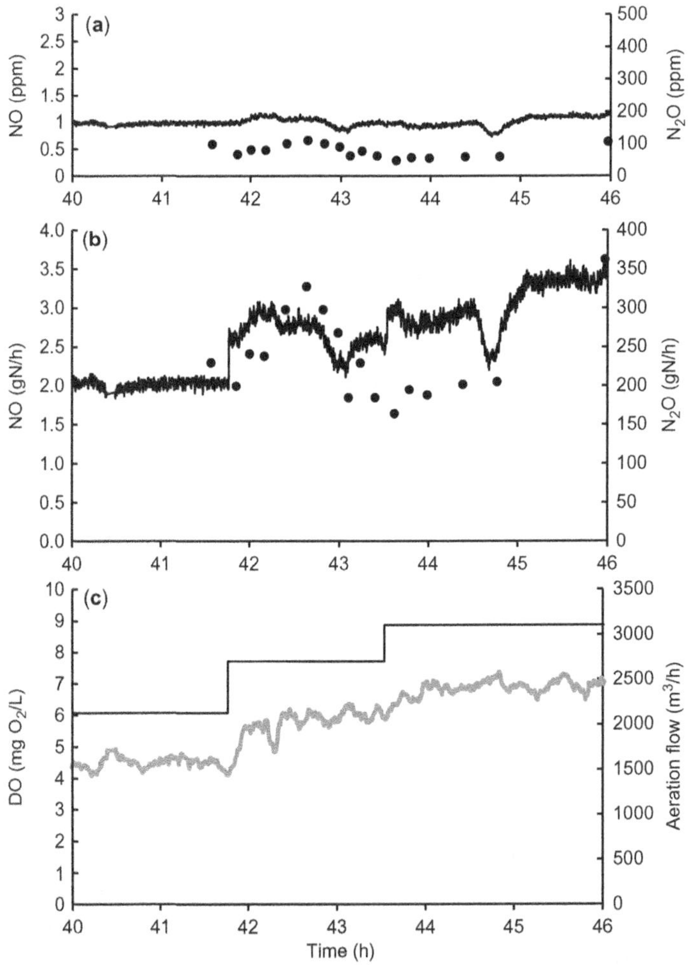

Figure 10.4 A NO (black line) and N_2O (•) concentrations in the off-gas of the single-stage nitritation-anammox reactor during variations in the aeration rates. 2B. NO (black line) and N_2O (•) emission loads. 2C. Aeration flow rate (black) and dissolved oxygen concentration (gray line). For more details see Kampschreur *et al.* 2009b.

10.4.2 Effects of long-term measures on GHG emission

Removing more organic carbon from the influent by pre-sedimentation or application of the two-stage AB-process, will lead to less CO_2-emission due to a decrease of aeration energy that is needed. Additionally, electricity can be generated, if the organic carbon is digested and methane is converted to electric power or if the sludge is incinerated in a biomass power plant.

However, it can lead to increased emission of GHGs other than CO_2 in two ways:

(1) It will lead to a decreased COD/N ratio in the denitrification stage, which can lead to increased N_2O emission during denitrification.
(2) Anaerobic digestion of more organic carbon will lead to increased CH_4 emission from the WWTP, which was clearly observed at full-scale WWTPs.

The effect of novel technologies like supercritical gasification of wastewater sludge on the total GHG footprint is yet unknown. In this case, it is also important to consider the impact on the total GHG footprint, including CO_2, CH_4 and N_2O.

N_2O emission also occurs during the nitritation-anammox process (Kampschreur *et al.* 2008), however the emission can probably be decreased by application of a single-stage process instead of a two-stage process (Weissenbacher *et al.* 2010). A single-stage process is observed to emit less N_2O due to prevention of nitrite accumulation. Because application of the nitritation-anammox process on reject water also decreases the energy consumption of aeration, and thereby the CO_2 emission, the overall GHG impact seemed to be positive in a full-scale nitritation-anammox reactor (Weissenbacher *et al.* 2010). More full-scale measurements are needed to further analyse the overall GHG-effect of application of the nitritation-anammox process.

10.5 CONCLUSIONS

Initiatives to reduce energy consumption at WWTPs can lead to a significant reduction in the GHG footprint due to reduced emission of CO_2 from fossil fuel consumption for energy generation. This fossil fuel derived CO_2 is an important factor in the overall GHG emission of a WWTP.

Short term measures to save energy often focus on aeration energy. It is very important in this case to prevent that a different aeration strategy leads to conditions in the activated sludge compartment that promote N_2O emission, like increased nitrite or low oxygen concentration.

WWTPs will increasingly be converted to energy recovery plants capable of harvesting the potential energy available in the organic carbon in wastewater. In this respect there is a risk of increased N_2O emission from the activated sludge process because of a low COD/N-ratio. Also CH_4 will be increasingly emitted due to sludge digestion and subsequent emission from the sludge handling and biogas combustion processes and reject water. The effect of application of more innovative sludge treatment technologies on N_2O and CH_4 emission is not yet known.

Clearly, to decrease both the energy consumption and the greenhouse gas emission simultaneously, the effect of measures to save energy on the emission of greenhouse gasses other than CO_2 (thus N_2O and CH_4) should be taken into consideration as well. This means that it is the responsibility of the WWTPs to measure also other GHGs than CO_2 when processes are changed at WWTPs to reduce the GHG emission.

At this moment there are not enough full-scale data available to adequately predict the effect of an energy reduction measure on the emission of other GHGs than CO_2.

REFERENCES

Ahn J.-H., Kim S., Park H., Pagilla K. and Chandran K. (2010). N_2O emissions from activated sludge 2008–2009: results of a nationwide monitoring survey in the United States. *Environmental Science and Technology*, **44**(12), 4505–4511.

Böhnke B. (1977). Das Adsorptions-Belebungsverfahren (The AB-type activated sludge treatment). *Korrespondenz Abwasser* (Wastewater Correspondence), 24. Jahrg., 2/77.

Colliver B. B. and Stephenson T. (2000). Production of nitrogen oxide and dinitrogen oxide by autotrophic nitrifiers. *Biotechnol. Adv*, **18**(3), 219–232.

Czepiel P. M., Crill P. M. and Harriss R. C. (1993). Methane emissions from municipal wastewater treatment processes. *Environmental Science and Technology*, **27**, 2472–2477.

Guisasola A., Haas de D., Keller J. and Yuan Z. (2008). Methane formation in sewer systems. *Water Research*, **42**(6–7), 1421–1430.

IPCC (2001). *Climate Change 2001: The Scientific Basis. Contribution of Working Group I to the Third Assessment Report of the Intergovernmental Panel on Climate Change.* [J. T. Houghton, Y. Ding, D. J. Griggs, M. Noguer, P. J. van der Linden, X. Dai, K. Maskell, C. A. Johnson (eds.)].Cambridge University Press, Cambridge, United Kingdom and New York, NY, USA, 881 pp.

Jetten M. S. M., Horn S. J. and van Loosdrecht M. C. M. (1997). Towards a more sustainable municipal wastewater treatment system. *Water Science and Technology*, **35**(9), 171–180.

Kampschreur M. J., Temmink H., Kleerebezem R., Jetten M. S. M. and van Loosdrecht M. C. M. (2009a). Nitrous oxide emission during wastewater treatment. *Water Research*, **43**(17), 4093–4103.

Kampschreur M. J., Poldermans R., Kleerebezem R., van der Star W. R. L., Haarhuis R., Abma W. R., Jetten M. S. M. and van Loosdrecht M. C. M. (2009b). Emission of nitrous oxide and nitric oxide from a full-scale single-stage nitritation-anammox reactor. *Water Science and Technology*, **60**(12), 3211–3217.

Kampschreur M. J., van der Star W. R. L., Wielders H. A., Mulder J. W., Jetten M. S. M. and van Loosdrecht M. C. M. (2008). Dynamics of nitric oxide and nitrous oxide emission during full-scale reject water treatment. *Water Research*, **42**(3), 812–826.

Keller J. and Hartley K. (2003). Greenhouse gas production in wastewater treatment: process selection is the major factor. *Water Science and Technology*, **47**(12), 43–48.

Matsumura Y., Minowa T., Potic B., Kersten S. R., Prins W., van Swaaij W. P., van de Beld B., Elliott D. C., Neuenschwander G. G., Kruse A. and Antal M. J. Jr. (2005). Biomass gasification in near- and supercritical water: status and prospects. *Biomass and Bioenergy*, **29**, 269–292.

STOWA (2010). *Emissies van broeikasgassen van RWZI's* (Greenhouse Gas Emission From Wastewater Treatment Plants), E. Van Voorthuizen, M. Van Leusden, A. Visser, J. Kruit, M. J. Kampschreur, U. van Dongen and M. C. M. van Loosdrecht, Published by STOWA, Amersfoort, the Netherlands. pp. 90., ISBN 9789057734618, with English summary.

van Dongen U., Jetten M. S. M. and van Loosdrecht M. C. M. (2001). The SHARON-Anammox process for treatment of ammonium rich wastewater. *Water Science and Technology*, **44**(1), 153–160.

van Nieuwenhuijzen A. F., Graaf J. H. J. M., van der Kampschreur M. J. and Mels A. R. (2004). Particle related fractionation and characterisation of municipal wastewater. *Water Science and Technology*, **50**(12), 125–132.

Weissenbacher N., Takacs I., Murthy S., Fuerhacker M. and Wett B. (2010). Gaseous nitrogen and carbon emissions from a full-scale deammonification plant. *Water Environment Research*, **82**(2), 169–175.

Wett B., Buchauer K. and Fimml C. (2007). Energy self-sufficiency as a feasible concept for wastewater treatment systems. Proc. IWA Leading Edge Technology Conference, Singapore, Asian Water, Sept. 2007, 21–24.

Yu R., Kampschreur M. J., van Loosdrecht M. C. M. and Chandran K. (2010). Mechanisms and specific directionality of autotrophic nitrous oxide and nitric oxide generation during transient anoxia. *Environmental Science and Technology*, **44**(4), 1313–1319.

Part 3

Energy Footprint of Water Reuse

By The Editors

FOREWORD

The ambitious goals of sustainable development and achieving a zero net carbon and pollution emission footprint call for a new holistic approach to the management of the water cycle with an increased role for water reuse. With the further growth of megacities and increasing efforts to optimise energy efficiency, water recycling is of growing interest and will take a leading role in the future of sustainable water reuse.

Part 3 of this book "Energy Footprint of Water Reuse", (e.g. Chapters 11 to 15), highlights several sustainable approaches to water reuse with emphasis on energy requirements and carbon footprint.

Decentralised or semi-centralised water distribution systems are more efficient for future cities when water reuse is inevitably considered. Water supply can be tailored to match water demand more closely in centralised water infrastructures, adapting water quality to the given use. With substituting fresh water with appropriately treated recycled water, it is possible to save 30–50% of domestic water demand. The implementation of the semi-centralised approach to urban infrastructures would not only allow substantial water and energy saving, but would also provide new opportunities for energy recovery from biowaste throughout biogas production and for the use of nutrients in sewage sludge to produce biosolids for agriculture and other uses.

The energy demand of wastewater or greywater treatment depends on the water quality requirements, the selected treatment process and the plant capacity. In any case, the energy consumption of water reclamation is significantly lower than that of desalination, and represents only a fraction of the energy intensities for water supply, treatment, and distribution.

Several case studies and data from full-scale operations were provided in order to better understand the implications of energy footprint for water reuse systems, including traditional applications such as agricultural irrigation and energy intensive systems such as indirect potable reuse using advanced water treatment technologies. For example, reverse osmosis (RO) can typically constitute the most expensive component of wastewater tertiary treatment. Approximately half of the total treatment costs in advanced water reuse systems results from the RO system. However, the successful experience of the Groundwater Replenishment System (GWRS) project in Orange County, California, indicates that the energy footprint of advanced treatment could be more efficient compared to other water supply alternatives, for example water transportation at long distance or desalination. The implementation of this indirect potable reuse project clearly demonstrated not only the relevance of water reuse for sustainable development in a context of water scarcity, but also provided relevant information on health safety of high-quality recycled water and strategies for public outreach, education and communication.

More intensive water treatment has a higher environmental impact in terms of carbon footprint. For example the state of the art advanced water reclamation based on high-tech, energy intensive technologies has a carbon footprint five times higher than the conventional water reclamation processes. Although carbon footprint provides a valuable criterion to assess sustainability of industrial products and processes, other environmental impact factors have to be taken into consideration for the assessment of water reuse projects, such as fresh water depletion, reduced pollution, preservation of biodiversity and lower toxicity.

Chapter 11

Semizentral Germany: energy self-sufficient infrastructure systems for livable cities of the future

Susanne Bieker, Carola Zeig and Peter Cornel

11.1 CHALLENGES OF THE FUTURE

While some countries in Europe are suffering from decreasing population rates with all their consequences for urban and rural areas, large parts of the world face highly increasing population rates in urban areas. Figure 11.1 shows the "Speed of Change": it illustrates the hourly increase in population of the fasted growing cities of the world. Among these, the urban region of Shanghai shows a "medium" increase rate, that is a surplus of more than 280,000 inhabitants per year – an average large city in European benchmarks. And this happens, as stated by Daigger (Chapter 1), in all development areas around the world. While in 1975, there were only 200 cities with more than one million inhabitants, presently there are about 400, with a clear emphasis in Asian countries. Not long ago, we passed the threshold of 7,000,000,000 people living on earth. More than 50% are already living in cities and in 2030, this number will have increased to more than 60%, equaling almost 5 billion people – related to a world population of 8.32 billon people in 2030 (UNDP 2011; UN report 2007).

Figure 11.1 The speed of urban change, hourly increase in population (Bieker 2009, adapted from Burdett & Rhode 2007)

All these people need water, produce wastewater, and need energy for daily life, transportation, communication, and infrastructure operation: water extraction, transfer, treatment and distribution, wastewater transportation and treatment, and so on. The intensity of the linkage between water and energy is also described by Cornel *et al.* 2011 and Novotny (Chapter 3).

Rising living standards additionally increase the required amounts and capacities: taking a bath instead of a shower, cleaning the car, watering gardens and golf courses, consuming more meat, and causing additional water demand in agriculture are only some examples.

Worldwide, urban growth and increasing consumption have tremendous effects on the infrastructure of supply, treatment, and disposal of water, wastewater, and solid waste. The environment is seriously stressed by deficient or missing wastewater and waste treatment plants.

11.1.1 New infrastructure solutions needed to cope with urban growth

To improve the access to water and sanitation, two main objectives are to be pursued. On the one hand, the environment has to be protected from pollution; on the other hand, wastage and overexploitation of resources have to be reduced to a minimum. Today, this has been recognized worldwide. Especially the formulation of the Millennium Development Goals (MDG) reflects the increased awareness of this necessity. In the context of Goal 7 ("Ensure environmental sustainability"), the following targets are relevant:

- Target 9: "reverse the loss of environmental resources" and
- Target 10: "halve, by 2015, the proportion of the population without sustainable access to safe drinking water and basic sanitation".

To achieve these targets, recommendations have been formulated, for example, the identification of new water sources, such as rainwater and reclaimed water (UN 2008; Lenton *et al.* 2005). With regards to urban areas, this inevitably requires intra-urban water reuse (see Chapter 1). The applicability of respective technologies has to be assessed context-related. Besides ecological aspects of water reuse, there are strong economic arguments for reclaiming water: the increasing water demand on the one hand and limited amounts of natural resources (groundwater, surface water) on the other hand inevitably lead to water scarcity in densely populated urban regions. Rising prices and a high energy demand for tap water purification from alternative resources (e.g. seawater by desalination) are direct consequences that can only be avoided by severely reducing the needed water amount via the reuse of main water fractions, thereby additionally saving energy for conditioning.

However, it is obvious that this objective is barely realizable with conventional (linear) centralised supply and disposal systems as they have been implemented in industrialized countries in the 20th century (Novotny, Chapter 3). Huge technical units are able to treat the sewage of millions of urban residents, but only in case the grid system is adequate, regularly operated and maintained and the number of inhabitants to be supplied is more or less steady. However, this is not urban reality, which instead is affected by an increasing population of some dozen up to several hundred people a day, as shown in Figure 11.1. The principal characteristic of centralised supply and disposal systems is the huge capital commitment for a fixed-scale system, whereby the scale is only an estimated figure in the moment of investment and therefore a high risk. In other words: the upscaling of centralised solutions to the 'Asian scale' leads to huge facilities of several million capita units that need a several years' notice for planning, approval, and realization. This lead time is characterized by change, possibly fast growth, but above all from uncertainty. In fact, we do not know what the world will look like in five years. In general, prognoses work out well enough, however in detail and regarding specific circumstances, such as the investment in a pipe and sewage-system that connects several million people, the uncertainty is considerably higher.

Because of these obvious shortcomings of centralised pipe and sewerage systems, alternative solutions have been discussed worldwide, such as decentralised sanitation systems for on-site treatment of wastewater on household level. These systems might be appropriate within areas characterized by low population densities, however not within densely populated urban areas where limited space for on-site treatment facilities is the decisive factor. Still, household-based solutions, such as compost toilets, household-based urine separation, and rainwater collection, have already been proposed as possible solutions. These systems provide valuable information, however in urban areas with high population densities, aspects such as (monitoring) quality standards and surveillance, hygiene, maintenance, and performance will question their widespread use as stand-alone solution.

Another challenge, coming up predominantly in fast growing developing countries, is the rising amount of sewage sludge, resulting from increasing treatment capacities. At the same time, fast growing urban areas have to focus increasing amounts of residual waste. For both challenges there are no suitable and reliable solutions, especially in urban regions with high population growth rates and densities and therefore lacking space.

To overcome the shortcomings of centralised systems on the one hand and household-based decentralised systems on the other hand, the 'Semizentral' approach is being developed.

Semizentral represents an approach of district-based infrastructure development, able to 'deliver' infrastructure facilities at the time needed and of the required size, effectively on demand. Semicentralised infrastructures are based on district levels, that is, scales of 20,000 to 100,000 people.

However, Semizentral does not only stand for a system's scale. Simply transferring sectored centralised supply and treatment systems from industrialized countries to fast growing urban regions in newly industrialized and developing countries seems not appropriate: Intra-urban water reuse fosters relatively compact structures, avoiding the substantial transports of wastewater out of the city borders for treatment and the treated water back into the city for reuse. Therefore, the move from conventional (linear) centralised systems towards semicentralised solutions is a decision towards integrative planning and proceeding of technical infrastructure (Böhm *et al.* 2006) as well as closed-loop material and energy flow (circular urban metabolism systems, Novotny, Chapter 3).

11.1.2 Strategies to cope with the water challenge

Philosophically speaking, there are four theoretical strategies to cope with rising water needs in growing cities (Figure 11.2).

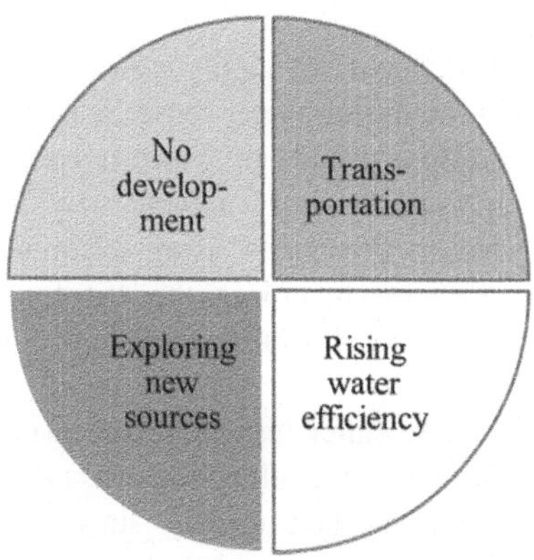

Figure 11.2 Strategies to cope with the water challenge

No development in areas where water is scarce

"No development" is a simple postulation, however especially in fast growing urban regions fairly unenforceable for two reasons: urban development is linked to economic development and economic development takes place where advantageous conditions and circumstances are given: industrial bases, mobility hubs, and of course governmental aids and subsidies. Especially in developing and threshold countries, market prizes for resources and energy are highly subsidized so that local or regional deposit and availability for example of water are no crucial factors for the choice of location.

Furthermore, due to the quantity of people and the population density within fast growing urban areas the inhabitants' demand usually exceeds the availability of natural resources in most of these areas: assuming the water demand to be 150 liters per capita and day and adding industrial and agricultural needs, urban agglomerations of 2.5 or more million people usually transcend the natural limits. Therefore, alternatives need to be found.

Transportation from areas with a surplus of water

The strategy of transporting water into densely populated urban areas is a very common approach. This way, however, the challenge of water scarcity is not solved, but only relocated. At the same time, transport infrastructures are needed, causing high investment, energy and maintenance costs. This is aggravated by the fact that the longer the distance of transportation (often more than 100 km, up to more than 1,000 km), the higher the water loss rates and the less efficient the systems. Furthermore, environmental impacts, such as lowering groundwater levels, salt intrusion, physical water scarcity, loss of biodiversity – to name but a few – affects regions, where in the past physical water scarcity has not been a topic. Therefore, this approach does not seem to be sustainable either.

Exploring new sources by desalination of seawater (in coastal regions)

In coastal areas desalination of seawater is used for water production. Although seawater is almost unlimited desalination has some major drawbacks. First to mention is the rather high energy consumption in the range of 2.5 to 4.5 kWh/m^3 product. Secondly, waste materials such as concentrated salt water and used chemicals have to be disposed, too. Both aspects cause considerable operation costs. In case of sea discharge, ecological aspects such as contamination have also to be noted (Grimm *et al.* 2008). Anyhow, desalination is becoming a more often used option as the technology of desalination improves.

Rising water efficiency in urban use

To cope with the balance between water demand and supply, especially in densely populated urban areas, an integrated water resource management, including water reuse, is advisable. The reduction of the potable water demand can be achieved by using methods of water reuse. Wastewater consists of more than 99.5% water and therefore is an appropriate and valuable new water resource. Depending on the subsequent application, wastewater has to be treated using adequate methods to reach the respective requirements (Cornel *et al.* 2011), thereby regarding the 'fit for purpose' aspect. Applications of non-potable intra-urban reuse, such as toilet flushing, irrigation, street cleaning, firefighting and air conditioning are of main interest.

The objective of intra-urban water reuse (after respective adequate treatment) is to preserve regional drinking water resources. One precondition for unobjectionable and successful water reuse in the long term is the consistent compliance with defined quality standards (Cornel *et al.* 2011). The US Environmental Protection Agency (EPA, 2004) has listed wastewater quality standards for defined wastewater constituents and their impact on wastewater reuse systems.

Besides technical solutions and parameter definitions, water reuse is – although for hygiene matters technically not a challenge any more – still an implementation challenge due to lacking acceptance. "Wastewater reuse" – the term implies hygienic challenges, risk and quality doubts for potential users. The only way to enable efficient and large-scale intra-urban water reuse will be to raise awareness for the need to save water (including the implementation of tariff systems) and the realization of lighthouse projects (e.g. Water Factory 21, NEWater or Yannawa, see §11.2.2) in order to show and prove the necessity and reliability of intra-urban reuse systems. Only the acceptance of the users, of house owners and renters, will enable a far more efficient use of water in terms of "fit for purpose" in the future.

11.2 THE SEMIZENTRAL APPROACH

Semizentral focuses on rising water efficiency in urban use by integrating energy and material flows. Instead of separate linear supply and disposal systems, the integration enables to build and close loops of these flows. Semicentralised Supply and Treatment Systems are circular systems that are flexible in the treatment technology of the different flows on the one hand and the design capacity or scale on the other hand. This is enhanced by their modular design.

11.2.1 Integrated energy and material flows for improved efficiency

Semicentralised Supply and Treatment Systems focus on integrated material and energy flow management, including treatment facilities for greywater, wastewater, biowaste, and sewage sludge as well as the energy sector: gaining energy from waste materials, reducing carbon emissions, the consumption of fossil fuels, and landfill capacities. Urban districts are the spatial scale that offers flexible solutions. Each district is provided with its own Semicentralised Supply and Treatment Center (STC). All material flows of the connected district are collected and transported to the STC. Wastewater from households is no longer collected as one flow, but separated into streams with different loads of pollutants: blackwater and greywater.

The separate collection and treatment of the greywater stream offers two opportunities: the lower pollution facilitates reuse options, from the energetic as well as the acceptance point of view. Technically, the reuse of blackwater is possible as well, even for purposes requiring relatively high water quality. However, the energy demand for biological treatment, as well as the "feel good factor" in the context of intra-urban water reuse should not be underestimated. For both factors, greywater is the preferable option. By converting greywater into reusable non-potable water for the exclusive use of toilet flushing, about 30% of the daily water demand can be saved, without any changes in comfort for customers or bathroom facilities. Of course, double collection and distribution pipes have to be provided for. Separately collected greywater also enables the opportunity to recover the inherent heat more efficiently as its temperature is higher compared to mixed wastewater. The heat can be used, for example, for preheating water, in order to reduce energy needs for water heating (Bieker *et al.* 2010).

The building complex of the STC comprises up to three main treatment modules (Figure 11.3).

The treatment of the entire wastewater is arranged in two modules of the STC, module A (greywater treatment) and module B (blackwater treatment). The third module (module C) is called 'Energy Center'. Here, the carbon-neutral energy resource is generated: biogas. The anaerobic treatment of sewage sludge (resulting from modules A and B) and sorted biowaste generates more biogas than needed for a self-sufficient operation of the whole STC, even a surplus can be produced. Additional potentials lie in the thermal energy within the STC, generated from the warm greywater flow and the so-called "waste heat" from power generation by the conversion of biogas into electric energy (Bieker *et al.* 2011).

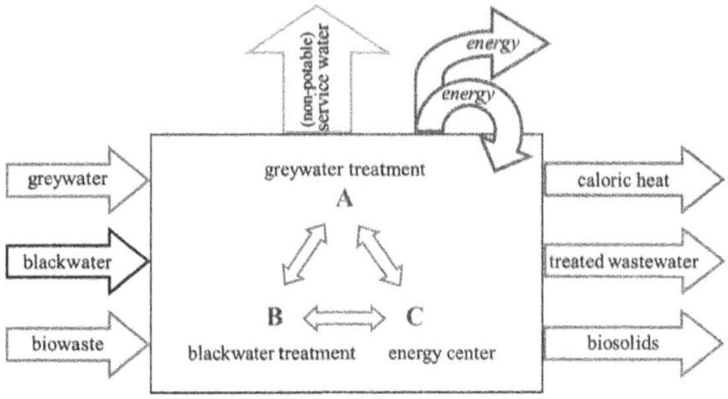

Figure 11.3 Principal material flows within a semicentralised supply and treatment center (STC) (Bieker *et al.*, 2010, modified)

Generally, all modules can be designed flexibly for variable scales to assure the optimal treatment of the accumulated wastewater and biowaste streams of the according district (Figure 11.3). This way, as mentioned above, the spatial scale of the Semizentral approach follows a district-oriented solution, where each district operates its own STC. The modular design allows:

- shorter planning and construction periods resulting in lower planning and construction costs
- improved conditions in the acquisition of plant components via framework agreements
- lower investment and maintenance costs of treatment facilities
- flexible deployment of personnel due to identical operating procedures

Finally, the energy production from waste and sludge improves the carbon footprint of the overall water, wastewater, waste, and sludge treatment: The energy used in the Semicentralised Water and Energy Center is (nearly exclusively) gained from organic material, the sewage sludge, as well as biowaste and residuals (see Section 11.2.3.3). Using the biogas out of this sludge and waste, not only the energy bill is reduced to a minimum, but also the CO_2 balance of the overall system is improved significantly. Ongoing research is going to clarify the carbon dioxide balance in more detail.

11.2.2 A matter of scale

The scale of integrated Semicentralised Supply and Treatment Systems follows the principle "as small as possible – as big as necessary" (Böhm *et al.* 2006), coping with the ambivalence of economic sustainability as well as social and ecological interests. Basic calculations assume a scale of at least several thousand people to achieve reasonable costs for anaerobic treatment. Corresponding to "as big as necessary": in order to use economies of scale, specific scale thresholds need to be considered.

To minimize the distance between the location of water treatment and households [as the source of grey- and blackwater as well as the location of consumption of (non-potable) service water] the STC is to be located very close to urban housing areas. This physical closeness implies the necessity of being "as small as possible": The contiguity of housing on the one hand and treatment modules on the other hand implies a small footprint of the STC due to land costs and development pressure. Therefore, compact treatment methods are of interest. Secondly, emissions need to be reduced to a minimum, odor as well as noise and air pollution. These requirements can be met best by subsurface or inhouse solutions. These "encapsulated" solutions for wastewater treatment already exist, for example inhouse solutions for membrane bioreactors in Porlock (UK), biofilter plants in France, Switzerland, and Germany, or even for larger-scale activated sludge plants, for example, the Yannawa Wastewater Plant in Bangkok or the NEWater factory in Singapore. Besides, small footprints are not only of interest in fast growing urban regions in China: the wastewater treatments plant in Helsinki (Finland) and Stockholm (Sweden) are built underground and within natural rock. Reasons are, for example, lacking space and climate conditions. Summarizing the mentioned arguments: the best-fit size is an individual scale, considering local circumstances and conditions and therefore cannot be given in a defined number of connected residents.

Key elements of the integrated Semicentralised Approach and the "encapsulated" underground solution for the STC are residential reuse of water on the one hand and integrated treatment of biowaste and sewage sludge on the other hand. It is not intended to determine certain treatment technologies – the choice has to consider the specific circumstances of the respective implementation case.

Additionally, the treatment of tap water for producing high-quality potable water is one possible option within the Semicentralised Approach that however is not included in the mentioned energy balance – resulting in energy self-sufficiency (Bieker *et al.* 2011).

11.2.3 A matter of flexibility: the construction kit

In order to realize the best-possible case-adapted solution for each implementation case and to ensure reasonable costs at the same time, a compromise of standardization and tailor-made solutions is required. The idea is to offer standardized modules within the STC, that is, to offer a kind of construction kit from which the modules can be chosen, thereby regarding the individual needs and requirements of each specific implementation case.

11.2.3.1 Module A: greywater treatment

The greywater is located in module A, where greywater arising from showers and washing machines is collected and treated. The resulting (non potable) service water can be used for toilet flushing or even higher-quality purposes, depending on the treatment methods and the implemented multi-barrier-approach on the one hand and on the acceptance of water reuse in the implementation area on the other hand. The module includes the pumping station, greywater treatment with upstream screening and the processing system for the recirculation of treated greywater as well as the disinfection device. The treatment consists of organics removal only, as the nutrient concentrations of greywater are assumed to be low. The temperature of the incoming greywater is elevated due to hot water from laundry and bathrooms. Pros and cons of different treatment processes are described by Chang (2010). Regarding the energy demand, sequencing batch reactors (SBR) and aerated biofilters are superior compared to membrane bioreactors (MBR). The electrical energy demand is in the range of 9 kWh/cap · yr for SBRs and up to 18 kWh/cap · yr using the MBR process. MBRs, of course, result in better effluent quality in respect of suspended solids and hygienic parameters and might be the best choice when considering water reuse (Meda *et al.* Chapter 14). Due to the higher temperature of greywater, compared to mixed wastewater, the efficiency of heat recovery will likely be increased. The caloric energy recovery of MBRs ranges from about 117 to 131 kWh/cap · yr (Figure 11.4). Further areas, such as depot, machine shop, electrotechnics, and blower installation for the MBR are also integrated in this module.

11.2.3.2 Module B: blackwater treatment

Due to the higher pollution, the separate treatment of blackwater from kitchen sinks and toilets is preferable. Therefore, module B is designed to treat blackwater as well as process water from module C, thereby considering the required effluent standard that usually includes nutrient elimination. Module B provides for mechanical treatment, consisting of inlet pumping station, screening system, grit separator, primary treatment, and the blackwater treatment unit. Due to higher concentrations of pollutants in comparison to greywater, the SBR process is chosen. Besides carbon removal, nutrient removal has to be considered. Within this module, excess water from module C that contains a considerable amount of ammonia nitrogen has to be treated as well. The electrical energy demand is about 20 kWh/cap · yr for the SBR process which is significantly higher for blackwater than for greywater treatment, as mentioned above. The reason is the higher pollution load that requires a higher treatment capacity and therefore more energy for aeration. The purpose of module C is to ensure reliable blackwater treatment in order to enable its reuse in street cleaning, irrigation or other non-potable reuse options, at least a discharge into receiving water bodies or stormwater sewerages.

11.2.3.3 Module C: energy center – waste to energy

Module C includes sludge and biowaste treatment with thickening, digestion, and dewatering as well as biogas generation. The anaerobic digestion of sewage sludge and sorted biowaste can be operated under mesophilic (37°C) or thermophilic (55°C) conditions. On the one hand, thermophilic digestion has a higher heat demand than mesophilic digestion; therefore the requirements for the thermal insulation are more comprehensive. On the other hand, thermophilic digestion implicates lower retention times and good disinfection effects – resulting in a higher biosolid quality. These advantages are the reason to favor thermophilic digestion within the STC. Anaerobic digestion of organic substrate produces biogas that is removed from the digester continuously.

The energy center features the possibility to increase the biogas yield via the co-treatment of sewage sludge (resulting from grey- and blackwater treatment) and biowaste. The biowaste fraction of the overall amount of domestic waste is much higher in developing and threshold countries than in industrialized countries. For example, in China, the organic fraction ranges between 50% and 65% (Rhode, 2007). In combination with the fact that landfilling is the main disposal strategy and that pretreatment generally only comprises manual sorting with the purpose of collect recyclables (Retech, 2009), the emission potential of landfills is alarmingly high.

Regarding the power generation from biogas, the unit consisting of gas turbine, gas tanks, boiler plant and gas flare purification is designed to ensure operation stability and reliability, both in case of heat deficiency (gas turbine) and in case of a surplus in gas production. Energy flow calculations indicate a power generation of around 300 Wh/cap · d respectively 110 kWh/cap · d whereof one third is used for operating the energy center, one third for operating the water treatment units, and the last third might be surplus and reserve (Figure 11.4). The "waste" heat from power generation via biogas is more than sufficient to compensate the needs of the digester.

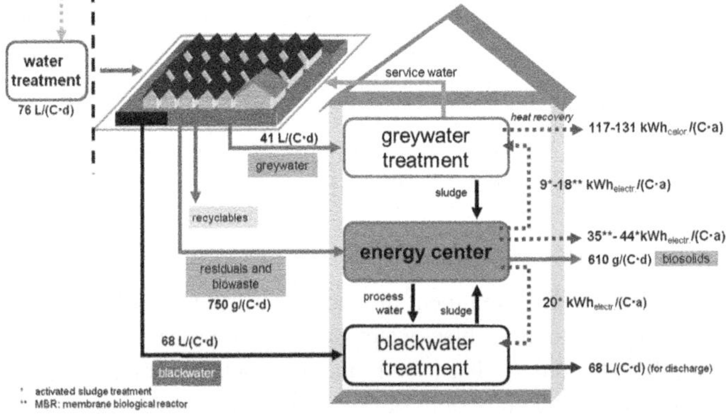

Figure 11.4 Energy flows of a semicentralised supply and treatment center for 20,000 inhabitants (Cornel *et al.* 2011, modified)

This way, the electrical and thermal energy yield from sewage sludge and biowaste combined with the thermal energy yield form greywater enables the energy self-sufficient operation of the integrated STC (Bieker *et al.* 2011). Stabilized biosolids can be used in agriculture, horticulture or similar applications of phosphorus and micronutrient reuse.

The treatment of the exhaust air is dimensioned to cover all exhaust air streams of the STC and is carried out via biofilters. Via an accumulative system the streams are transported to the exhaust air purification unit.

During post-treatment, the digested sludge is dewatered and the filtrate is intended to be processed in module B. The sludge dewatering unit consists of centrifuges for sludge thickening and a chamber filter press in order to provide a total solid content above 40%, necessary for landfill disposal. Furthermore, there is the possibility – after further treatment – for commercial use of biosolids as soil conditioner.

11.3 THE NEXT LEVEL – NEXUS OF SCALE AND FLEXIBILITY

The Semicentralised Supply and Treatment Approach offers the possibility to integrate different supply and treatment streams: In a selected new urban area wastewater streams are collected separately, treated greywater is used for toilet flushing, and organic waste is co-treated with sewage sludge.

Due to the limited spatial capacities within rapidly growing urban areas, footprints as small as possible are favored and the construction design of the STC has to be adjusted to the city structure. This way, the living quality of the neighboring residential areas will not be affected, thus increasing the acceptance of the facilities to be integrated into the residential district.

In order to design integrated semicentralised systems, a change in the perception of supply and disposal systems is requisite. Flexibility and adaptability to respective boundary conditions play a decisive role in semicentralised systems. It is possible to design an STC for two neighboring districts or to include potential expansions for the integration of future urban areas already in the planning of the first STC (Figure 11.5).

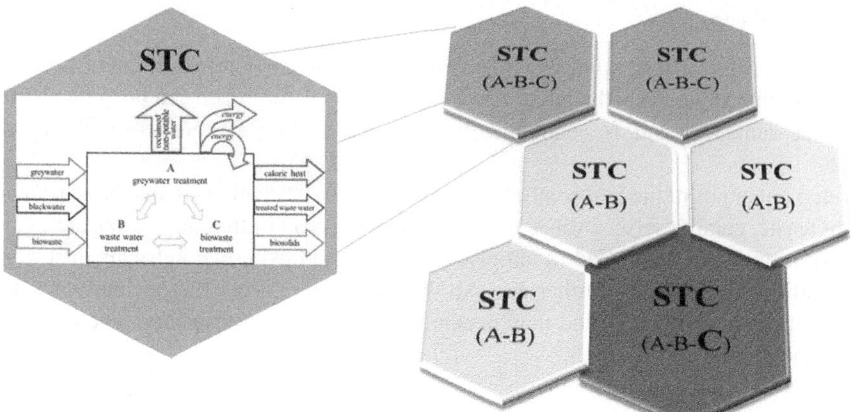

Figure 11.5 Integrated Semicentralised Supply and Treatment Centers: left: the principle – one STC for each district; right: different scales of STC, development-related expansion of housing and infrastructure

The modular design of STCs allows for growing capacities in case of additional catchment areas to be integrated. Furthermore, the modules can be duplicated and individually combined according to the respective requirements. For example, it is possible to operate an STC just for grey- and blackwater treatment. In this variant, the accumulating sewage sludge is transported to the STC of the neighboring district where it is treated in the energy center (module C) that accordingly has been designed for larger capacities From the economic point of view, there are many advantages to be gained in comparison to conventional design and construction techniques.

Usually, treatment modules of centralised wastewater treatment plants need extensive footprints. With integrated semicentralised systems, the design of treatment modules favors new construction designs with small footprints. Semicentralised plants that – in comparison to central plants – are designed to service significantly smaller catchment areas, allow for the construction of multistory treatment units. This way, compact and encased facilities are involved that can be optimally adjusted to the given needs and offer improved service reliability.

11.4 CONCLUSIONS

Up to this point, we can conclude that the application of the integrated semicentralised supply and treatment system reduces the freshwater amount substantially – by one third – and therefore:

- Conserves natural resources (or offers the opportunity to supply one third more people with the same amount of water),
- Reduces the arising daily wastewater amount by one third,
- Reduces energy needs for water purification, water transport, and wastewater treatment,
- Reduces system scales and transport distances and therefore water losses,
- Offers the opportunity of reusing thermal energy from separately collected greywater flows.

Additionally, waste and sewage sludge are used for biogas production – biogas that:

- Substitutes fossil fuels, and therefore,
- Reduces carbon emissions in energy production,
- Offers the opportunity of using the stabilized output fractions as soil conditioner, thereby reducing the need of landfill capacities.

Moreover – and this is the crucial point when leaving behind exclusive research interests and entering the field of implementation strategies – the integrated Semicentralised Approach can be implemented stepwise, considers market demands, and is economically competitive with sectored centralised infrastructure solutions.

The illustrated potentials of the Semizentral Approach can be summarized in one basic principle: recover resources from waste, gaining service water (fit for purpose) instead of treating wastewater, gaining biogas and thus energy from biowaste instead of simply depositing, and using the nutrients in sewage sludge to produce and establish biosolids for further use. All this is possible, if existing technologies are combined in an intelligent manner in order to find new perspectives of resource efficiency where we need it most: in the Cities of the Future.

REFERENCES

Bieker S. (2009). *Semizentrale Ver- und Entsorgungssysteme – ein neuer Ansatz für schnell wachsende urbane Räume in Entwicklungs- und Schwellenländern?* (Semicentralised Supply and Treatment Systems – a new approach for fast growing urban areas in developing countries?) In: *WAR-Schriftenreihe 200, Förderverein des Instituts WAR* (IWAR-Schriftenreihe 200, booster club IWAR), Technische Universität Darmstadt, Darmstadt, Germany.

Bieker S., Cornel P. and Wagner M. (2010). Semicentralized supply and treatment systems: integrated infrastructure solutions for fast growing urban areas. *Water Science & Technology*, **61**(11), 2905–2913.

Bieker S., Zeig C. and Cornel P. (2011). Semicentralized Infrastructures – 'Construction kit' for Flexible, Resource-saving and Energy Self-sufficient Supply and Treatment Solutions. *IWA International Conference, Cities of the Future Xían: Technologies for Integrated Urban Water Management*, Sept 15–19, 2011, Xían, China.

Böhm H. R., Bieker S. and Selz A. (2006). Semi-centralized Infrastructure Systems: Integrated Systems for a Sustainable Development of Fast Growing Urban Areas. *The Twelfth Annual International Sustainable Development Research Conference*, University of Hong Kong, 978-962-7589-27-3, Hong Kong, China.

Burdett R. and Rode P. (2007). The urban age project. In: The Endless City, R. Burdett and D. Sudjic (eds), Phaidon, London, England.

Chang Y. (2010). Greywater Treatment within Semi-centralized Supply and Treatment Systems by the Example of the Peoples Republic of China. PhD thesis, Technische Universität Darmstadt, Germany.

Cornel P., Meda A. and Bieker S. (2011). Wastewater as a source of energy, nutrients, and service water. In: Treatise on Water Science, P. Wilderer (ed.), Academic Press, Oxford. vol. 4, pp. 337–375.

Daigger G. (2012). Designing and Implementing Urban Water and Resource Management Systems Which recover Water, Energy and Nutrients, elsewhere in this book.

EPA (2004). Guidelines for Water Reuse. EPA/625/R-04/108, US Environmental Protection Agency.

Grimm V., Glauner C., Eicknbusch H. and Zweck A. (2008). *Wasserknappheit & Technologie, Übersichtsstudie, Zukünftige Technologien Consulting des VDI–Technologiezentrums* [Water Scarcity & Technology. Survey: Future Technologies Consulting of VDI–Technologiezentrum], Düsseldorf, Germany.

Lenton R., Wright A. M. and Lewis K. (2005). Health, Dignity and Development: What Will it Take? UN Millenium Project, Earthscan, London, Sterling, Novotny, Urban Metabolism.

Meda A., Henkel J., Chang Y. and Cornel P. (2012). Comparison of Processes for Greywater Treatment for Urban Water Reuse: Energy Consumption and Footprint, elsewhere in this book.

Novotny V. (2012) Water and Energy Link in the Cities of the Future – Achieving Net Zero Carbon and Pollution Emissions Footprint, elsewhere in this book.

Retech. (2009). http://www.retech-germany.net/files/pdf_dokumente/application/pdf/lp_vr_china_091126.pdf (accessed 14 December 2011).

Rhode C. (2007). *Milchsäurefermentation von biogenen Abfällen* [lactic Acid Fermentation of Biowaste]. In: *WAR-Schriftenreihe 186, Förderverein des Instituts WAR* (IWAR-Schriftenreihe 186, booster club IWAR). Technische Universität Darmstadt, Darmstadt, Germany.

UN report (2007). For humanity's sake, developing world must prepare for soaring urbanization. Report by UN Population Fund with additional reporting by Lisa Schlein and Joe De Capua (VoA) as well as Sven Krüger (City Mayors), 28 June 2007.

UN report (2008). The Millennium Development Goals Report, New York, USA.

UNDP (2011). Sustainability and Equity: A Better Future for All. Human Development Report 2011, published for the United Nations Development Program (UNDP), ISBN: 9780230363311, Communications Development Incorporated, Washington, DC, USA.

Chapter 12

Groundwater Replenishment System – energy usage implications

Mehul V. Patel

12.1 INTRODUCTION

In 1933 the Orange County Water District (OCWD) was formed in Fountain Valley, California to maintain the quality and quantity of Orange County's groundwater basin. Orange County's groundwater basin (the Basin) currently provides potable water for over 2.5 million residents of North and Central Orange County. The Basin's primary water supply is obtained from the Santa Ana River, which is recharged into the Basin. Supplemental sources are imported from the Colorado River and State Water Project via the Metropolitan Water District (MWD).

OCWD has seen significant decreases of water flow in the Santa Ana River. Much of this is a result of dry weather conditions and water conservation and recycling upstream. Orange County receives an average on only 330 to 380 mm (13 to 15 inches) of rainfall annually. Legal mandates to protect endangered fish in the Delta, have greatly reduced the delivery of imported water supplies from Northern California to Southern California. Frequent drought conditions in the Western United States and competing demands from six other western states, Mexico and several Native American tribes have caused supply reductions of water imported from the Colorado River. North and central Orange County relies on the Delta and the Colorado River to provide approximately 40% of the region's water supply needs. Historically, OCWD has been able to purchase water from these sources to recharge the groundwater basin, but these water supplies are not always available and will be more difficult to rely on in the future.

Faced with these challenges, and an expected population growth of more than 300,000 people by 2035 in north and central Orange County, OCWD has to find new ways to maximize the annual sustainable yield of the groundwater basin. Consequently, water recycling has become an essential element of local water supply.

To address the issue of seawater intrusion, in 1976 OCWD began recycling Orange County Sanitation District (OCSD) wastewater at a facility known as Water Factory 21. Product water from this facility was injected along the Talbert Gap via a series of injection wells to form a hydraulic barrier. This barrier was used to block the migration of seawater into the main groundwater aquifer. Since early 1990s, it was evident that Water Factory 21 was reaching the end of its useful life. During the latter years, 18,930 m^3/d (5 million gallon per day, MGD) of injection water were produced, which was blended with other potable and deep well water sources to achieve a total of 56,780 m^3/d (15 MGD) of injection water for the Barrier.

In 2008 OCWD completed the Groundwater Replenishment System (GWRS) project to replace the Water Factory 21 facility which had reached the end of its useful life and was under-sized for current seawater intrusion barrier needs. This project included a 265,000 m^3/d (70 MGD) Advanced Water Purification Facility (AWPF) to purify secondary effluent for use in an existing seawater intrusion barrier and for aquifer replenishment by means of a 22.5 km (14 miles) distribution pipeline (Figure 12.1). The GWRS is the largest indirect potable reuse project in the United States and one of the largest in the world.

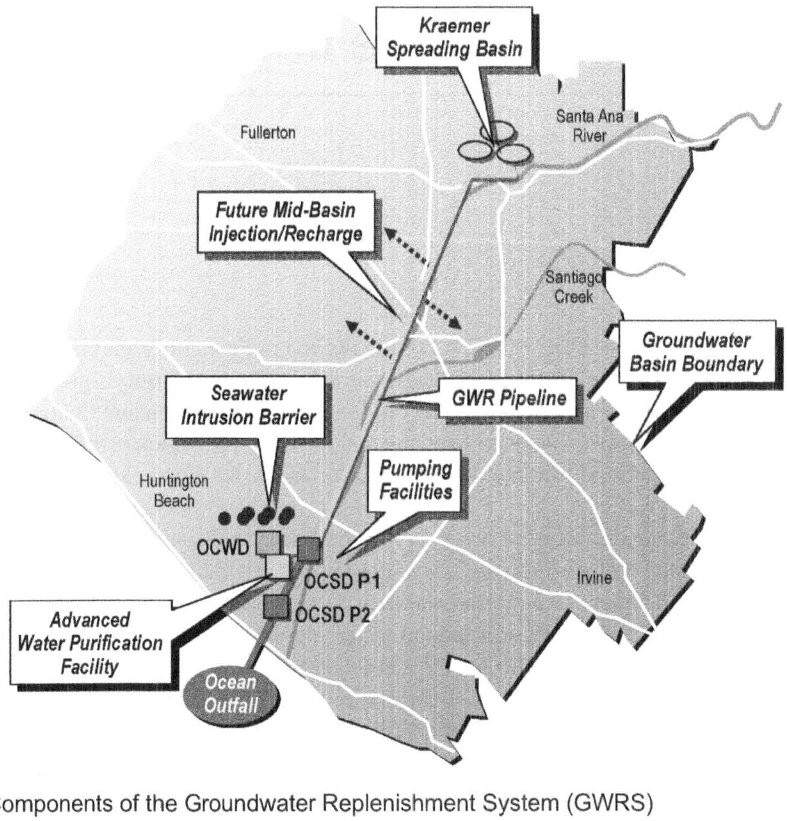

Figure 12.1 Components of the Groundwater Replenishment System (GWRS)

The California Department of Public Health (CDPH) requires stringent product water quality, especially in regards to total nitrogen and total organic carbon. CDPH requirements call for total nitrogen of less than 5 mg/L and total organic carbon of less than 0.5 mg/L. These stringent standards drive the treatment facility toward the use of advanced membrane technology. The AWPF product water is roughly equally distributed to the seawater intrusion barrier and recharge basins.

The final product water from the AWPF is tested regularly for hundreds of constituents including contaminants of emerging concern. In addition, the water quality is tested throughout the groundwater basin on a regular basis. This includes testing at areas closest to the introduction point of the recycled water. This allows OCWD to determine if the use of recycled water has any detrimental affect on the overall water quality of the groundwater basin. To date there have been no detrimental affects by the use of this highly treated recycled water.

12.2 DESIGN OF THE ADVANCED WATER PURIFICATION FACILITY

The Advanced Water Purification Facility, AWPF (Figure 12.2), is the major component, the heart, of the Groundwater Replenishment System. The treatment train includes the following processes: microfiltration (MF), reverse osmosis (RO), and advanced oxidation process (AOP) using ultraviolet light (UV) and hydrogen peroxide (H_2O_2).

Figure 12.2 View of the advanced water purification facility of the GWRS in Orange County

A detailed plant process diagram is shown in Figure 12.3, including water quality sampling locations. The first treatment processes of the AWPF are the pre-treatment systems including enhanced source control and clarified secondary treatment, fine screening, and chloramination. Clarified secondary effluent from OCSD Wastewater Treatment Plant N°1 (WWTP N°1) travels by gravity to the screening facility and to the MF system.

The screening facility has four rotating 2-mm gravity screens to remove larger particles from the secondary effluent ahead of the MF process. Upstream screening is vital to MF System performance and is required as part of the MF membrane manufacturer warranty conditions.

During normal operation, the MF system requires upstream chloramination, which helps to extend the interval between full MF chemical cleanings and decreases the rate of biofouling in the first stage of the RO. Chlorine in the form of sodium hypochlorite is mixed into the screened secondary effluent, forming chloramines from the abundant ammonia in the secondary effluent. Chloramines are necessary to minimize biological fouling of the MF membranes without oxidizing the membranes, which are made of polypropylene and are susceptible to oxidation by free chlorine.

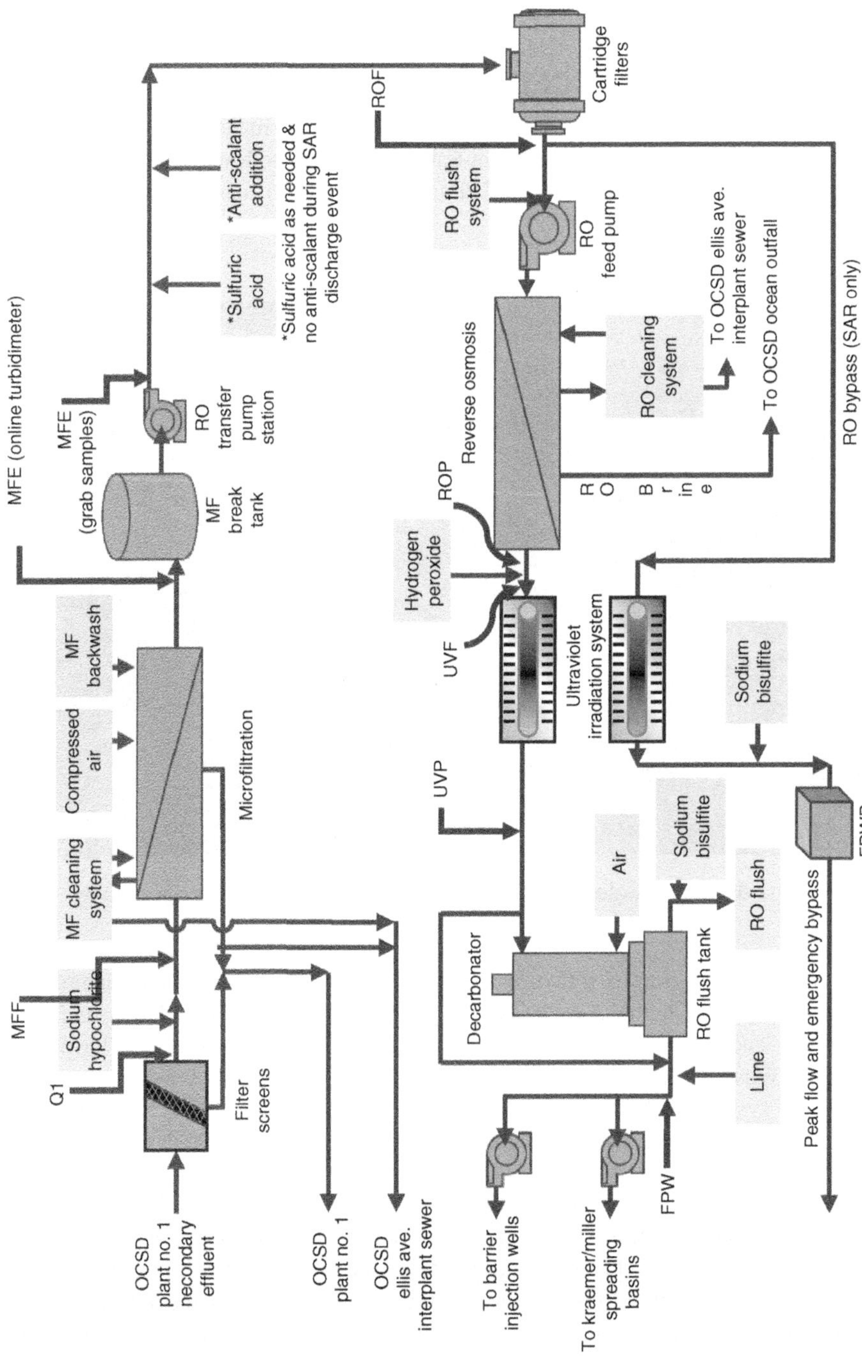

Figure 12.3 AWPF process flow diagram with sample locations

The MF system treats the screened secondary effluent to a filtrate water quality that has no significant particulate materials as indicated by low turbidity (<0.2 NTU) and a 15-minute Silt Density Index (SDI) below 3 at a suitable cleaning interval. The MF system consists of 26 Siemens/Memcor CMF-S units (cells). The units are grouped into 3 trains of 8 cells each and a fourth train of 2 cells. Each cell contains 608 hollow fiber membrane modules, and is capable of holding up to 684 modules. The hollow fiber pore size is 0.2 μm, and the recovery rate is between 88% and 90%. The backwash from the MF system is returned to the primary clarifiers at OCSD WWTP N°1. The MF system has an initial design capacity of 325,545 m^3/d (86 MGD) at an instantaneous design flux of 0.83 $m^3/m^2 \cdot d$ (20.4 gallons per square foot per day, gfd).

The second component of the AWPF process is the RO system (Figure 12.4). The RO system includes the RO transfer pump station, RO pre-treatment chemical addition, cartridge filtration, high pressure membrane feed pumps, RO treatment trains, flushing systems and clean-in-place systems.

Figure 12.4. Views of the reverse osmosis building and the RO treatment trains

The RO transfer pump station pumps MF effluent from the MF filtrate tank through the cartridge filters to the RO high pressure membrane feed pumps. The pre-treatment chemical addition includes facilities for the addition of sulfuric acid and a proprietary threshold inhibitor to the MF filtrate for pH adjustment and for prevention of sparingly soluble salt precipitation on the membranes. Cartridge filters with a nominal 10 μm rating are located upstream of the RO membranes to ensure suspended solids removal that may be introduced through pipeline residuals or impurities in the chemical treatment of the filtrate, thereby providing an additional level of protection for the RO membrane elements.

After the cartridge filters, the in-line high-pressure pumps boost the pressure of the pre-treated MF effluent to the RO treatment trains. There are a total of 15 RO treatment units, with 150 vessels per unit (Figure 12.4). The units are arranged in five groups of three with 14 units on line at any one time. The 15th unit serves as a standby unit. Each vessel in a unit houses 7 membrane elements. Each unit has a rated capacity of 18,930 m^3/d (5 MGD), operating over a range of flows from 15,140 to 18,930 m^3/d (4 to 5 MGD). Each unit operates at a recovery of 85% with all concentrate piping sized to accommodate a recovery as low as 80%. The reject stream (concentrate) is sent to OCSD where it is combined with secondary effluent and disposed of in the OCSD ocean outfall. The RO permeate is sent to the UV/AOP system for further treatment.

The final components of the AWPF are the post-treatment systems. The post-treatment facilities include advanced oxidation (UV treatment with hydrogen peroxide), decarbonation, lime stabilization and product water pumping. Direct photolysis (UV alone) provides disinfection and is also effective in reducing levels of n-Nitrosodimethylamine (NDMA), a low molecular weight organics (LMWO) not fully removed by RO.

Since direct photolysis is not effective for all LMWOs, approximately 3 mg/L hydrogen peroxide (H_2O_2) will be added upstream of the UV system for hydroxyl radical generation. These hydroxyl radicals are more effective for reduction of contaminants such as 1,4-dioxane.

The AWPF UV facility consists of nine trains; each train is designed to treat 33,120 m^3/d (8.75 MGD). The UV system is designed to provide a 4-log reduction of viruses and a 1.2 log reduction of NDMA. The UV system will utilize LPHO lamp technology to treat the RO permeate. After treatment by the RO system and the UV system, the product water still has a high carbon dioxide content created by the pH reduction ahead of the RO system.

The AWPF utilizes a decarbonation system consisting of blowers and pack towers to partially strip the carbon dioxide from the water to an optimal level. After partial decarbonation, the UV product is stabilized with hydrated lime. The stabilized product water is then be pumped to the seawater intrusion barrier and to groundwater recharge basins located 22.5 km (14 miles) away from the plant site to supplement other sources.

12.3 WATER QUALITY

The feed water quality to the GWRS is typical of what is shown in Table 12.1 (water quality sampling designations are shown in Figure 12.3). The primary measure of the plant's performance is based on total organic carbon (TOC less than 0.5 mg/L/recycled water content), total nitrogen (less than 5 mg/L), TDS and NDMA (less than 10 ng/L). These parameters give an indication of the overall plant performance, especially in regard to the RO and AOP processes.

The AWTF also removes emerging organic parameters that are not currently regulated in California, but may be considered for potential future regulations. The emerging parameters most often found in wastewater include pharmaceuticals (acetaminophen, ibuprofen, caffeine, carbamazapine, gemfibrozil, primidone, sulfamethoxazole), flame retardants (TCEP), and pesticides (DEET, diuron, triclosan). None of these chemicals, nor any other organic micropollutants tested, are found in measurable concentrations after AOP treatment.

12.4 ENERGY CONSUMPTION

The GWRS uses an energy intensive treatment process. The high level of treatment can only be achieved by using advanced technologies. These technologies are highly automated and include the need for frequent preventative and reactive maintenance.

The heart of the treatment process is reverse osmosis. This process is also the most energy intensive process of the AWPF. The RO process energy consumption is directly rated to level of pre-treatment and the dissolved content concentration in the water.

The power consumption of each treatment process is measured by a series of power monitors. The power monitoring system allows for plant personnel to determine the energy usage by specific treatment process. Figure 12.5 shows the breakdown of energy requirements by major treatment process in the AWPF based on actual data compiled. The data in the pie chart is in terms of percentage, as well as in kilowatt hours per cubic meter (kWh/m^3) and per million gallons (kWh/MG). The data shown in Figure 12.5 is based on a yearly production rate of 74,996,000 m^3/yr (19,812 MG or 60,800 acre feet).

As stated earlier it is clear the RO process is the main source of power consumption: 48.2% or 0.53 kWh/m^3 (2014.5 kWh/MG). Consequently, the optimization of the RO process is the primary focus by OCWD operations personnel.

Table 12.1 Water quality data for AWPF (sampling location is indicated in Figure 12.3).

Parameter name	Units	Q1	MFF	MFE	ROF	ROP	UVP	FPW	Permit
Electrical conductivity	µm/cm	1649.83	1723.44[1]	1715.00	1769.14[1]	42.44[1]	54.82	88.72[1]	N/A
Total dissolved solids	mg/L	949.77	na	na	1012.53	20.22	22.49	45.37	500[2]
Suspended solids	mg/L	8.94	7.10	2.82	na	na	na	na	N/A
Turbidity	NTU	3.43	4.804[1]	0.088[1]	0.184[1]	0.03[1]	na	0.156[1]	≤0.2/≤0.5
Ultraviolet transmittance	%	na	na	62.18	na	97.71	na	99.40	>90
pH	units	7.63	7.42[1]	7.49	6.65[1]	5.52[1]	5.43	8.59[1]	6–9
Total hardness (as $CaCO_3$)	mg/L	317.17	na	na	299.17	<1	<1	24.61	240[2]
Calcium	mg/L	84.64	na	na	80.76	<1	<1	9.69	N/A
Magnesium	mg/L	25.68	na	na	23.68	<1	<1	<1	N/A
Sodium	mg/L	217.83	na	na	204.42	6.49	7.09	6.48	45
Potassium	mg/L	17.34	na	na	16.88	0.41	0.44	0.36	N/A
Bromide	mg/L	na	na	na	na	na	na	<0.1	N/A
Chloride	mg/L	258.08	na	na	230.67	4.00	4.53	4.75	55
Sulfate	mg/L	231.42	na	na	285.92	<0.5	<0.5	<0.5	100
Hydrogen Peroxide	mg/L	na	na	na	na	na	2.61	2.51	N/A
Bicarbonate (as $CaCO_3$)	mg/L	na	na	na	200.42	13.29	12.09	34.73	N/A
Nitrate Nitrogen	mg/L	2.79	na	na	2.48	0.28	0.42	0.35	3[2]
Nitrite Nitrogen	mg/L	0.57	na	na	0.47	<0.002	0.005	0.06	N/A
Ammonia Nitrogen	mg/L	20.53	na	na	21.15	1.25	na	1.24	N/A
Organic Nitrogen	mg/L	2.26	na	na	1.27	<0.1	na	0.08	N/A
Total Nitrogen	mg/L	25.79	na	na	25.35	na	na	1.75	5
Phosphate Phosphorus	mg/L	0.57	na	na	na	na	na	<0.01	N/A
Iron	µg/L	451.25	na	na	113.76	<1	<1	2.98	300
Manganese	µg/L	42.13	na	na	41.86	<1	1.88	<1	50
Aluminum	µg/L	18.55	na	na	10.83	2.35	2.98	7.27	200[2]
Arsenic	µg/L	1.58	na	na	1.37	<1	<1	<1	10
Barium	µg/L	27.21	na	na	23.97	<1	<1	<1	1000
Boron	µg/L	0.37	na	na	0.38	0.24	0.22	0.23	1

(Continued)

Table 12.1 Water quality data for AWPF (sampling location is indicated in Figure 12.3) (*Continued*).

Parameter name	Units	Q1	MFF	MFE	ROF	ROP	UVP	FPW	Permit
Cadmium	μg/L	<1	na	na	<1	<1	<1	<1	5
Chromium	μg/L	<1	na	na	<1	<1	<1	<1	50
Copper	μg/L	5.88	na	na	9.28	<1	<1	<1	1000[2]
Cyanide	μg/L	9.3[3]	6.7[3]	na	12.24[3]	<5	na	<5	150
Fluoride	mg/L	0.93	na	na	na	na	na	<0.1	2
Lead	μg/L	<1	na	na	<1	<1	<1	<1	15
Mercury	μg/L	0.35	na	na	0.33	<0.1	<0.1	<0.1	2
Nickel	μg/L	6.94	na	na	6.83	<1	<1	<1	100
Selenium	μg/L	2.58	na	na	2.37	<1	<1	<1	50
Silica	mg/L	22.18	na	na	22.73	<1	<1	<1	N/A
Silver	μg/L	<1	na	na	<1	<1	<1	<1	100
Zinc	μg/L	20.07	na	na	28.31	2.08	1.33	0.78	5000
N-nitrosodimethylamine	μg/L	49.42	na	na	71.00	35.00	<2	<2	10
1,4-Dioxane	μg/L	2.19	na	na	na	na	<1	<1	3
Total Trihalomethanes	μg/L	na	na	na	2.00	1.20	na	0.14	80
Dibromoacetic acid	μg/L	na	na	na	<1	<1	na	<1	60,total HAA5
Dichloroacetic acid	μg/L	na	na	na	13.59	1.70	na	<1	60,total HAA5
Monobromoacetic acid	μg/L	na	na	na	<1	<1	na	<1	60,total HAA5
Monochloroacetic acid	μg/L	na	na	na	<1	<1	na	<1	60,total HAA5
Trichloroacetic acid	μg/L	na	na	na	9.70	<1	<1	<1	60,total HAA5
Apparent color (unfiltered)	units	na	na	na	35.00	<3	na	<3	15
Total Organic Carbon (unfiltered)	mg/L	13.59	14.95	na	10.13[1]	0.10[1]	na	0.19	0.5–0.7[2]
Surfactants	mg/L	0.20	na	na	0.25	<0.02	na	<0.02	0.5
Total Coliform, MPN/100 mL		1 633 010	38 419	<2	na	<2	<2	<2	2.2
Fecal Coliform, MPN/100 mL		532 663	5 904	<2	na	<2	<2	<2	N/A

[1]On-line average
[2]See App. A for more information
[3]Cyanide detected due to lab processing issue.
Q1, Secondary Effluent (AWPF influ); MFF, Microfiltration Feed; MFE, Microfiltration Effluent; na, Not analyzed
ROP, Reverse Osmosis Product; UVP, Ultraviolet UV/AOP Product; FPW, Finished Product Water; N/A, Not applicable

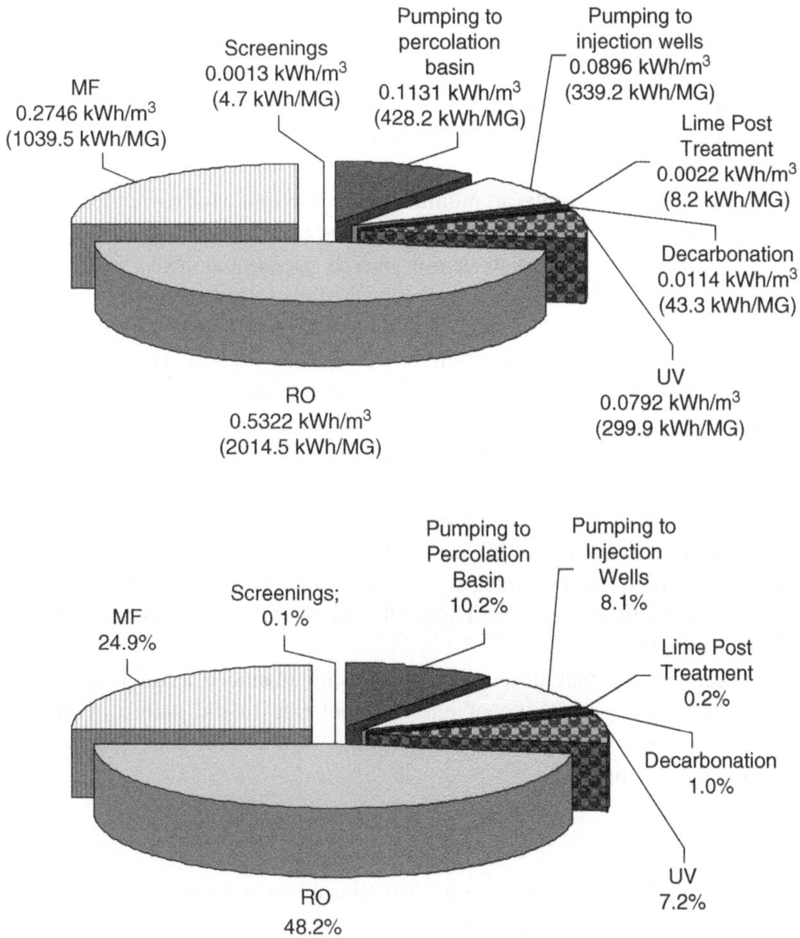

Figure 12.5 Energy consumption by treatment process for the AWPF

The driver for chemical cleanings of the reverse osmosis is the increase in total feed pressure requirements. A dramatic rise in feed pressure due to accelerated membrane fouling can have a large affect on the AWPF overall energy consumption and plant O&M costs. It is the RO system energy consumption that tends to govern the criteria for determination of the proper time to perform chemical cleanings.

However, for the MF process the cleaning interval is well defined in terms of a set terminal transmembrane pressure (TMP) of 0.83 bar (12 psi). The MF process has a set range of operating pressure between 0.21 and 0.83 bar (3 and 12 psi). The MF system also has a control system that shuts down the process once the terminal TMP of 0.83 bar (12 psi) is reached forcing a full chemical cleaning.

12.5 CONCLUSIONS

The Advanced Water Purification Facility of the Groundwater Replenishment System in Orange County is a large and complicated treatment facility that is able to take secondary effluent and produce purified water that exceeds drinking water standards. The ability of the system to remove multiple types of contaminants helps to ensure that the product water discharged from the facility will in itself not cause degradation to the Orange County groundwater aquifer.

The most significant operational cost relates to the energy consumption required for the AWPF process. The treatment technologies used and the large scale of the facility make energy consumption of key importance to plant optimization.

The AWPF is equipped with state of the art monitoring devices in the form of power monitors. The power monitoring capability allows OCWD to track energy usage by treatment process. This in turn lends itself to possible optimization of treatment process to save in overall operational costs.

The reverse osmosis portion of the AWPF is by far the greatest source of energy consumption among the treatment processes used. As such, a key strategy in reduction of energy consumption is to maximize filtrate water quality from the upstream microfiltration process. This strategy is implemented by operation the microfiltration system with the highest possible level of membrane fiber integrity. All possible membrane fiber or mechanical seal issues are addressed immediately and closely monitored by the microfiltration system on line integrity monitoring system or pressure decay test. The other key strategy implemented is to control the level of biofouling on the first stage of the RO system. As biofouling increases the energy demand of the RO increases and the ability to remove the fouling layer becomes less over time. OCWD attempts to control biofouling by carrying a continuous chloramine residual in the RO feed while ensuring a high level of integrity to the upstream cartridge filtration system. These two strategies are crucial for targeting the highest possible energy efficiency on the most energy intensive process used in the GWRS project.

It is important to keep in mind that all reuse facilities using advanced technologies such as membranes and advanced oxidation require a significant energy input due to the nature of the equipment used. Understanding the relationship between energy and water production can lead to a more successful implementation of any reuse scheme.

REFERENCES

Burris D. (2011). Groundwater Replenishment System. 2010 Annual Report, Orange County Water District, Fountain Valley, CA.

Chalmers B. and Mehul P. (2009). A Drought-proof Supply: Implementing an Advanced Recycled Water Plant. *Proc. 2009 AWWA Annual Conference and Exhibition*, San Diego, CA.

Patel M. (2010). Why Operating Membrane Plants Can Drive You Nuts. *Proc. of the 2010 AMTA Annual Conference*, San Diego, CA.

Chapter 13

Comparative study of carbon footprint of water reuse

Laura Delgado, Yvan Poussade and Pere Aguiló

13.1 INTRODUCTION

Water reuse has been identified as an effective new source of water which allows reducing the demand on available surface and ground water supplies and can replace or be complementary to other high-energy demanding sources such as desalination. Water reuse has also other beneficial impacts such as the reduction of water pollution from wastewater discharge (Sala and Serra, 2004; Asano *et al.* 2007).

However, water recycling may have also detrimental impacts which should be studied. This concern has resulted in the introduction of environmental risk management support tools for decision making. Life Cycle Analysis (LCA) methodology is currently used for estimating the most relevant environmental impacts of potable water supplies (Vince *et al.* 2008), wastewater treatments (Pasqualino *et al.* 2009) and reclamation treatments (Stokes and Horvath, 2006). Global warming is one of the key environmental indicators considered in a LCA and, in addition, it is projected to have a significant impact on water resources and their management (IPCC, 2008). It is therefore important to understand the contribution of water reuse on GHG emissions, in order to identify and encourage environmental improvements against climate change.

Water reuse guidelines and regulations can vary from one country to another and result in requirements for water quality and treatment technologies that are very dependent on the final product use, ranging from industrial reuse, pasture irrigation, recreational bathing, indirect or even direct potable reuse applications. Most developed countries have established conservatively low risk guidelines or standards based on a high technology/high-cost approach, such as in California (USEPA, 2004). As quality requirements become more stringent, advanced technologies that consume more energy and other resources are required.

The objective of this chapter is to assess the major contributors to greenhouse gas emissions, that is carbon footprint, of water reclamation according to the grade of water quality achieved. Two different water reuse schemes are discussed: a basic reclamation treatment in Barcelona, Spain and an advanced reclamation treatment in South East Queensland, Australia.

13.3 DESCRIPTION OF WATER REUSE FACILITIES

Spain and Australia are two leading countries by the amount of recycled water produced (GWI, 2010). The two facilities presented in this chapter are part of regional integrated water resources management plans: El Prat de Llobregat Water Reclamation and Reuse Project in Barcelona (Spain) and the Western Corridor Recycled Water Project, Australia.

13.3.1 El Prat de Llobregat water reclamation plant

This water facility has a design flow of 300,000 m^3/d, making it one of the largest reuse plants in the world (GWI, 2010). The basic reclamation process treatment of the El Prat de Llobregat Water Reclamation Plant (WRP) has been in operation since 2006 and includes (Figure 13.1, Mujeriego *et al.* 2008):

- Equalization tanks for the secondary effluent and a pumping station.
- ACTIDisk® process: Actiflo ballasted clarification and lamella settling followed by microfiltration (MF) through 10 μm mesh Hydrotech Discfilter.
- Disinfection by UV light and sodium hypochlorite dosage.

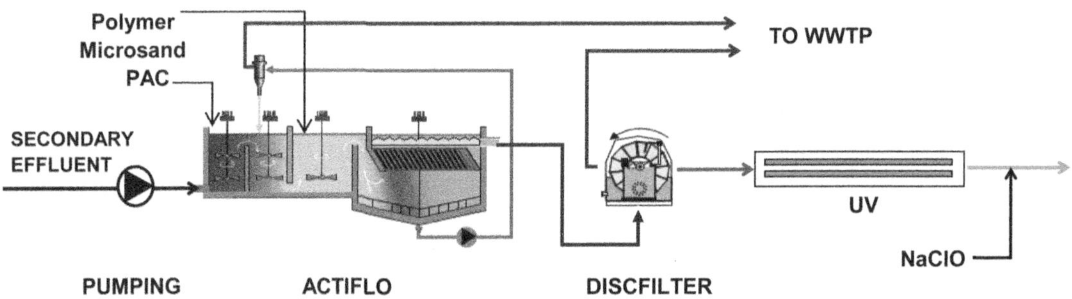

Figure 13.1 El Prat de Llobregat WRP process flow scheme

Actiflo is a coagulation, flocculation and sedimentation technology which uses microsand as a ballasting agent in the flocculation step. The addition of microsand as a ballast promotes the formation of high density flocs and results in a high rate settling technology. Chemicals used in the Actiflo include polyaluminium chloride (PAC) and anionic polymer. The microsand is recycled into the system whereas sludge is withdrawn from the Actiflo. Clarified water is filtered through 10 μm Discfilters and subsequently disinfected by means of UV light and sodium hypochlorite. Backwash water of Discfilters and Actiflo sludge are sent to the primary treatment of the wastewater treatment plant.

Sludge from the reclamation treatment accounts for 6–7% of the total amount of sludge (dry solids) produced by the WWTP. In order to include the impact of this waste flow on the carbon footprint analysis, the sludge process line of El Prat WWTP has been evaluated and a 6% of its carbon footprint is allocated to the water recycling plant. The sludge treatment scheme consists of a dewatering step by means of thickeners and centrifuges, anaerobic treatment, further centrifugation and thermal drying. A combined heat and power system (CHP) uses biogas from the anaerobic digester and natural gas. Supplied electricity and heat are used on site, which allows for a reduction of approximately 50% in the total energy consumption. Part of the dewatered sludge (25% DS) is directly used in agriculture whereas most of it is further dried (90% DS) and finally used as fuel in cement industries (65%) or allocated for direct agricultural uses or composting. Figure 13.2 shows the sludge treatment scheme.

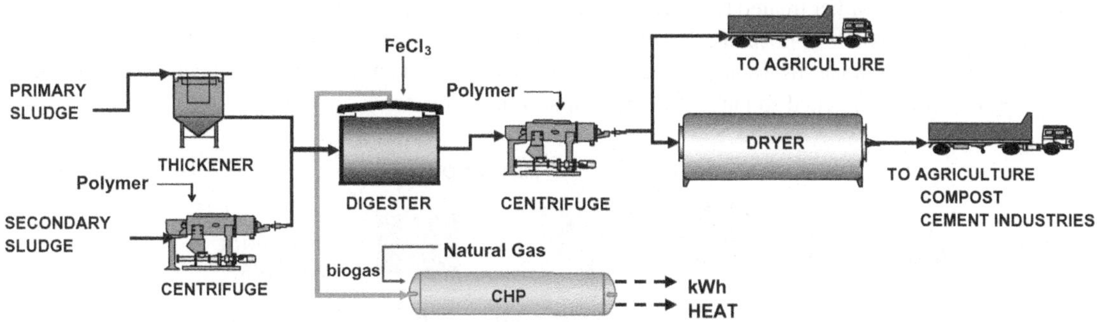

Figure 13.2 El Prat de Llobregat WWTP sludge treatment scheme

13.3.2 Bundamba advanced water recycling plant

Bundamba Advanced Water Recycling Plant (AWRP) is one of the three advanced water treatment facilities of the Western Corridor Recycled Water Project. It is designed to treat effluent from four different wastewater treatment plants and has a capacity of 66,000 m^3/day. It is based on the following treatment process units (Srinivas *et al.* 2009) (Figure 13.3):

- Chloramination and pretreatment by coagulation and settling.
- Microfiltration (MF) and reverse osmosis (RO) filtration.
- Advanced oxidation (AOP) using hydrogen peroxide and UV light.
- Post treatment by remineralisation and final chlorination with sodium hypochlorite.

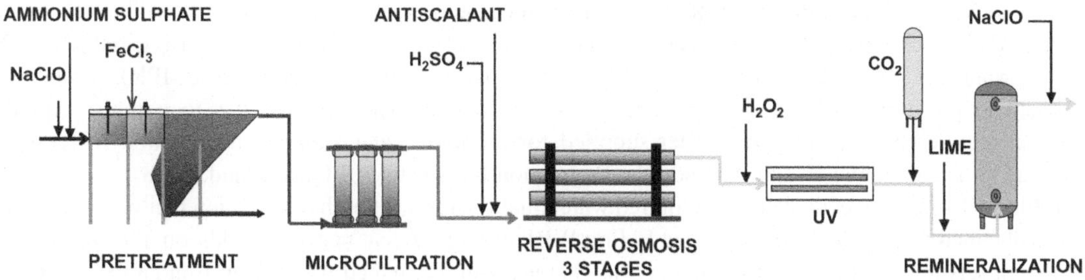

Figure 13.3 Bundamba AWTP process flow scheme

Chloramination is done with ammonium sulphate and sodium hypochlorite and it is aimed to limit biofouling on the RO membranes. Ferric chloride is added for phosphate precipitation and plate settlers are used for water clarification. Microfiltration, with 92% recovery, provides an effective pre-treatment for RO. A three-stage RO filtration with large diameter (18″) membrane elements is implemented with a typical 85% recovery. Citric acid, sodium hypochlorite and caustic soda are used for membrane systems chemical cleanings. RO permeate is further treated by advanced oxidation combining hydrogen peroxide and UV light to remove any trace organic that may have gone through the membranes and is subsequently treated with lime and carbon dioxide for stabilization. Finally sodium hypochlorite is added to maintain a disinfectant residual in the distribution system.

RO concentrate is further treated to reduce nutrients, before discharge in the Brisbane River, by means of nitrification with MBBR™, phosphate precipitation with ferric chloride and denitrification in anoxic filters with methanol. Other waste flows are collected and treated by thickening and centrifugation, as shown in Figure 13.4. Sludge is trucked to landfill for final disposal (bioreactor).

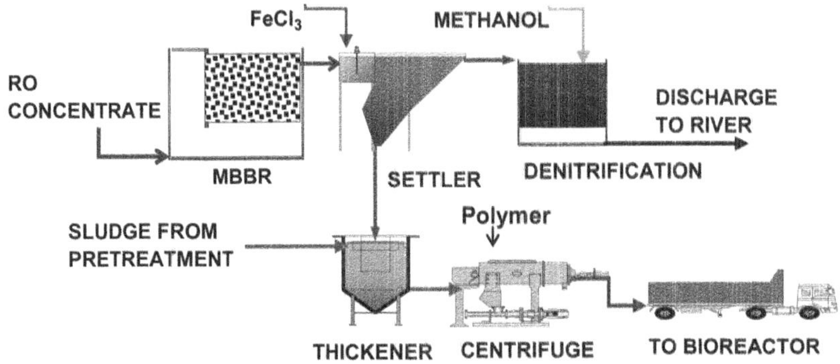

Figure 13.4 Bundamba AWTP sludge process scheme

13.3.3 Water quality requirements and uses

Recycled water from El Prat de Llobregat WRP consistently complies with the reuse limits established by Spanish Regulation RD 1620/2007 for the following uses: landscape irrigation, natural wetlands restoration and in-stream river flow substitution. It also provides a resource for other applications requiring higher quality. Two facilities further treat the basic reclaimed water for other purposes: an electrodialysis reversal demineralization plant to remove dissolved solids for agricultural irrigation and a reverse osmosis facility to achieve the water quality required for injection into a seawater intrusion barrier.

The AWRP scheme in Australia produces purified recycled water for industrial use (predominantly cooling water for power stations) and potentially irrigation and indirect potable reuse (IPR). It has been designed to produce only one quality of water so the default water quality has to comply with the stringent requirements for recycled water supplied for drinking water supply augmentation (IPR) at all times, as set in the Public Health Amendment Regulation No.1 (State of Queensland, 2008).

Table 13.1 summarizes some of the required water quality criteria at both reclamation facilities and actual values obtained from follow-up analysis at El Prat WRP during a 3 year period. In addition, the Queensland regulation sets maximum limits for another 388 chemicals in the treated water from Bundamba AWRP.

Table 13.1 Reclaimed water quality at El Prat WRP and Bundamba AWRP.

Parameter	Unit	El Prat WRP		Bundamba AWRP
		Limit RD 1620/2007	P90	Limit
Microbiological criteria				
Escherichia coli	cfu/100 mL	<100	<2	ND
Clostridium perfringens	cfu/100 mL			ND

(Continued)

Table 13.1 Reclaimed water quality at El Prat WRP and Bundamba AWRP (*Continued*).

Parameter	Unit	El Prat WRP		Bundamba AWRP
		Limit RD 1620/2007	P90	Limit
F-RNA bacteriophage	pfu/100 mL			ND
Somatic coliphage	pfu/100 mL			ND
Nematodes (pathogenic helminthes)	ova/10 L	<1	<1	
Legionella	cfu/L	<100	<100	
Physical and chemical criteria				
Turbidity	NTU	<10	1.6	<5
Total Suspended Solids	mg/L	<20	5	NA
Total Dissolved Solids	mg/L			<250
N_{tot}	mg N/L	<10 (P50)	6.9 (P50)	<1.5 (<1.2 annual P50)

*ND – not detected.

13.4 METHODOLOGY

13.4.1 Carbon footprint

A carbon footprint (CF) is the total amount of greenhouse gas (GHG) emissions caused directly and indirectly by an individual, organization, event or product and evaluated in kg of CO_2 equivalent. It is a cradle-to-grave analysis and it may be considered as a simplified Life Cycle Analysis (LCA) focused on greenhouse gas emissions and fossil fuels scarcity.

The methodology applied for carbon accounting in this study is a "bottom up" procedure developed by Veolia Water Solutions & Technologies based on the Bilan Carbone®, a methodology developed by ADEME (French Agency for the Environment and Energy Management), which is compatible with the major international standards ISO 14064 and GHG Protocol. In accordance with these standards, three emission scopes are considered in the assessment:

- Scope 1: direct emissions from sources within the boundary of an organisation, such as fuel combustion.
- Scope 2: indirect emissions from the consumption of purchased electricity, steam or heat produced by another organisation.
- Scope 3: all other indirect emissions that are consequence of the project but are not from sources owned or controlled by the organisation, such as chemicals production and freight.

The procedure follows the emission factor-based method described in the IPPC Guidelines (IPCC, 2006) "combination of activity data with coefficients which quantify the emissions or removals per unit activity". It can be formulated using the equation $E = \Sigma(A^*EF)$, where E represents the emissions in kg CO_2eq, *A* represents the activity data (e.g. fuel consumption) and EF represents the emission factor (expressed as a specific value in kg CO_2eq/kg or kg CO_2eq/kWh).

Figure 13.5 depicts all the items covered by the inventory in a water treatment plant.

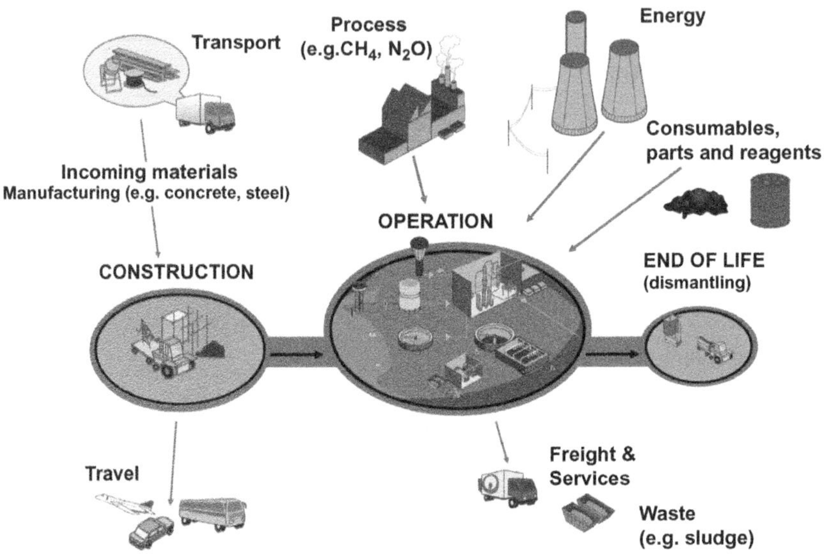

Figure 13.5. Water treatment plant activities covered by a carbon footprint

Scope of study

In order to elaborate a consistent comparison of both facilities, the same boundaries and variables have been considered in this study. Only the operational carbon footprint has been assessed. Site specific operating data for a 12-month period have been provided by the operators: the period of study for El Prat de Llobregat WRP extends from January 2009 to December 2009 and for Bundamba AWRP from July 2009 to June 2010. Main activity data consist of energy consumption and production, chemicals use and freight, as well as sludge production and disposal. An estimation of the GHG emissions resulting from the replacement of reverse osmosis and microfiltration membranes has been included in Bundamba CF.

Perimeter

The perimeter considered is shown in Figures 13.1 to 13.4 and it includes the water and sludge treatment lines of both water reclamation plants. Energy for pumping water to and from the facilities is not taken into account as it is site specific.

Emission factors

Emission factors have been selected from different databases: ADEME for freight EF, Ecoinvent v2.2 and specific VWS database for chemicals EF, the International Energy Agency Report for electricity EF (IEA, 2009) and IPCC Guidelines for process EF. The electricity EF depends on the electricity production mix of a country and consequently, the emissions per kWh will vary from year to year and from country to country. The same EFs have been applied and the world average electricity EF from the IEA (0.5 kg CO_2eq/kWh) has been chosen as default value.

Uncertainty

The accuracy of this methodology depends on the EFs and activity data used. EFs are associated with a default degree of uncertainty so each carbon footprint assessment is provided with an overall uncertainty.

Process emissions

Process emissions in water treatment plants are the greenhouse gases directly emitted during water or sludge treatment. CO_2 emissions from aerobic digestion, CH_4 from anaerobic processes and N_2O as a by-product of nitrification – denitrification reactions are all direct fugitive emissions from wastewater treatment plants. CO_2 emissions resulting from the digestion of biological waste are not taken into account in a carbon footprint assessment because they are considered as part of the "normal" biological cycle of carbon. CH_4 and N_2O have a much higher global warming potential (GWP) than CO_2, respectively 25 kg CO_2eq/kg CH_4 and 298 kg CO_2eq/kg N_2O (IPCC, 2007) so these emissions should be considered carefully.

Process emissions from water reclamation treatments are negligible since they are mainly physical-chemical treatments. However, process emissions from sludge treatment and disposal might be significant. These emissions have been estimated according to IPCC default values (IPCC, 2006) and the following hypothesis:

- An average 1% biogas leakage from the anaerobic digestion in El Prat is taken into account, but no measured data was available.
- N_2O emissions from the MBBR^{TM} treatment in Bundamba are estimated according to the method described in the National Greenhouse and Energy Reporting Measurement Technical Guidelines (Commonwealth of Australia, 2010).
- Bundamba sludge disposal releases no uncontrolled emissions since sludge is disposed offsite in a bioreactor, where methane is captured and used to generate electricity.

Avoided emissions

Some processes used for sludge management at both facilities contribute to a reduction of the carbon footprint: biogas is used for energy production at El Prat WWTP, sludge from Bundamba produces biogas in a bioreactor and sludge from El Prat replaces either mineral fertilizer in agriculture or fuel and raw material in cement industries. However, in the absence of more detailed data, avoided emissions have not been considered in this case study.

13.5 COMPARISON OF CARBON FOOTPRINTS

Figure 13.6 shows the global carbon footprint of the two water reclamation plants and the contribution of each activity according to the selected emission factors.

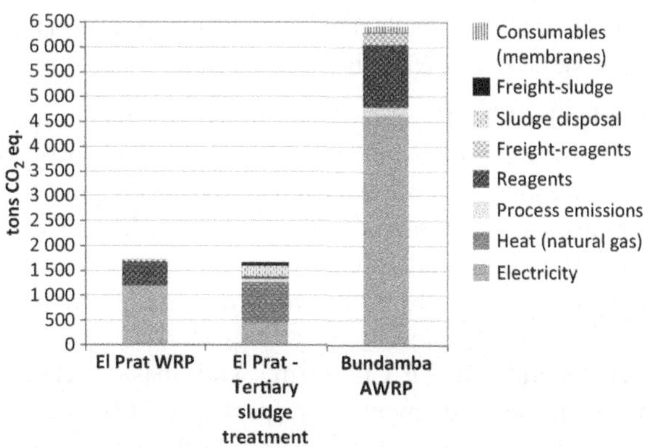

Figure 13.6 Overall greenhouse gas emissions in 12 months operation

Energy consumption represents the highest impact on GHG emissions for both basic and advanced water reclamation facilities, accounting for 69% of GHG at El Prat WRP and 72% at Bundamba AWRP. Chemicals production (28% and 19%) and chemicals freight (3% and 4%) represent the second and third largest impacts, respectively.

It is important to notice that tertiary sludge treatment from El Prat WRP has a similar CF, around 1,700 ton CO_2eq, than the water treatment line itself. The main contributors in the sludge process line are natural gas consumption (50%) and electricity (26%). Process emissions due to sludge treatment and disposal in agriculture and cement industries make up to 17%.

Focusing on energy demand, El Prat basic water reclamation process has a specific energy demand of 0.1 kWh/m^3 but, if energy requirements of the tertiary sludge treatment line (electricity and natural gas) are taken into account, this value amounts up to 0.28 kWh/m^3. On the other hand, the electricity demand at Bundamba AWRP is 1.14 kWh/m^3, due principally to the energy intensive processes of microfiltration (21% of the global) and RO (53%), as shown in Figure 13.7.

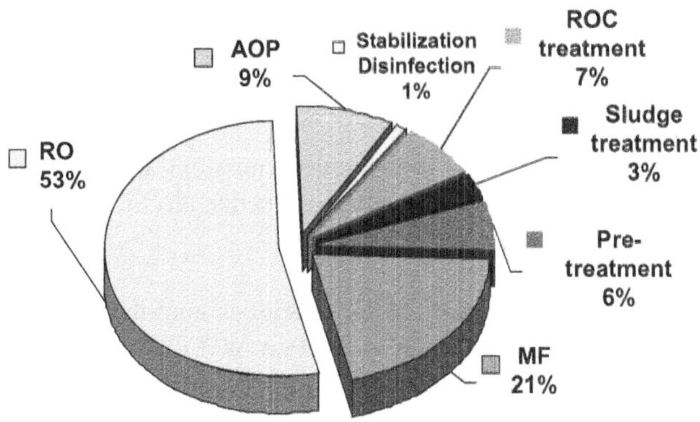

Figure 13.7 Energy breakdown at Bundamba AWRP

Figure 13.8 compares the GHG emissions from Bundamba AWRP if a different electricity EF was applied. Three electricity generation mixes are compared:

(a) France (EF = 0.085 kg CO_2eq/kWh) which mainly relies on nuclear power plants;
(b) Spain (EF = 0.35 kg CO_2eq/kWh), where electricity produced from natural gas cogeneration plants and renewable energies are displacing coal and fuel power plants;
(c) Australia (EF = 0.921 kg CO_2eq/kWh), that heavily relies on coal thermal power plants.

In these countries, the Bundamba AWTP carbon footprint would range from 2,500 to 10,200 ton CO_2eq, and emissions due to electricity consumption would represent respectively 30%, 60% and 82% of the global assessment. Therefore, GHG emissions from water reclamation plants relying on relatively energy intensive processes such as RO membrane separation are heavily affected by the way electricity is produced.

To compare the carbon footprint independently on plant capacity, the ratio GHG emissions per cubic metre of recycled water is shown in Figure 13.9 (the world average electricity EF of 0.5 kg CO_2eq/kWh from the IEA is used).

Advanced recycled water generates 802 g CO_2eq/m^3 while basic reclaimed water results in 142 g CO_2eq/m^3. The uncertainty of these assessments is respectively 16% and 17%. Hence, a basic water reclamation treatment plant generates around 80% less GHG emissions than an advanced water reclamation plant.

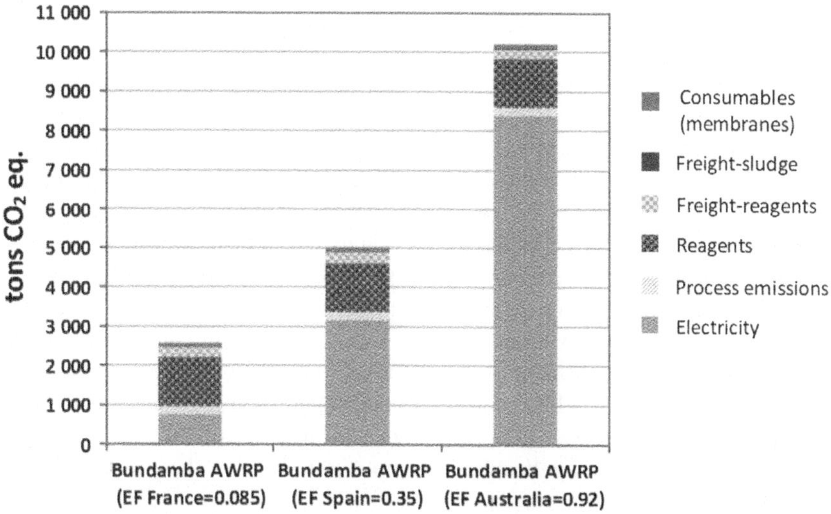

Figure 13.8 Bundamba carbon footprint breakdown with different electricity EF

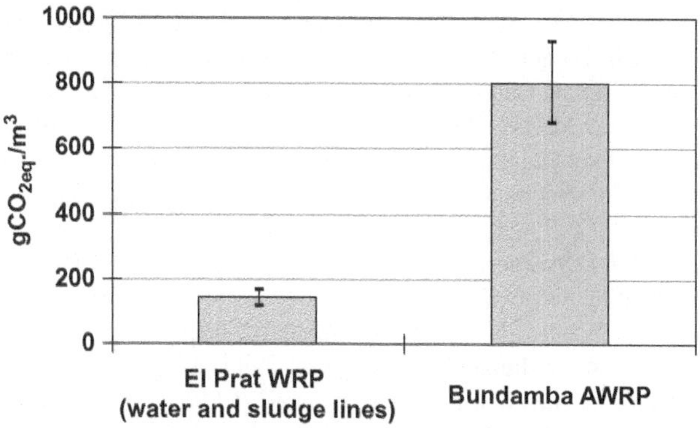

Figure 13.9 Specific greenhouse gas emissions per cubic metre of reclaimed water

In order to integrate these results in a comparative CF study with different water supply sources, the local context should be considered. At the same time, it is important to evaluate the CF of the whole water cycle and not the water industry alone, as demonstrated in a report issued by the CSIRO in 2008, which estimated that the total energy use by some of Australians' largest water utilities represented less than 15% of the energy used for residential water heating (Kenway *et al.* 2008).

13.6 CONCLUSIONS

Current technologies for water reuse are able to satisfy any quality standard, as shown in these two state-of-the-art facilities. As expected, more intense water treatments have a higher environmental impact in terms of carbon footprint. This case study illustrates that the advanced water reclamation treatment based on high-tech, energy intensive technologies, has a carbon footprint 5 times higher than the basic water reclamation process. Specifically, estimated GHG emissions are 802 g CO_2eq/m^3 and 142 g CO_2eq/m^3, respectively.

Energy consumption is the most important contributor in both basic and advanced water reclamation, accounting for 70% of the total CF, when an average electricity EF of 0.5 kg CO_2eq/kWh is applied. Hence, the way electricity is produced in each country or region has a considerable weight on the CF of water reuse. To minimize this impact, the use of renewable or less carbon-emitting energy sources is a key issue.

Although carbon footprint provides a valuable criterion to assess sustainability of water reuse projects, other environmental impact factors accounting for fresh water depletion and pollution are necessary to provide additional information on water resources management. For this reason, Veolia Environnement Research & Innovation is currently developing a water footprint indicator called Water Impact Index that provides a more comprehensive assessment of the effects of human activity on water resources.

The water reclamation industry will be required to contribute to greater sustainability and encourage environmental improvements. However, process design and intensity of treatment in water reclamation must be driven first and foremost by health protection concerns, as they remain key factors when implementing any type of water reuse project.

REFERENCES

Asano T., Burton F. L., Leverenz H. L., Tsuchihashi R. and Tchobanoglous G. (2007). Water Reuse – Issues, Technologies, and Applications. Metcalf & Eddy/WECOM, McGraw Hill, New York.

Commonwealth of Australia (2010). Technical Guidelines for the Estimation of Greenhouse Gas Emissions by Facilities in Australia. Department of Climate Change and Energy Efficiency. Canberra.

GWI (2010). Municipal Water Reuse Markets 2010. Global Water Intelligence. Oxford.

IEA (2009). CO_2 Emissions from Fuel Combustion. Highlights. International Energy Agency. Paris.

IPCC (2006). IPCC Guidelines for National Greenhouse Gas Inventories. National Greenhouse Gas Inventories Programme. IGES, Japan.

IPCC (2007). Climate Change 2007. Synthesis Report. Contribution of Working Groups I, II and III to the Fourth Assessment Report on the IPCC. Geneva.

IPCC (2008). Climate Change and Water. IPCC Technical Paper VI. IPCC Secretariat, Geneva.

Kenway S., Priestley A., Cook S., Seo S., Inman M., Gregory A. and Hall M. (2008). Energy use in the provision and consumption of urban water in Australia and New Zealand. *CSIRO: Water for a Healthy Country National Research Flagship*, ISBN: 978 0 643 09616 5. CSIRO Australia and Water Services Association of Australia.

Mujeriego R., Compte J., Cazurra T. and Gullón M. (2008). The water reclamation and reuse project of El Prat de Llobregat, Barcelona, Spain. *Water Sci. Tech.*, **57**(4), 567–574.

Pasqualino J., Meneses M., Abella M. and Castells F. (2009). LCA as a decision support tool for the environmental improvement of the operation of a municipal wastewater treatment plant. *Environ. Sci. Tech.*, **43**, 3300–3307.

Sala L. and Serra M. (2004). Towards sustainability in water recycling. *Water Sci. Tech.*, **50**(2), 1–7.

Srinivas V., O'Toole G., Bates J. and Waer M. (2009). Operational results of a fast track Australian Advanced Municipal Effluent Recycling Facility. IDA World Congress, Dubai.

State of Queensland (2008). Public Health Amendment Regulation (No. 1).

Stokes J. and Horvath A. (2006). Life cycle energy assessment of alternative water supply systems. *International Journal of Life Cycle Assessment*, **11**(5), 335–343.

U.S. EPA (2004). Guidelines for Water Reuse. EPA/625/R-04/108, September 2004. Office of Water, Washington, D.C., USA.

Vince F., Aoustin E., Bréant P. and Marechal F. (2008). LCA tool for the environmental evaluation of potable water production. *Desalination*, **220**, 37–56.

Chapter 14

Comparison of processes for greywater treatment for urban water reuse: energy consumption and footprint

Alessandro Meda, Jochen Henkel, Yue Chang and Peter Cornel

14.1 INTRODUCTION

Greywater is an excellent resource for service water in intra-urban reuse. With substituting fresh water with appropriately treated greywater, it is possible to save 30–50% of fresh water at household level (Cornel, 2007). In the literature, slightly different definitions for greywater can be found. Greywater gained solely from showers/bathing and washing machines is called low-strength greywater or 'light' greywater (Nolde, 1999; Bieker *et al.* 2010). This definition excludes in particular the heavier polluted wastewater originating from kitchen sinks. Greywater characteristics are highly variable and depend on living standards, population structures, customs and habits, water installations, and the degree of water abundance (Morel & Diener, 2006). Reviews of current literature data can be found in FBR[1] (2005), DWA (2008), Li *et al.* (2009), Chang (2010), Meda & Cornel (2010).

Variation ranges of the relevant quality parameters of "light" greywater are given in Table 14.1. The huge variation range of the quality parameters which can be found in the literature makes clear that there is no "typical" composition of greywater, even of 'light' greywater, so that the expected composition must be carefully measured and evaluated in each practical case. Apart from the extreme values, greywater contains fewer pollutants in comparison to mixed domestic wastewater, for example, 12% of nitrogen and 47% of BOD_5 load (Morel & Diener, 2006). In addition, due to its origin, it has a higher water temperature. These features facilitate treatment. Thus, greywater is an excellent resource for service (recycled) water in intra-urban reuse (Meda & Cornel, 2010).

Applicability and suitability of technologies for greywater treatment have to be assessed under different points of view, such as treatment efficiency, energy demand, reliability, reactor volume, chemicals consumption, and so on. Based on the research project "Semicentralized Supply and Treatment Systems for Urban Areas in China – Part II" (BMBF, 2009), investigations of greywater treatment with respect to intra-urban reuse were carried out. The following treatment techniques were tested: BAF (Biological Aerated Filter), MBR (Membrane Bioreactor), and SBR (Sequencing Batch Reactor). Further information on the implementation of a greywater treatment and reuse system within an integrated,

[1]Association for process water and rainwater utilization

semicentralized supply and treatment infrastructure for urban areas can be found in BMBF (2009), as well as in Bieker *et al.* (2010).

Table 14.1 Literature data and measured greywater characteristics in this study and in Shanghai and Qingdao, PR China (BMBF, 2009; Meda & Cornel, 2010).

Type of greywater	COD mg/L	BOD_5 mg/L	TSS mg/L	TN mg/L	TP mg/L	Anionic surfactants mg/L	Total coliforms n/L	Turbidity NTU
Greywater (Shanghai)	430	186	285	38	–	72		
Greywater from shower (Shanghai and Qingdao)	270–280	50–130	–	21–23	<1	10–12		
Literature data ('light' greywater)	41–1,100	48–472	7–1,475	1.1–45	0.11–171*			
This study (synthetic greywater)	300 ± 50	155 ± 25	65 ± 30	18 ± 5	1.7 ± 1	60 ± 20	10^7	–
Chinese technical standard GB/T 18920–2002	–	≤10	–	–	–	≤1	<3	≤5

*More than 171 mg/L for greywater from laundry (Meda and Cornel 2010).

14.2 METHODOLOGY

Within this chapter, selected results of three biological techniques tested for the treatment of 'light' greywater are presented focusing on the variation of design parameters that are crucial for biological processes, such as sludge residence time (SRT), volume loads and F/M ratio.

The tests were carried out using synthetic 'light' greywater, prepared by mixing commercial personal care products and detergents (shower gel, shampoo, toothpaste, body lotion, washing powder, fabric softener, etc.), according to average specific consumption data in Germany (Chang, 2010). In order to simulate microbial contamination and dirt particles such as grease, dandruff, hairs, and sand, 3% of raw wastewater from a municipal treatment plant was added. Lack of nutrients seems to be distinctive for synthetic greywater, as found by Jefferson *et al.* (2001a). In contrast, real greywater, even less polluted 'light' greywater, contains sufficient nitrogen and phosphorus to serve as nutrients for microbial growth, as reported in several studies (Friedler *et al.* 2005; Liu *et al.* 2005; Eriksson *et al.* 2007; Knerr *et al.* 2007; Hernández Leal *et al.* 2007; Merz *et al.* 2007). No limitation of microbial growth or poor effluent quality due to the lack of nutrients was reported for biological treatment of low-strength greywater. In order to avoid a lack of nutrients, in this study, urea and potassium dihydrogenphosphate (KH_2PO_4) were added. The resulting COD:N:P ratio of about 100:6:0.6 is realistic for real 'light' greywater as reported in the quoted literature.

The concentrations of the relevant parameters in synthetic greywater measured during the investigation period are listed in Table 14.1.

These concentrations are in the range of the reported literature data as well as in the range of data for real greywater measured in the PR China (BMBF 2009), see Table 14.1. Therefore, they are considered to be representative for an "average" 'light' greywater. Only the concentration of anionic surfactants seems high (60 ± 20 mg/L) compared to real greywater. However, reported values differ in analytical methods and definitions. Quite often they are not regarded at all.

The pilot-scale setup (Figure 14.1) consisted of a unit for preparing synthetic greywater and four biological treatment units: a biological aerated sand filter (Huber SE), two membrane bioreactors (hollow fiber module from ITT Water & Wastewater AB, plate module from Huber SE) with a common storage tank, as well as a sequencing batch reactor (ITT Water & Wastewater AB) with its separated storage tank. The biological reactors are designed for biological oxidation of organics. Nutrient removal was not intended since nutrient concentrations in greywater usually are low[2]. The plants were operated in a way to meet the Chinese technical standard for process water quality for urban miscellaneous (non-potable) water reuse (GB/T 18920–2002, China), as depicted in Table 14.1. This standard is one of the most restrictive for urban non-potable water reuse (Meda & Cornel, 2010)

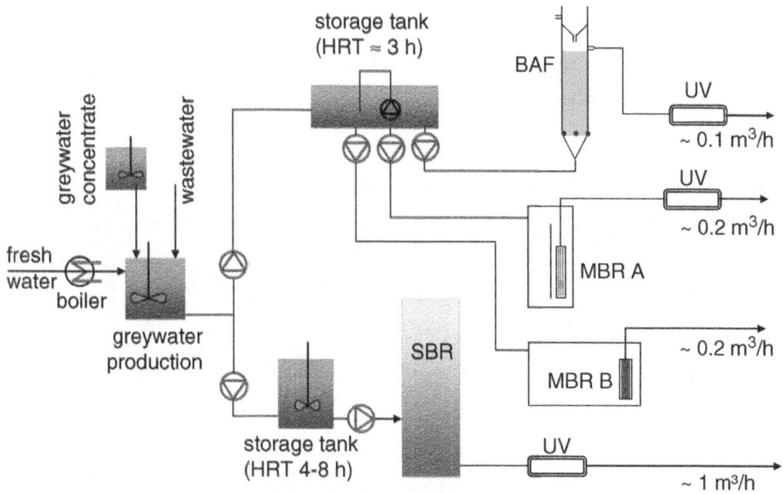

Figure 14.1 Scheme of the pilot plant for greywater treatment

14.3 EFFICIENCY OF GREYWATER TREATMENT

The results of the pilot plant experiments show that all tested treatment processes are adequate to meet the very stringent Chinese limit values for greywater in intra-urban water reuse. Selected results on the performance of the investigated processes are described below. More detailed results on the single processes are given in BMBF (2009), Cornel *et al.* (2009), Henkel *et al.* (2009), Chang (2010), & Meda & Cornel (2010).

14.3.1 BAF

The start-up of the biofilter until achieving a stable biological degradation took about two months. Initial operational problems occurred due to clogging of the filter bed and foaming at the top of the biofilter caused by stripping non-degraded surfactants. After the sand circulation had been optimized, the biofilter was able to cope with the TSS present in the synthetic greywater (in average 65 ± 30 mg/L with peaks up to 100 mg/L).

[2]Greywater contains around 10% of the load of wastewater from toilets and kitchen sinks (see also Table 14.1) (BMBF, 2009).

It was found that the effluent concentrations of COD, BOD_5, and anionic surfactants decreased with decreasing organic loads. The required effluent standards were met at an organic load of 8.8 kg $COD/m^3 \cdot d$ (effluent concentrations of COD (41 ± 7 mg/L), BOD_5 (5.4 ± 1 mg/L), and anionic surfactants (0.6 ± 3 mg/L) and lower. At higher loads, the anionic surfactant concentration in the effluent increased (Figure 14.2) due to insufficient degradation. In addition, the TSS increased due to the higher required air flow rates necessary to maintain a sufficient oxygen concentration. The NH_4-N concentration was approximately 10 mg/L. With a reduction of the organic load below 5 kg $COD/m^3 \cdot d$, a complete elimination of ammonium (nitrification) and a reduction of the effluent COD (from 41 to 21 mg/L), BOD_5 (from 5.4 to 1.6 mg/L) and TSS (from 8 to 4 mg/L) concentration were achieved. Anionic surfactants were not further reduced. The effluent values fulfilled even the very strict Chinese quality standard for greywater reuse (GB/T 18920–2002). Depending on the ammonium influent concentration and effluent requirements, nitrification may not be necessary, since a certain portion of ammonium (2–2.5% referred to the eliminated COD) is already removed from the greywater by physiological incorporation into biomass and removed with the backwash water.

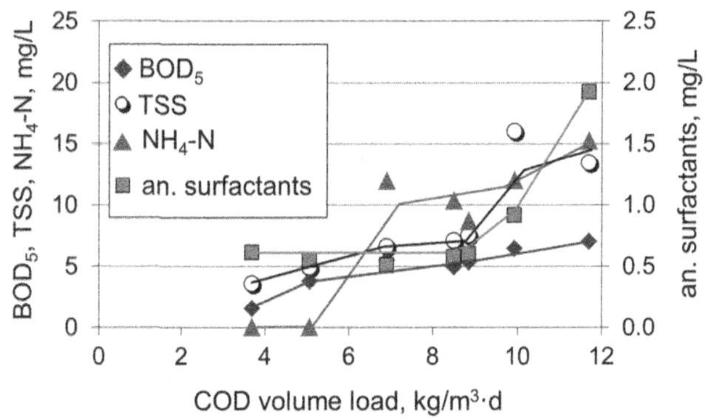

Figure 14.2 Effluent quality of BAF in dependence of the volume load

The loading rates are much higher than other published values for the organic load of biofilters for greywater treatment, for example, 1.09 ± 0.73 $kg/m^3 \cdot d$ (Jefferson *et al.* 2001b), 1.5 $kg/m^3 \cdot d$ (Laine, 2001), and 0.8 $kg/m^3 \cdot d$ (Franci Gonçalves *et al.* 2007). The established organic load is in the same range as for biofilters for full-scale treatment of municipal wastewater (8–10 $kg/m^3 \cdot d$, according to Metcalf & Eddy, 2003). For a given organic load, no influence of the filtration velocity on the effluent quality was found. Concentrations of total coliforms and *E. coli* in greywater (BAF influent) were approx. $1.0 \cdot 10^7$/100 mL and $6.3 \cdot 10^5$/100 mL, respectively. The BAF reduced the microbial contamination approximately 2 orders of magnitude within 200 cm bed height. Nevertheless, the microbial contamination of the BAF effluent is still too high and does not meet the quality requirements for urban reuse (faecal coliform <3/L for toilet flushing, irrigation, and washing purposes in China (GB/T 18920–2002), ND/100 mL for unrestricted reuse in USA (Asano *et al.* 2007), making UV disinfection necessary. Total coliforms and *E.coli* were not detected in any sample of the UV effluent.

14.3.2 MBR

Two membrane bioreactors, one with a hollow fiber module (reactor A, airlift design), the other with a plate module (reactor B), were investigated, using different experimental setups (Henkel *et al.* 2009). The first setup used a baffled membrane bioreactor equipped with only one fine bubble aeration unit. The operation mode was 9 min of filtration followed by 1 min of backwash. Once per week, chemical backwash was performed with hypochlorite. The second reactor was equipped with a coarse-bubble unit for crossflow aeration and a fine-bubble unit for biological aeration and mixing. The operation mode was 9 min filtration followed by 1 min relaxation time. It was not possible to apply backwash measures on the plate modules, and consequently no chemical cleaning was performed.

During the test series lasting three to five weeks each, the sludge parameters of reactor A and reactor B were almost identical (Table 14.2).

Table 14.2 Parameters analyzed in the pilot plants (± indicates standard deviation).

Source for sampling	Parameter	Unit	Reactor A	Reactor B
Effluent Parameters	COD	mg/L	10 (±2)	10 (±2)
	Anionic surfactants	mg/L	0.3 (±0.1)	0.3 (±0.1)
	pH	–	7.6 (±0.1)	7.7 (±0.1)
	E.coli	UFC/100 mL	0	0
	Total coliforms	n/100 mL	1 (±1)	1 (±1)
Sludge Parameters	Loss on ignition	%	40 (±7)	37(±5)
	SVI	mL/g	39 (±6)	34 (±6)
	CST	s	6.9 (±0.7)	7.0 (±0.7)

In contrast to wastewater sludge from other membrane bioreactors operating at high SRT, the greywater sludge was characterized by very low ignition loss (<40%), and very good settling and filtration behavior (sludge volume index, SVI < 40 mL/g, capillary suction time CST < 8 s). This is attributed to the high ratio of filterable solids (FS) to the COD concentration in the influent (FS/COD ~ 0.33), caused by abrasives in toothpastes (e.g. powdered white mica) or silicates in detergents used for water softening (e.g. zeolite). As expected, the anionic surfactant concentration in the influent was high and the COD concentration was low compared to real wastewater. Approximately 95% of the COD load and 99% of the anionic surfactants load were removed by both bioreactors. Coliform bacteria occurred only sporadically and *E. coli* were not detected at all in the effluent.

Membrane bioreactor B was successfully operated without chemical cleaning as long as the SRT was higher than 25 d. At lower SRT, fouling increased and the membranes had to be cleaned externally after one month of operation.

The configuration of reactor A (airlift design, only one aeration device) seems to be a promising configuration, especially for greywater treatment. The fine-bubble crossflow device was sufficient to prevent membrane fouling and to ensure the necessary biological oxygen uptake rate at the same time. Consequently, lower airflow rates were required compared to reactor B, which was equipped with two aeration units (crossflow/biological aeration). However here as well, a SRT of 25 d is recommended, since at lower SRT chemical cleaning have to be intensified. This conclusion is consistent with observation of several authors (Chang *et al.* 2002; Estrada-Arriaga & Mijaylova, 2011).

14.3.3 SBR

The tests with the sequencing batch reactor (SBR) included four main phases, using different water temperatures and two MLSS concentrations (7 and 4.5 kg/m^3, Table 14.3).

Table 14.3 Test phases of SBR and analytical results.

Source for sampling	Parameter	Unit	Phase 1	Phase 2	Phase 3	Phase 4
Input parameters and test conditions	MLSS (average)	g/L	7		4.5	
	F/M ratio	kg BOD_5/kg MLSS · d	0.14		0.06	
	BOD_5 load	kg BOD_5/d	4.1		2.2	
	Volume load	kg $BOD_5/m^3 \cdot d$	0.48		0.26	
	Water temperature	°C	16	25	26	20
Effluent parameters	BOD_5	mg/L	5 (±1)	8 (±2)	2 (±1)	3 (±2)
	COD	mg/L	39 (±5)	40 (±7)	23 (±5)	30 (±5)
	Anionic surfactants	mg/L	0.5 (±0.2)	0.7 (±0.2)	0.4 (±0.2)	0.6 (±0.3)
	Turbidity	NTU	32 (±20)	23 (±5)	8 (±3)	21 (±10)

After adaptation, the effluent concentrations of BOD_5 and anionic surfactants fell below the required limit values as given in the Chinese reuse standard. Degradation of BOD_5 and anionic surfactants were 95% and 99% in average. However, suspended solids and subsequently high turbidity values have been challenging. Due to high mechanical stress, caused by the centrifugal pump used for feeding and recirculation and by the jet aerator, the turbidity by far exceeded the given limit value of 5 NTU. By adding polyamine as flocculation aid (10 mg/L), the requirements were met (turbidity 4 ± 1 NTU, COD 22 ± 2 mg/L). Since the reduction of the microbial contamination by SBR treatment proved insufficient, UV disinfection is mandatory to meet the quality standards for reuse.

14.4 COMPARISON OF PERFORMANCE AND ENERGY DEMAND OF GREYWATER TREATMENT TECHNOLOGIES

In order to compare the three tested techniques, a full-size greywater treatment plant for effluent reuse was designed, based on the results of the pilot plant and including the three systems. The investigation was expanded by including the conventional activated sludge process (CAS), enabling a broader view on existing techniques. Thereby, identical input data and boundary conditions (specific greywater amount: 41 L/cap · d, greywater concentration of COD 300 mg/L, BOD_5 120 mg/L, TSS 250 mg/L, TN 7 mg/L and TP 4 mg/L, respectively, were applied.

Due to strong daily variations of greywater flow, a buffer tank is necessary to equalize the influent. For CAS and MBR, a separate buffer tank was not included since the reactor itself includes sufficient buffer volume. Disinfection is also indispensable due to high hygienic safety requirements.

The details of reactor volume, energy demand, and so on. of the different processes in comparison are summarized in Table 14.4.

Table 14.4 Comparison of reactor volume and energy demand for greywater treatment processes.

Parameter	Unit	BAF	MBR	SBR	CAS
Reactor volume (net)	L/cap	2.5–3.7	15	18	6[1]
Influent buffer tank volume	L/cap	3.7–7.8	–	8	–
Required total volume	L/cap	7.4–10.3	15	26	16[2]
Energy demand	kWh/m^3	0.16 – 0.24	0.5 – 0.7	0.1–0.3	0.11 – 0.22
	kWh/cap · yr	2.4 – 3.6	7.5 – 10.5	1.5–4 5	1.9–3.3

[1]volume of the aeration tank

[2]including primary and secondary clarification, no influent buffer

The BAF requires the smallest specific volume in comparison to the other processes. Depending on the design of the influent buffer tank, a total volume of approx. 7.5 L/cap is sufficient. The reactor volume of the MBR and SBR is much bigger than for BAF and CAS. For MBR, despite higher MLSS concentration, this is caused by the SRT of 25 d, required to reduce the fouling potential of the membrane. It has to be noted that in all tested processes, the total volume for greywater treatment is very low compared to the volume for full municipal wastewater treatment (e.g. 250–350 L/cap for aeration and sedimentation tanks with CAS for nutrient elimination as typical in Western Europe). This feature makes greywater treatment a compact and low-footprint process that can be encased, and therefore, easily integrated in the urban context.

The energy consumption is caused mainly by aeration and strongly depends on the efficiency of the aeration system. SBR and CAS processes will achieve the lowest energy consumption of approx. 0.1 kWh/m^3 with an efficient aeration system. This value is based on the very low energy consumption for aeration caused by choosing a low SRT (3–4 days) for carbon removal without nitrification. The energy demand of the MBR process is much higher than for the other processes due to the crossflow aeration of the membrane modules in order to reduce permeability losses by fouling. However, it has to be noted that, while considering the higher energy consumption, the MBR process produces the best effluent quality: the effluent is free of suspended solids and bacteria and is immediately useable for all purposes of intra-urban non-potable reuse.

Besides the energy consumption, the consumption of chemicals has to be considered when comparing the processes. For the MBR process, cleaning chemicals, such as sodium hypochlorite and citric acid, are required for membrane cleaning. The SBR process might require some polymer dosage in order to ensure low suspended solids effluent concentrations. In contrast, BAF as well as the conventional activated sludge process do not need any chemicals. This is a clear advantage in terms of operating costs and sustainability.

14.5 CONCLUSIONS

The energy demand for greywater treatment has been calculated to be 0.1–0.5 kWh/m^3, depending on the process. This is only a fraction of the energy intensities for water supply, treatment, and distribution, for example, 0.5 to 7 kWh/m^3 in California, 0.6 to 1.4 kWh/m^3 in Germany (Cornel *et al.* 2011).

Therefore, reuse of treated greywater can contribute to the preservation of valuable freshwater resources and, at the same time save electric power, especially when freshwater has to be transported over long distances or elaborate treatment is required.

Acknowledgement

The authors would like to thank the German Federal Ministry for Education and Research for financing the project "Semicentralized Supply and Treatment Systems for urban areas in China – Part II" (02WD0607)

REFERENCES

Asano T., Burton F. L., Leverenz H. L., Tsuchihashi R. and Tchobanoglous G. (2007). Water Reuse Issues, Technologies, and Applications. McGraw-Hill, New York.

Bieker S., Cornel P. and Wagner M. (2010). Semicentralized supply and treatment systems – integrated infrastructure solutions for fast growing urban areas. *Water Science and Technology*, **61**(11), 2905–2913.

BMBF (2009). Semizentrale Ver- und Entsorgungssysteme für urbane Räume Chinas (Semicentralized Supply and Disposal Systems for Urban Areas in China). Subproject 2, project code 02WD0607, Final Report. German Ministery for Education and Research (BMBF) (in German).

Chang Y. (2010). Greywater Treatment within Semi-Centralized Supply and Treatment Systems by the Example of the Peoples Republic of China. PhD thesis, Technische Universität Darmstadt, Germany.

Chang I. S., Le Clech P., Jefferson B. and Judd S. (2002). Membrane fouling in membrane bioreactors for wastewater treatment. *Journal of Environmental Engineering*, **128**(11), 1018–1029.

Cornel P. (2007). Semi-centralized supply and treatment systems for fast growing urban areas. In: Water Supply and Sanitation for All, H. Huber, P. Wilderer and S. Paris, (eds), IWA Publishing, London, UK, pp. 211–222.

Cornel P., Meda A. and Bieker S. (2011). Wastewater as a source of energy, nutrients and service water. In: Treatise on Water Science, P. Wilderer (ed.), Elsevier Verlag, Oxford, UK.

Cornel P., Wagner M., Meda A. and Bieker S. (2009). Semicentralized Supply and Treatment Systems – Integrated Solutions for Fast Growing Urban Areas, *Proc. 7th IWA International Conference on Water Reclamation & Reuse*. Brisbane, Australia, 21–25 September 2009.

DWA (2008). *Neuartige Sanitärsysteme* (Innovative Sanitation Systems), DWA (ed.) (German Association for Water, Wastewater and Waste). Hennef, Germany. (in German).

Eriksson E., Yan X., Lundsbye M., Madsen T. S., Andersen H. R. and Ledin A. (2007). Variation in Grey Wastewater Quality Reused for Toilet Flushing. *Proc. 6th IWA Specialty Conference on Wastewater Reclamation and Reuse of Sustainability*, Antwerp, Belgium, 9–12 October 2007.

Estrada-Arriaga E. B. and Mijaylova P. N. (2011). Influence of operational parameters (sludge retention time and hydraulic residence time) on the removal of estrogens by membrane bioreactor. *Environmental Science and Pollution Research*, **18**(7), 1121–1128.

FBR (2005). *Grauwasser-Recycling, Planungsgrundlagen und Betriebshinweise* (fbr information sheet H 201 (Greywater Recycling – Planning fundamentals and operation information), Fbr-Hinweiseblatt H 201, Fachvereinigung Betriebs- und Regenwassernutzung e.V. (fbr), Darmstadt, 2005 http://www.fbr.de/publikation/H201_Grauwasser.pdf, (accessed June 2010) (in German).

Franci Gonçalves R., Bazzarella B. and Cassini S. T. (2007). Graywater Reuse in Edifications Using Combined Anaerobic/Aerobic Compact Treatment System. *Proc. 6th IWA Specialist Conference on Wastewater Reclamation & Reuse for Sustainability*, Antwerp, Belgium, 9–12 October 2007.

Friedle E., Kovali R. and Galil N. I. (2005). On-site greywater treatment and reuse in multi-storey buildings. *Water Science and Technology*, **51**(10), 187–194.

GB/T 18920–2002 (2003). Quality Standard of Intra-Urban Water Reuse in China, GB/T 18920–2002, Ministry of Housing and Urban-Rural Development, P. R. China, enacted in 2003.

Henkel J., Lemac M., Wagner M. and Cornel P. (2009). Oxygen transfer in membrane bioreactors treating synthetic greywater. *Water Research*, **43**(6), 1711–1719.

Hernández Leal L., Zeeman G., Temmink H. and Buisman C. (2007). Characterisation and biological treatment of greywater. *Water Science & Technology*, **56**(5), 193–200.

Jefferson B., Burgess J. E., Pichon A., Harkness J. and Judd S. (2001a). Nutrient addition to enhance biological treatment of greywater. *Water Research*, **35**(11), 2702–2710.

Jefferson B., Laine A. L., Stephenson T. and Judd S. (2001b). Advanced biological unit processes for domestic water recycling. *Water Science and Technology*, **43**(10), 211–218.

Knerr H., Engelhardt M., Hansen J., Sagawe G. and Steinmetz H. (2007). The Project KOMPLETT – A New Approach for Closed Loop Urban Water Cycles, *Proc. 2nd International Conference SmallWat07 on Wastewater Treatment in Small Units*, Sevilla, Spain, 11–15 November 2007.

Laine A. T. (2001). Technologies for Greywater Recycling in Buildings. PhD thesis, Cranfield Unniversity, UK.

Li F., Wichmann K. and Otterpohl R. (2009). Review of the technological approaches for grey water treatment and reuses. *Science of the Total Environment*, **407**(11), 3431–3610.

Liu R., Huang H., Chen L., Wen X. and Qian Y. (2005). Operational performance of a submerged membrane bioreactor for reclamation of bath wastewater. *Process Biochem*, **40**(1), 125–30.

Meda A. and Cornel P. (2010). Greywater treatment with Biological Aerated Filter (BAF) for urban water reuse. *Water Science & Technology: Water Supply*, **10**(6), 907–913.

Merz C., Scheumann R., Hamouri B. E. and Kraume M. (2007). Membrane bioreactor technology for the treatment of greywater from a sports and leisure club. *Desalination*, **215**(1–3), 37–43.

Metcalf and Eddy (2003). Wastewater engineering. In: Treatment and Reuse, G. Tchobanoglous, F. Burton and H. D. Stensel (eds), 4th edn, McGraw-Hill Publishing Co., New York, pp. 887–982.

Morel A. and Diener S. (2006). Grey Water Management in Low and Middle-Income Countries. Water and Sanitation in Developing Countries (Sandec). Swiss Federal institute of Aquatic Science and Technology, Eawag, Dübendorf, Switzerland, 2006. http://www.sandec.ch (accessed November 2008)

Nolde E. (1999). Greywater reuse systems for toilet flushing in multi-storey buildings – over ten years experience in Berlin. *Urban Water*, **1**, 275–284.

Chapter 15

Sustainable approaches to water reuse systems with emphasis on energy requirements

Roumiana Hranova

15.1 INTRODUCTION

The system approach to the analysis of wastewater systems requires consideration of all elements of the system as separate units, as well as, the performance of the system as a whole and its impact on the surrounding environment. Wastewater systems are closely related to urban planning solutions and are dependent on urban population density and the urban development pattern. The size of residential plots, the type of buildings and residential, commercial or institutional units will influence the quantity, variations and quality of the generated wastewater. In the vast majority of medium and large cities or towns, the wastewater system is centralized, for example it comprises of a bulk gravity sewer network with one or several treatment plants, depending on the size of the town and topography. This approach requires relatively large amounts of clean water to washout the waste and to transport it at a long distance. The main objective of the system is to collect and convey the wastewater from buildings to a place, suitable for discharge into natural water bodies, after an appropriate level of treatment in a centralized facility, which would guarantee a good environmental status. These systems are bulk, robust and expensive; it is very difficult and costly to alter or renovate them after they are constructed.

The application of decentralized wastewater systems for medium and large towns is gaining momentum more recently as a recommendable alternative to ever increasing centralized systems (Speers, 2007; Heaney, 2007). The advantages and disadvantages of this approach are discussed in Hranova (2008). Decentralization can be executed at different spatial levels of magnitude: at on-site level – individual buildings; at neighbourhood level – a cluster of several buildings or townhouse complex; at sub-urban level – including a whole district, sub-urban area or satellite townships (Saywell *et al.* 2007). Decentralized solutions at on-site or neighbourhood level would require the application of close-loop technologies, where most of the treated effluent should be reused at a relatively close distance to the source of generation because no suitable discharge point might be available. Thus, water reuse becomes an essential part of the decentralized solution.

In any case of implementation of centralized or decentralized wastewater systems, the water reuse system involves the reuse facility and the wastewater system becomes a part (a sub-system) of the whole water management scheme. Reuse alternatives will vary based on the amount of treated water, space availability, climatic and geological conditions and public acceptance. The most common reuse option suitable for warm and semi-arid climates is the irrigation of agricultural fields or landscaping (Metcalf & Eddy, 2003).

Sustainability is a widely accepted term, describing requirements and conditions regarding the future of human development. Sustainable development is discussed in three core dimensions: environmental, social and economic sustainability (Adams, 2006). Pintér *et al.* (2005) includes one additional dimension related to institutional framework and capacity, as well as the integration of economic and environmental indicators. The assessment of the level of sustainability in the process of development of a country, and the achievements in this direction, is based on a number of well-defined measurable indicators under each one of the different dimensions, determined at national level (Indicators for sustainable development, 2001).

The indicators relevant for this publication are:

- Related to water resources management:
  - Environmental dimension: water quantity – annual withdrawal of ground and surface water as a percent of total available water; concentration of BOD and fecal coliforms in fresh water bodies.
  - Social dimension: sanitation – percent of population with adequate sewage disposal facilities; drinking water – population with access to safe drinking water.
- Related to energy use:
  - Economic dimension: annual energy consumption per capita – share of consumption of renewable energy resources; intensity of energy use (energy use per unit GDP) with specific indicators for the commercial and services sector, manufacturing sector, residential sector and the transportation sector.
  - Environmental dimension: indirectly, the energy use indicators are related to the environmental indicators – emissions of green house gases (GHG) and consumption of ozone depletion substances, as well as to the social dimension by improving public health conditions and mitigating the climate change impact.

Water related systems are consuming significant amount of energy mainly due to pumping and treatment. The application of reliable, safe and economically viable water reuse systems would contribute to the improvement of the national sustainability indicators by reducing the water quantity abstracted from the natural environment and by reducing the quantity of effluents discharged to the environment, correspondingly reducing pollution indicators of this environment.

The application of a system approach to water reuse practice would allow for the assessment and optimization of the process, where energy requirements could be limited in order to find solutions, which are not only viable technically and economically but also efficient in terms of energy consumption. This chapter aims to discuss aspects of the application of water reuse systems in terms of their environmental, social and economic sustainability. Emphasis is given on the inclusion of energy use in the analysis of the system, which is illustrated by a specific case.

15.2 ENVIRONMENTAL SUSTAINABILITY

The major objective of wastewater systems is the protection of public health and the environment from the harmful substances contained in wastewater. They could be seen as a tool/structure for achieving sustainability indicators as related to water quality of natural (fresh) water resources in urban or rural population centers. The achievement of this goal is based on the application of water quality legal and regulatory framework, which defines and describes the criteria and requirements to be achieved, in order to prevent pollution and obtain a treated effluent of acceptable quality. If such guidelines are not available locally, international documents and sources might be used (Anderson *et al.* 2001; WHO, 2006). In the case of water reuse, different standards, criteria and guidelines for each specific type of beneficial use would be required.

Centralized wastewater systems have been developed based on effluent orientated standards and regulations. Usually, these are well defined in order to provide protection of the receiving water body. Decentralized solutions, where different types of water recycling might be applied, would require a more diversified and specified regulatory basis. In both cases, if such standards are not available, if they are available but not enough explicit, or if they are not enforced in practice, public health and environmental risks and hazards would be higher, leading to occasional epidemic outbreaks or pollution of the fresh water resources. Such risks would increase considerably in decentralized wastewater systems, because they would contain a large number of small elements, each one a potential pollution source. The owner would be responsible for the proper functioning of the system. Incompetence or deliberate applications of unreliable or technically inferior systems might lead to unsustainable applications.

Therefore, the sustainable application of decentralized systems would require different type of control and enforcement of the regulatory basis, including not only institutional control, but also auditing of effluent quality and specialized services to maintain a large number of small systems. In this direction, the implementation of standardized wastewater treatment units, approved by the controlling institution, could improve the reliability of the treatment process.

Considering the indicators for sustainability mentioned above, a regulatory basis, which is oriented towards public health and environmental protection only does not contribute fully to a sustainable practice. Minimizing the energy use of the system and the application of renewable sources of energy are additional requirements for sustainable solutions. A holistic assessment of different alternatives during the design stage, including new and innovative technologies, would allow developing optimal systems, which are cost effective, provide environmental and public health protection and use minimal amounts of energy.

15.3 TECHNICAL VIABILITY

For sustainable applications, the technical solutions chosen to convey, treat and dispose or reuse the effluent must be robust, reliable and should provide the required level of treatment. This section discusses several technical aspects, which are basic requirements for the sound and reliable functioning of water reuse systems. The reuse subsystem is the limiting one, which will determine the objectives and requirements of the wastewater subsystem. In other words, the type of beneficial reuse of the effluent and the corresponding technical structure to support it would define the objectives and the structure of the wastewater reuse system.

Sustainable water reuse applications would require the consideration of the following points during the planning stage, the design process and the operation/maintenance (O&M) of the facilities involved:

- Is the generated effluent capable to satisfy the needs of the reuse facility in terms of quantity and quality?
- Are all by-products disposed off in a sustainable way?
- Is the technology cost effective and energy efficient?
- Are all conditions in place for a reliable O&M process?

15.3.1 Meeting water quantity requirements

In order to provide a sustainable water reuse system, the correct estimation of the effluent quantity to be generated in the area under consideration is essential. Quantitative assessments of the generated wastewater have two aspects: the first is the correct estimation of the daily wastewater volumes; the second is related to the variability of the generated amounts. Quantitative assessments of generated wastewater volumes are based on the unit flow rate per capita estimated to be supplied to the premises. The correct and detailed estimation of the expected daily amounts would consider all sources of wastewater in the area, connected to the sewerage. It could be recommended that the management institutions convey regular assessments of the actual consumption rates and their variability in order to provide reliable figures for planning and design purposes. The generated wastewater quantities depend on the urban development pattern, the population density and the established consumption practices. These parameters differ depending on the country or locality and for this reason location specific data would be necessary in order to estimate correctly the expected wastewater flow rates.

The wastewater quantities estimated as average daily flows are subject to diurnal and seasonal variations. The diurnal variations reflect the daily use pattern and are represented by the peak hour coefficients used in the design of water and wastewater structures. In residential areas the maximum hourly flow rate is recorded in the morning and evening hours; in institutional buildings (offices) – during the lunch break. Typically, low wastewater volumes, as one in decentralized solutions, are subject to large variations, such that 40 to 50% of the daily consumption might be generated in two hours for selected cases of single housing on-site solutions.

Larger population results in lower variation of the daily flow. The correct estimation of the variability pattern, for each specific type of water use, would result in a correct design of the conveyance structures and treatment units. The seasonal variations are related to larger consumption volumes during the hot months of the year. However, previous studies (Hranova & Kgomotso, 2007) show that at the scale of the City of Gaborone (Botswana), wastewater flows do not show significant seasonal variations. This can be explained by the fact that relatively low water volumes are used to irrigate loans or gardens, which are the major consumption activity during the hot season. At smaller scale such as residential areas, the effluent stream is generally regular and continuous, but in shopping centres, touristic complexes and schools, the volume of the effluent might be considerably reduced during periods of low activity. From this point of view, centralized wastewater systems provide a more reliable supply of effluent with lower diurnal and seasonal variation, compared to decentralized solutions.

A sustainable wastewater reuse solution would require matching the requested volumes for proper functioning of the reuse subsystem, considering the variability of the flows as mentioned above. If the effluent generated cannot be reused immediately, this variability must be accounted for by means of appropriate storage facility or disposal in the natural environment. Therefore, the choice of reuse alternative and corresponding facilities would depend on the analysis and correct estimation of the available effluent quantity and its variability.

15.3.2 Meeting water quality requirements

In order to achieve sustainable solutions regarding the wastewater subsystem, both the sewerage and wastewater treatment subsystems must be reliable, but the reliability of the treatment unit is essential in order to provide effluents with adequate quality. The conveyance subsystem might be based on gravity or vacuum principles; the last is more expensive and difficult to design, maintain and operate, compared to gravity systems. In addition, vacuum sewer systems require a considerable amount of energy for their functioning. However, they might be an appropriate option in special cases of rocky terrain or high groundwater tables, where shallow laying of pipes might be required. If such conveyance subsystem is applied, it should be coupled with treatment plant at the location of the vacuum station (neighbourhood level), in order to avoid a secondary pumping of the sewage.

Small bore systems, also known as solids-free sewers (Crites & Tchobanoglous, 1998) are a form of gravity sewers of small diameter and low depths. Their application and economic assessment should be analysed from a systematic point of view, including the necessity for septic tanks to remove the solids before the liquid is transported to its destination and the cost of maintaining the septic tank and the sewers.

The treatment subsystem may consist of a wide variety of treatment processes and units. The choice of an appropriate treatment technology should be based on specified criteria, considering general rules and characteristics of the technology in conjunction with local conditions. It is important to apply technologies that correspond to the economic and human resources status of the locality. In developing countries, a number of treatment plants, which are properly designed do not provide the expected quality of effluents because of lack of proper operation and maintenance practice.

15.4 SOCIAL AND ECONOMIC ASPECTS

The social acceptance of water reuse projects is one of the pillars of sustainability for the implementation of any project. It has been acknowledged that one of the barriers for a wider application of water reuse projects is the insufficient public education (Miller, 2006; Anderson *et al.* 2008). Therefore, all stakeholders' involvement, including the general public, should be considered and involved at each stage of the project's planning and implementation. This is of specific importance in cases where a relatively large number of people will be affected by the implementation of such projects or will use the treated effluent for different purposes.

Different forms of involvement should be envisaged, depending on the decisions to be made and the level of stakeholders' expertise. Public awareness, educational programs and other forms of interaction should be implemented as an indispensable part of the project's implementation. The public acceptance process could be supported by a proper economic analysis, which has always been a major consideration for decision-makers during the planning stage of water reuse projects.

The assessment of the projects is based on a cost-benefit analysis in most cases. It has been reported that often such type of analysis shows higher values of the project cost, as compared to its benefits. This is due to the fact that some of the water reuse benefits, such as the protection and recovery of the environment and the improved social conditions, usually are difficult to be presented in monetary terms and consequently are omitted from the analysis (Miller, 2006).

In order to improve the sustainability of water reuse projects, the following points must be considered:

- Economic analysis and assessment should be based on detailed design of the whole water reuse system and also should include its environment – the water supply system and the fresh water resources. Sustainability indicators under the social and environmental themes, such as percentage of population with adequate water supply and sanitation facilities and reducing the pollution and abstraction of fresh water resources, could be incorporated in the analysis in order to support the economic parameters.
- Different alternatives regarding the reuse option and the corresponding treatment facilities should be considered and the most cost effective and energy effective system, which meets the safety requirements and is technically viable, should be selected. This approach would contribute to improve the sustainability indicator "energy use".
- Provision of financial means to sustain the project throughout its live should be made, including the O&M costs.
- Market assessment forms an indispensable part of the economic analysis process. During this stage the potential customers of the wastewater reuse project will be identified, and their willingness to participate, use the reclaimed water and pay for it, will be assessed.

15.5 SYSTEM ANALYSIS INCLUDING ENERGY USE

The system analysis process, as related to water reuse projects, requires a holistic and multi-objective approach. It requires the assessment of the reuse system itself and its interrelation with the water supply system and the natural water environment. It should be based on models, which allow for the variation of different parameters to simulate different outcomes for different scenarios and also would allow for an optimization procedure in order to choose the best option. Such approach would allow the water reuse project to be technically viable, economically sound and energy efficient, as well as, would help to achieve public approval in a transparent and substantiated way.

Fagan *et al.* (2010) had performed a similar analysis for the development of the wastewater system of a new suburb in Melbourne, Australia, based on a dynamic modelling framework which allows the variation of different types of wastewater solutions, including decentralization of the wastewater system at different levels and different water reuse scenarios. The model outcomes provide information regarding the variation of nutrients loads, costs, energy consumption and corresponding GHG emissions under the different scenarios and allow choosing the best option.

Another example of system analysis, specifically related to the O&M of water supply and wastewater systems in Oslo, Norway (Venkatesh & Brattebøa, 2011), assessed the energy use, costs and environmental impacts of these systems, based on actual data from the municipality.

Similar assessment have been done at on-site level, as related to the water services in buildings, assessing different alternatives of water reuse (Hranova, 2009) and the water-energy nexus (Cheng, 2001; Hafez *et al.* 2008).

The development of such type of models, especially for centralized systems, require extensive input of efforts, time and costs, as well as, substantial amount of reliable data for the model calibration and validation. In this chapter such an extensive analysis has not be provided due to lack of enough reliable data, but the system analysis process is illustrated by the solution of a simplified case of water reuse, which shows how the incorporation of specific constraints (in this case limitation of the energy required for the project implementation) would influence the economic output, and correspondingly, the final decision regarding the type of water reuse application.

The description of this case study (Figure 15.1) is as follows:

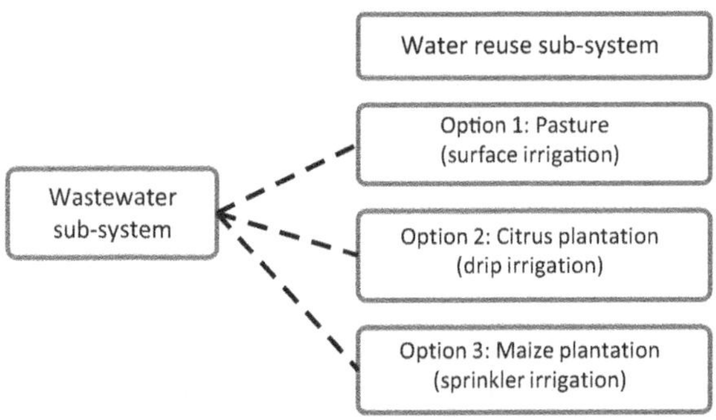

Figure 15.1 The water reuse system configuration

- A centralized wastewater system, including sewerage, treatment plant and the pumping and storage facility to the irrigation field, produces effluent (10 000 m^3/d), which has an appropriate quality for reuse for irrigation purposes, but does not meet the requirements for discharge into an ephemeral stream.

- It has been decided that the treatment plant and the irrigated field will be operated by the municipality, which has selected three viable options. The area available for the reuse option is 750 ha.
- It is necessary to decide which one of the options to be chosen as the most efficient one in terms of profit, energy saving and the use of all available effluent (closed loop approach).

15.5.1 Methodology

The modelling approach used for the solution of the problem is linear optimization (the Excel tool "Solver" has been used), where the economic profit of each option is maximized, subject to constraints as related to available irrigation water resources, area and energy use. The main variable is the area used for irrigation in each option. The input parameters are given in Table 15.1.

Table 15.1 Input parameters.

Parameter	Option 1	Option 2	Option 3
1. Water requirements – W (m^3/ha · yr)	5000	3000	6000
2. Energy requirements – E (kWh/ha)	0	30	85
3. Annual expenditure (cost/unit area)			
3.1 Capital cost (US$/ha)	3500	7700	4000
3.2 Annual capital cost (US$/ha · yr)	350	770	440
3.3 Annual operation cost (US$/ha · yr)			
3.3.1 Labour (US$/ha · yr)	90	1150	120
3.3.2 Operating expenditure (US$/ha · yr)	300	750	400
3.3.3 Pumping energy cost (US$/ha · yr)	0	3	8
3.4 Annualised cost (subtotal points 3.2 and 3.3)	740	2673	968
4. Annual revenue (US$/ha · yr)	2900	7500	3000
5. Annual profit (US$/ha · yr)	2160	4827	2032

The input parameters were calculated based on data typical for the conditions in Botswana and Zimbabwe (semi-arid climate) and the agricultural practices in these countries (Hranova, 2005), supplemented by data available free on the World Wide Web. The energy requirements (Eq. 1) estimated the pumping power and multiplied it by the duration of the irrigation procedures. Thus, only the energy required for pumping has been considered. A fixed value for head loss/ha for each option has been assigned, based on the type of irrigation system. The combined pump/motor efficiency assumed is 65%. Additional energy requirements, related to fuel and electricity used during the operation process, should also be considered in a more detailed assignment, but this has not been included in this case. In option 1 the energy requirement is 0, because water is delivered by gravity to the surface irrigation system.

$$P_i = \frac{\rho g H Q}{\eta} \tag{1}$$

where: P_i is the input power required (W); ρ is the fluid density (kg/m^3); g is the standard acceleration of gravity (9.80665 m/s^2); H is the energy Head added to the flow (m); Q is the flow rate (m^3/s); η is the efficiency of the pump plant as a decimal.

The economic assessment is based on expected profits from the three options, considering the total capital and O&M costs averaged per year (yr) and per unit area (ha). The capital costs have been transformed to annual costs as per Eq. 2, with assumed interest of 10% and a design life time of 25 years.

$$A = C\{i/[1 - (1 + i)^{-n}]\} \tag{2}$$

where: A is annual cost of investment; C is initial value (capital cost at present value); i is interest rate; n is design life time.

The economic analysis is based on a linear optimization technique (Denardo, 2002), which requires for each specific problem the formulation of a linear equation with "n" number of variables and "z" number of constraints, formulated as linear equations as well. The optimization procedure (performed by the Excel data analysis tool "Solver") leads to a well defined extreme (maximum or minimum). In this specific case, the optimum result is the profit maximization following the variation of the area of each option, under the specified constraints. The model formulation and the corresponding optimization procedure are case specific and might vary for different conditions and scenarios.

15.5.2 Analysis of water reuse options

The model developed for this case study is presented in Table 15.2. The constraints limit the use of water to the total available annual amount, the available area and the energy use. The solution has been repeated for different values of the energy use in order to assess the impact of the limitation on the profit.

Table 15.2 Model description.

Parameter
Model: Maximize ($2160\ X_1 + 4827\ X_2 + 2032\ X_3$)
Subject to the following constraints:
Water requirements: $5000\ W_1 + 3000\ W_2 + 6000\ W_3 \leq 3650000$
Area requirements: $X_1 + X_2 + X_3 \leq 750$
Energy requirements: $0\ E_1 + 30\ E_2 + 85\ E_3 \leq 1000$
$X_1;\ X_2;\ X_3 \geq 0$

Where: X_i – irrigated area for each option (ha)

It is evident that option 1, water reuse for pasture, is the best option as related to energy requirements, because of the gravity delivery of the water to the field. For an initial value of energy use of 1500 kWh/yr, the results show that the best solution leads to an annual profit of 1,753,350 US$, where option 1 and option 2 have an area of 700 ha and 50 ha, respectively.

Under this scenario, the total amount of available water is consumed and the energy requirements are minimized because the pasture irrigation does not require pumping. Further increase in the constraint of available energy (Figure 15.2) shows an increase in the profitability due to increase in the area for option 2, which is more profitable, and decrease of the area in option 1. However, this leads to reduction of the amount of water used (Figure 15.2b), thus requiring disposal of the excess water on land or its further treatment for discharge into a natural water body. The variations show a linear trend.

In all studied cases, the option 3 is excluded due to high energy requirements and relatively low profitability.

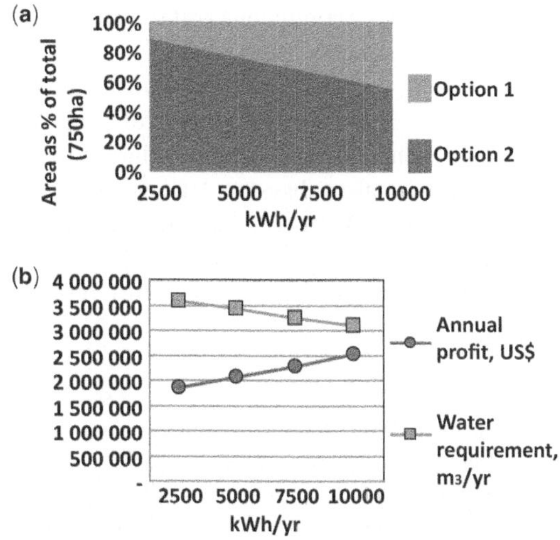

Figure 15.2 Variation of energy requirements and corresponding areas and profits: (a) area variations; (b) profit and water requirements variations

If the energy requirements are not limited, evidently the most profitable option would be chosen for the whole area in order to maximize the profit. Under such conditions, the profit will be 3,620,250 US$ with annual energy requirements of 22,500 kWh, and an annual amount of water requirement of 2,250,000 m^3, which represents 62% of the available water quantities.

In order to use the total available amount of water (close-loop approach), an increase in the area will be needed or additional options might be developed.

It should be noted that any change in the initial system description and input data would lead to a new model development with corresponding outputs.

This simplified example should be viewed as an illustration of the importance of the application of system analysis and optimization during the decision-making process of water reuse projects. A more substantial and detailed analysis must include the wastewater system with additional constraints, variables, and with a multi-objective approach; and also, would consider a non-linear variation of the energy requirements with the area. This would require a more complex model structure, more refined and detailed input data and a different optimization technique.

Nevertheless, this example illustrates the water-energy nexus by the choice of option 2 (pasture irrigation by gravity) as compared to option 3 (irrigation of citrus by drip irrigation). The drip irrigation technique requires about 30% less water as compared to conventional sprinkler or surface irrigation, which allows for larger areas to be irrigated with the same amount of water and also reduces the energy for pumping.

15.6 CONCLUSIONS

Sustainable applications of water reuse systems requires an integrated and multi-objective approach, where a system analysis process during the planning stage would support the decision-making process and make it more transparent and reliable. It is also important to follow this approach during the whole life cycle of the project, including the design and O&M phases. It requires the application of innovative technologies and computational methods and the correct formulation of the objectives, variables and constraints. In order to obtain reliable results, however, the need for continuous collection of sound and reliable data regarding the functioning of the system and its environments is essential.

The energy and water requirements, together with the protection of the public health and the environment, are the most important indicators of sustainability of water reuse projects. Therefore, they must be assessed in conjunction with economic considerations during the planning stage of the project. Usually, the reduction of water use in the reuse subsystem leads to reduction of the energy requirements thus leading to more sustainable reuse option. But limiting energy requirements might lead to considerable reduction of the expected profit of the reuse system. Therefore a balance between economic and sustainability objectives is essential to the success of water reuse projects.

REFERENCES

Adams W. M. (2006). The Future of Sustainability: Re-thinking Environment and Development in the Twenty-first Century. Report of the IUCN Renowned Thinkers Meeting, 29–31 January, 2006 @ http://cmsdata.iucn.org/downloads/iucn_future_of_sustanability.pdf (accessed 16 April 2010)

Anderson J., Adin A., Crook J., Davis C., Hultquist R., Jimenez-Cisneros B., Kennedy W., Sheikh B. and van der Merwe B. (2001). Climbing the ladder: a step by step approach to international guidelines for water recycling. *Water Sci. Tech.*, **10**(43), 1–8.

Anderson J., Baggett S., Jeffrey P., McPherson L., Marks J. and Rosenblum E. (2008). Public acceptance of water reuse. In: Water Reuse: An International Survey of Current Practice, Issues and Needs, B. Jimenez and T. Asano (eds), 1st edn, IWA Publishing, London, UK, 2008, pp. 332–350.

Cheng C. L. (2001). Study of the inter-relationship between water use and energy conservation for a building. *Energy and Buildings*, **34**(3), 261–266.

Crites R. and Tchobanoglous G. (1998). Small and Decentralized Wastewater Management Systems. WCB and McGraw-Hill, New York, USA, pp. 355–364.

Denardo E. V. (2002). The Science of Decision Making: A Problem Based Approach Using Excel. John Wiley & Sons, Inc., New York, USA.

Fagan J. E., Reuter M. A. and Langford K. J. (2010). Dynamic performance metrics to assess sustainability and cost effectiveness of integrated urban water systems. *Resources, Conservation and Recycling*, **54**, 719–736.

Hafez M. M., Bouman C. N., Van de Giesen N., Mushtaq S., Vlek P. and Khan S. (2008). Water reuse and cost-benefit of pumping at different spatial levels in a rice irrigation system in UPRIIS, Philippines. *Physics and Chemistry of the Earth*, **33**, 115–126.

Heaney J. P. (2007). Centralized and decentralized urban water, wastewater and storm water systems. In: Cities of the Future: Towards integrated sustainable water and landscape management, V. Novotny and P. Brown (eds), IWA Publishing, London, UK, pp. 236–250.

Hranova R. (2008). Integrated Applications of Decentralized Solutions to Municipal Wastewater Systems in Developing Countries with Arid Climates. *Proceedings of the 2nd IASTED International Conference – Africa WRM 2008-Science and technology for sustainable development*, Gaborone, Botswana.

Hranova R. (2009). Application of a system approach and evolutionary optimization technique in the practice of wastewater reuse. *Civil Engineering and Environmental Systems*, Taylor & Francis (online), London, UK, pp. 281–294.

Hranova R. and Kgomotso A. (2007). Statistical analysis of monitoring data from Gaborone wastewater treatment plant and implications on the effluent reuse practice. In: Water management Challenges in Global Change, B. Ulanicki, K. Vairavamoorthy, D. Butler, P. Bounds and F. Memon (eds), Taylor & Francis, London, UK.

Hranova R. and Manjonjo M. (2005). Sewage sludge disposal on land – impacts on surface water quality. In: Diffuse pollution of water resources – principles and case studies in the Southern African region, R. Hranova (ed.), Taylor and Francis, Balkema, 153–175.

Indicators of Sustainable Development: Guidelines and Methodologies (2001). UN, Division for Sustainable Development, http://www.un.org/esa/sustdev/publications/indisd-mg2001.pdf (accessed 29 August 2011)

Metcalf and Eddy (2003). Wastewater Engineering, Treatment and Reuse. 4th edn, McGraw-Hill Inc., New York, USA.

Miller W. G. (2006). Integrated concepts in water reuse: managing global water needs. *Desalination*, **187**, 65–75.

Pintér L., Hardi P. and Bartelmus P. (2005). Indicators of sustainable development – proposals for a way forward, UN division for sustainable development. Expert group meeting on Indicators of sustainable development, New York, 13–15 December 2005, UNDSD/EGM/ISD/2005/CRP.2 http://www.iisd.org/pdf/2005/measure_indicators_sd_way_forward.pdf (accessed 28 August 2011)

Saywell D., Evans B. and Somlyody L. (2007). Sanitation 21: Simple Approaches to Complex Sanitation. *Proceedings of the IWA Specialized Conference 'Advanced Sanitation'*, Aachen, Germany.

Speers A. (2007). Water and Cities – overcoming inertia and achieving a sustainable future. In: Cities of the Future: Towards Integrated Sustainable Water and Landscape Management, V. Novotny and P. Brown (eds), IWA Publishing, London, UK, pp. 18–32.

Venkatesh G. and Brattebøa H. (2011). Energy consumption, costs and environmental impacts for urban water cycle services: case study of Oslo (Norway). *Energy*, **36**, 792–800.

WHO (2006). Guidelines for the safe use of wastewater, excreta and gray water, Vol. **4**, http://www.who.int/water_sanitation_health/wastewater/en/ (accessed 10 March 2011)

Part 4

Energy Footprint of Alternative Water Resources

By The Editors

FOREWORD

Alternative resources such as desalinated water, rainwater, reclaimed wastewater and greywater are becoming reliable options for sustainable long-term water supply, as well as a viable tool to mitigate climate change impacts and solve the water energy nexus. However, the introduction of alternative resources in the urban water supply portfolios requires significant modifications of existing urban infrastructure, such as dual reticulation networks, use of more complex treatment technologies, and additional conveyance and energy management facilities. It is important to underline that alternative water resources are not in competition with each other or with conventional water supply sources, but in contrary, should be considered essential components of one integrated water management portfolio which balances their use depending on local conditions to improve water and energy management and efficiency and to secure reliable and sustainable long term water supply.

Part 4 of this book "Energy Footprint of Alternative Water Resources", (e.g. Chapters 16 to 20), provides case studies illustrating the economic and environmental viability of alternative water resources, breaking the dogmas associated with the perceived high energy use and costs of water supply alternatives such as desalination, water reuse and rainwater harvesting. This part of the book highlights their mutually beneficial roles as alternative water supplies in the context of integrated water portfolio and describes the state of the art methods, technologies and planning approaches for implementation of carbon-neutral water reuse, desalination and rainwater harvesting projects.

Until recently, use of seawater desalination was limited to desert-climate dominated regions of the world. Latest technological advances and associated decrease in water production costs and energy demand have expanded its use in coastal areas traditionally supplied with fresh water resources. Energy consumption of desalination is still higher than other water supply alternatives, but the implementation of high efficiency reverse osmosis membranes and energy recovery devices enabled to reduce specific energy consumption to 2.5–2.8 kWh/m^3, as well as to lower dramatically costs for fresh water production from saline water sources.

Forward looking water utilities consider desalination and water reuse as two integral components of the water cycle, with a number of synergies and mutual benefits in terms of water production capacity, product water quality and energy use. While at present the water production costs and energy use associated with large water reuse projects such as the Groundwater Replenishment System (GWRS) in Orange County (California), NEWater facilities (Singapore) and Western Corridor Recycling Scheme (Australia) are typically lower than the costs for production of desalinated water in the same utility service area, water utilities which have implemented such projects are also major stakeholders in the development of large desalination plants because of the drought-proof benefits which desalination brings to their diversified water portfolio.

The detailed case study of GWRS project in Orange County shows that specific energy cost of advanced treatment by microfiltration, reverse osmosis and advanced oxidation by UV and hydrogen peroxide requires 0.53 kWh/m^3 of energy, of which 48% is associated with reverse osmosis separation. A key strategy in reduction of energy consumption for this project was to maximize filtrate water quality from the upstream microfiltration process. With a daily production of 265,000 m^3/d, water recycling for groundwater recharge this project has become an essential component of Orange County's water supply portfolio.

Globally, rainwater harvesting (RWH) technologies are being more readily adopted, as the impetus for buildings to become more adaptable and resilient to climate change and population growth increases. At present, the majority of the existing RWH systems use pumping equipment and schemes, characterised by high energy consumption and carbon footprint. In recognition of the need to address these challenges, recently developed innovative gravity driven RWH systems, recover kinetic energy of the water flow or solar energy, and reduce the energy consumption and thereby improve the sustainability of rainwater harvesting.

The selection of the best option for water management with alternative water resources requires the development and application of a multi-criteria analysis, that is the triple bottom line approach, based on economic, environmental and social impacts. As demonstrated by several case studies, water scarcity index, climate change impacts, population dynamics, geographic location (coastal versus non-coastal areas), water and energy sources, existing infrastructure and especially water transportation and distribution systems are all evaluated and reflected into the makeup of well-balanced long term water supply portfolios.

As illustrated in this book section, life cycle analysis (LCA) has found numerous applications for assessment of the environmental impact of various products and processes, including water reuse and desalination. The comparison of the application of three different LCA methods (CML 2001, Ecoindicator 99, and Ecopoints 97) for seawater desalination plants has demonstrated the high influence of the energy supply source and the need to maximize in energy efficiency when planning such projects. However the practical implementation of existing LCA methodologies have some limitations when applied for the assessment of the environmental impact of alternative water supply resources. For example, the increased temperature and salinity of ambient seawater in the area of the plant outfall due to concentrate discharge of desalination plants, difference in terms of removal of pollutants in water recycling, and more globally the high subjectivity in the selection of the impact assessment method and the weighting factors.

Chapter 16

Energy use for seawater desalination – current status and future trends

Nikolay Voutchkov

16.1 INTRODUCTION

World's oceans contain over 97.2% of the planet's water resources. Because of the high salinity of ocean water and the significant costs associated with seawater desalination, most of the global water supply has traditionally come from fresh water sources – groundwater aquifers, rivers and lakes. Changing climate patterns observed worldwide over the past decade, combined with accelerating population growth pressures and limited availability of new and inexpensive fresh water supplies, are shifting water industry's attention to an emerging trend – seawater desalination.

Until recently, use of seawater desalination was limited to desert-climate dominated regions. Latest technological advances and associated decrease in water production costs and energy demand have expanded its use in coastal areas traditionally supplied with fresh water resources. At present, desalination plants provide approximately 1% of the world's drinking water supply. Potable water production capacity from saline sources (brackish and seawater) has been increasing exponentially over the past 15 years. In 2010, more than 15,000 desalination plants worldwide produced a total of 65.2 Mm^3/d (million cubic meters per day) of fresh water from seawater and brackish water (IDA, 2011). Approximately 60% of the installed plant production capacity is attributed to reverse osmosis (RO) membrane desalination plants and 34.8% by thermal distillation facilities. Currently, reverse osmosis is the most commonly used desalination technology because for most saline sources and applications, it yields fresh water at overall energy use and costs lower than other traditional alternative desalination technologies.

Salt separation from seawater requires a significant amount of energy to overcome the naturally occurring osmotic pressure exerted on the reverse osmosis membranes. This in turns makes seawater reverse osmosis (SWRO) desalination several times more energy intensive than conventional treatment of fresh water resources. Table 16.1 presents the energy use associated with various water supply alternatives.

Analysis of this data indicates that the energy needed for seawater desalination is approximately eight to ten times higher than that for production of potable water from conventional sources such as rivers, lakes, and fresh water aquifers. It should be pointed out however, that such resources are limited to less than 2.5% of the water available on the planet, and that in large urbanized areas of most developed countries worldwide traditional fresh water resources are near depletion, while new sources are not readily available to sustain long-term population growth, industrial development and quality of life.

Table 16.1 Energy use of various water supply alternatives.

Water supply alternative	Energy use (kWh/m^3)
Conventional treatment of surface water	0.2–0.4
Water reuse	0.5–1.0
Indirect potable reuse	1.5–2.0
Brackish water desalination	1.0–1.5
Desalination of Pacific Ocean water	2.5–4.0

As indicated in Table 16.1, energy use for water reuse is several times lower than that for seawater desalination. However, compared to desalination, water recycling does not create new fresh drinking water - it merely provides a more efficient use of the already available fresh water resources. Therefore, in most coastal urban areas worldwide both seawater desalination and water reuse are implemented in parallel and are viewed as integral parts of well-balanced and environmentally sustainable long-term water supply portfolio.

16.2 ENERGY USE FOR DESALINATION

16.2.1 Current status of energy use for seawater reverse osmosis (SWRO)

Table 16.2 provides typical ranges for cost of potable water production and energy use of reverse osmosis membrane systems of medium and large seawater desalination plants (i.e., plants with fresh water production capacity of 40,000 m^3/d, or more). This table is based on actual data from over 20 SWRO plants constructed between 2005 and 2010. As seen from Table 16.2, SWRO systems of best-in-class seawater desalination plants use between 2.5 and 2.8 kWh of electricity to produce 1 cubic meter of fresh water (kWh/m^3), while the industry average energy use is approximately 3.1 kWh/m^3. The typical cost for production of fresh drinking water from seawater at present is approximately US\$1.1/m^3. Energy expenditures typically contribute 25 to 40% of this cost depending on the unit power rate and the SWRO plant design, and equipment efficiency.

Table 16.2 Typical cost and energy use for medium and large size SWRO systems.

Classification	Cost of water (US\$/m^3)	SWRO system energy use (kWh/m^3)
Low-end bracket	0.5–0.8	2.5–2.8
Medium range	0.9–1.5	2.9–3.2
High-end bracket	1.6–3.0	3.3–4.0
Average	1.1	3.1

*SWRO: seawater reverse osmosis

It should be pointed out that the energy use presented in Table 16.2 only encompasses SWRO system operations, rather than that of the energy consumption of the entire seawater desalination plant. Usually, SWRO systems contribute between 65% and 80% of total desalination plant energy demand.

Figure 16.1 presents a breakdown of energy consumption within a typical seawater desalination plant using Pacific Ocean water of total dissolved solids (TDS) concentration of 33,500 mg/L (33.5 g/L) as a saline source water. In this illustrative example, the total plant energy use is 3.57 kWh/m^3 and the SWRO system's energy demand is 2.54 kWh/m^3 (71% of the total plant energy use).

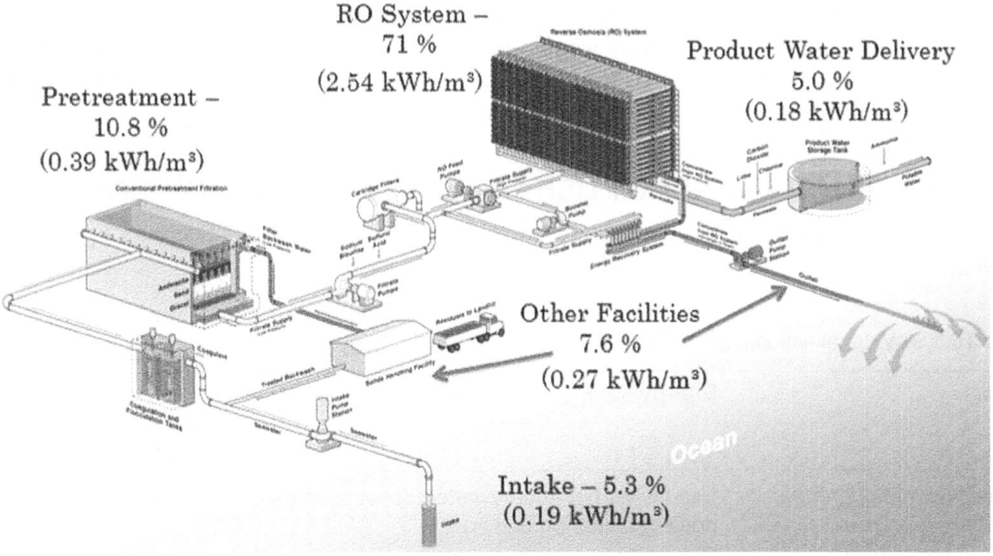

Figure 16.1 Energy use breakdown of typical SWRO desalination plant

Usually the desalination plant's source seawater pretreatment system is the second largest user of energy. However, for some plants, where the point of desalinated water delivery is located at a long distance and/or high elevation, the energy for product water conveyance could be higher than that for seawater pretreatment.

16.2.2 Minimum energy demand for SWRO desalination

The lowest theoretical energy consumption for desalination of seawater with a salt content of 33,500 mg/L and temperature of 25°C (i.e. typical Pacific Ocean Water) is 0.7 kWh/m^3. This energy use corresponds to condition of complete conversion of seawater into fresh water (100% recovery), which cannot be achieved in practical terms. For a more realistic 50% recovery, this minimum theoretical energy use would be 1.1 kWh/m^3. However, this energy consumption assessment assumes that all desalination plant equipment has 100% energy efficiency and all energy contained in the desalination plant concentrate is reused in the desalination process. Therefore, this energy threshold is the ideal theoretical minimum for seawater desalination.

Based on the systematic long-term testing of a full-scale state-of-the-art desalination system by the Affordable Desalination Collaboration (ADC) in the United States (US), the lowest energy use that could be achieved with actual state-of-the art highly efficient commercially available desalination equipment and RO membranes at the time of testing (years 2006–2007) was determined to be 1.58 kWh/m^3 (MacHarg *et al.* 2008). Such energy use was measured at RO system recovery of 42% and average SWRO membrane flux of 10.2 L/m^2 · h (lmh). The ADC testing was completed using Pacific Ocean seawater collected by an open ocean intake and pretreated by granular media pressure filters. The ADC study concluded however, that SWRO system operation at such low recovery and flux does not yield the lowest overall cost of water production at unit cost of energy of US$0.10/kWh used for life-cycle cost assessment.

Based on a detailed cost-benefit analysis, ADC researchers have determined that the "Most Affordable Point" of SWRO system design is at plant recovery of 48% and flux of 15.3 L/m^2 · h. At such operational conditions, the minimum SWRO system energy use was determined to be 2.0 kWh/m^3. It should be pointed out that the "Most Affordable Point" design would vary with unit cost of energy and the project-and-location specific construction and engineering costs. A detailed description of the ADC study scope, assumptions, test conditions, and results is presented elsewhere (MacHarg *et al.* 2008).

16.3 DESALINATION ENERGY USE FACTORS AND TECHNOLOGY TRENDS

Energy use for seawater reverse osmosis desalination varies in a wide range and depends upon a number of factors (Table 16.3).

Table 16.3 Key desalination plant energy use factors and technology trends.

Factor	Energy saving technology trends	Potential for energy savings (% of industry average)
Source water temperature	Use of warmer source water (collocation with power generation plants)	3 to 5%
Source water salinity	Use of lower-salinity source water or blend of seawater and brackish water	Over 50%
Membrane element energy and productivity losses	Use of higher productivity elements & lower-energy RO system configurations	5 to 15%
RO high pressure feed pump efficiency	Maximizing pump and motor efficiency by the use of large pumps serving several RO trains	5 to 10%
Recovery of energy from RO concentrate	Use of isobaric chamber type technologies	10 to 15%

16.3.1 Collocation of desalination and power plants – use of warmer source water

Warmer source seawater has lower viscosity and therefore lower osmotic pressure. As a result, desalinating warmer seawater requires less energy for salt separation than using seawater of ambient temperature. This potential energy reduction benefit could be applied by using warm water discharges from coastal power plants as source water for desalination. Coastal power generation plants often use seawater of ambient temperature for cooling of their electricity generation units. The cooling water discharged from a typical power generation station is usually 5 to 15°C warmer than the ambient ocean water. Taking under consideration that energy needed for salt separation is reduced with 3% to 5% for every 10°C of elevated seawater temperature, using warmer seawater can result in measurable energy reduction. Under a desalination plant – power station collocation configuration, the intake of the seawater desalination plant is connected to the discharge canal of the power plant to collect a portion of the cooling water of this plant for desalination (Figure 16.2).

The collocation configuration allows using the power plant cooling water both as source water for the seawater desalination plant and as blending water to reduce the salinity of the desalination plant concentrate prior to its discharge to the ocean.

Collocation has found a large-scale implementation at the 95,000 m^3/d Tampa Seawater Desalination Plant in the US, which has been in continuous operation since 2008, and is currently planned to be implemented at several seawater desalination plants in California, Cyprus and in other parts of the world.

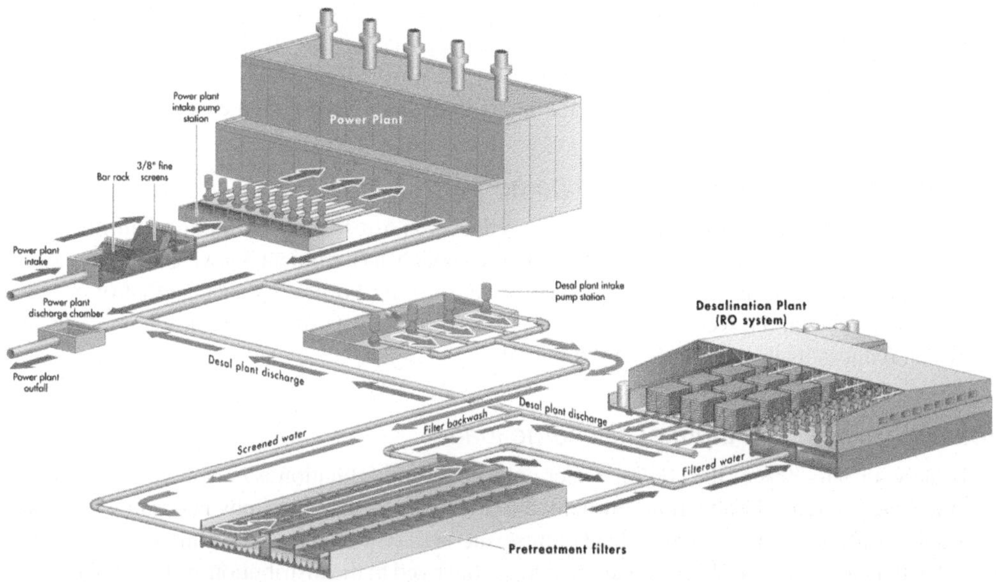

Figure 16.2 Typical configuration of collocated desalination plant

The collocation of SWRO desalination plants with existing once-through cooling coastal power plants could yield four key benefits:

(1) Reduced plant capital and operating costs;
(2) Dilution of desalination plant brine discharge;
(3) Decrease of power plant thermal discharge load;
(4) Accelerated dissipation of brine and thermal discharge loads.

Reduced plant capital and operating costs

Desalination plant capital cost is reduced by avoiding the construction of separate ocean intake and outfall – the plant collects source seawater for production of fresh water from the cooling water discharge of the power generation station and returns its brine to the same discharge downstream of the point of intake.

Usually, coastal power plants with once-trough cooling systems use large volumes of seawater. Because the power plant intake seawater has to pass through the small diameter tubes (typically 10-mm or less) of the plant condensers to cool them, the plant discharge cooling water is already screened through bar racks and fine screens similar to these used at surface water intake desalination plants. Therefore, a desalination plant which intake is connected to the discharge outfall of a power plant usually does not require the construction of a separate intake structure, intake pipeline and screening facilities (bar-racks and fine screens).

Since the construction cost of a new surface water intake structure for a desalination plant is typically 5% to 30% of the total plant construction expenditure, the power plant collocation could yield significant construction cost savings.

As a result of the desalination plant collocation, power costs could be further decreased by avoiding the need for using the power grid and the associated fees for power transmission to the desalination plant. Typically, the electricity tariff (unit power cost) structure includes two components: fees for power production and for power grid transmission. Often, the power transmission grid portion of the tariff is 30% to 50% of the total unit power cost. By connecting the desalination plant directly to the power plant electricity generation equipment, the grid transmission portion of the power fees could be substantially reduced or completely avoided, thereby further reducing the overall seawater desalination cost.

Dilution of desalination plant brine discharge

The salinity of the desalination plant discharge is reduced as a result of the mixing and dilution of the membrane concentrate with the power plant discharge, which has ambient seawater salinity. As a rule, the salinity concentration of the desalination plant brine is 1.5 to 2 times higher than that of ocean water. In the case of collocation, the RO brine is diluted by the cooling water discharge of the power generation station, which reduces its impact on the environment.

Decrease of power plant thermal discharge load

Approximately 45% to 50% of the cooling water collected for desalination is converted to drinking water. The cooling water used for desalination is usually 3 to 5°C warmer than the ambient ocean water and under the collocation configuration, the thermal load of this water will be directed to the drinking water distribution system rather than to the ocean. While the thermal load conveyed to the distribution system would not have a significant impact on its function, discharging this load to the ocean would have a negative impact on marine environment. Therefore, collocation will result in reduction of the total thermal pollution caused by the power plant discharge.

Accelerated dissipation of brine and thermal discharge loads

Desalination plant brine is heavier than the ocean water and therefore, has negative buoyancy (i.e. sinks to the ocean bottom after discharge). When it reaches the bottom, the brine tends to accumulate and may cause negative impact on aquatic species of low salinity tolerance inhabiting the ocean bottom in the vicinity of the discharge.

Power plant discharge has a positive buoyancy (i.e. the thermal plume of the power plant discharge floats to the surface of the ocean). This thermal plume dissipates heat very slowly because it only mixes with the shallow surface top layer of the ocean.

When the negatively buoyant brine and the positively buoyant cooling water mix before they enter the ocean their mixing has a mutually beneficial effect on buoyancy. The brine is lifted upwards by the warm cooling water discharge, which prevents its negative impact on the benthic flora. The thermal load of the cooling water is pushed downwards by the heavier brine, which reduces the surface area of the thermal plume and mixes it with the entire depth of the water column engaging more ambient ocean water in contact with the heat and dissipating it faster.

Adverse impacts and constraints of collocation

It should be pointed out that using warmer water for desalination could have certain disadvantages associated with the accelerated bio-growth of marine bacteria on the surface of the SWRO membrane elements which could result in the need for more frequent membrane cleaning, especially, if the source seawater temperature is already higher than 25°C.

Therefore, collocation is more likely to be feasible for locations where the ambient seawater is relatively cold (ocean temperatures of 18°C or lower).

16.3.2 Using lower salinity source water to reduce energy consumption

In reverse osmosis membrane desalination systems energy demand for salt separation is proportional to the salinity of the source water. Therefore, desalination of lower salinity source water results in lower energy demand for drinking water production. From this prospective, desalinating brackish water is preferable if such saline water source is readily available. If brackish water sources at a given location are not adequate to produce a desired volume of fresh water, then the available brackish water could be blended with seawater to reduce the source water salinity of a seawater desalination plant and thereby to decrease the overall energy used for desalination. While this approach is not commonly practiced at present, it holds significant potential benefits under the right circumstances.

Besides brackish source water, concentrate from brackish water desalination plants (desalter brine) could also be used as feed water to a seawater desalination plant in order to reduce feed water salinity (Figure 16.3). Such approach has already found practical implementation at a 10,000 m^3/d desalination plant located in City of Eilat, Israel and is under consideration for implementation in Orange County and San Diego County, California, USA.

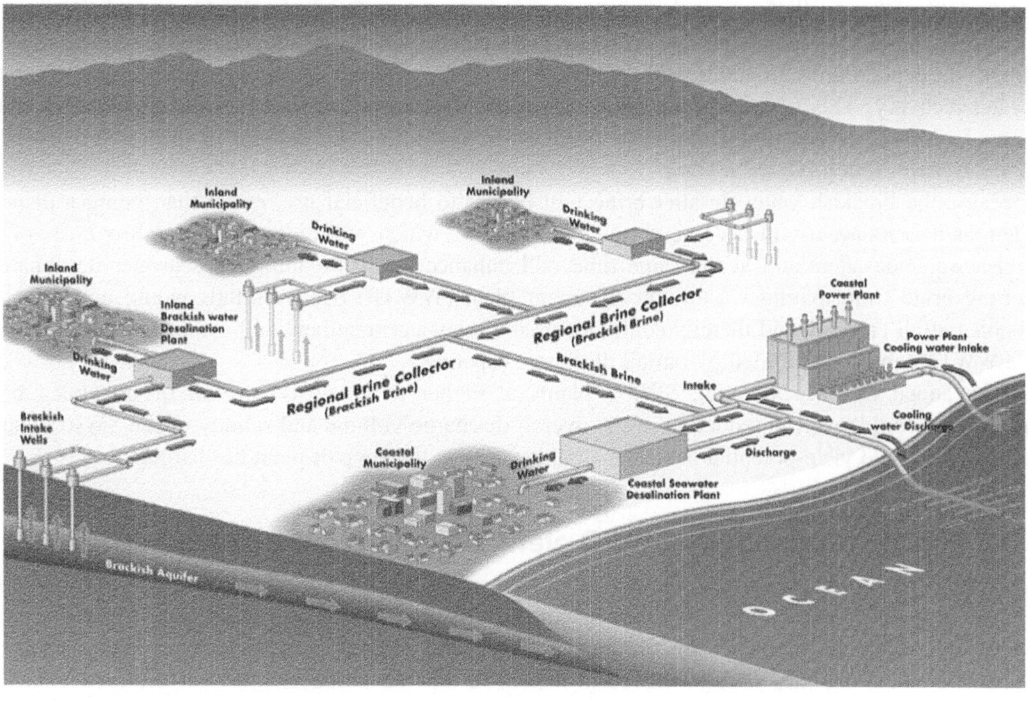

Figure 16.3 Integrated Inland Desalter Brine Disposal and Seawater Desalination

The key components of such regional integrated desalination system include: (1) inland brackish water desalination plants, (2) regional brine interceptor/collector, and (3) centralized coastal seawater desalination plant. The purpose of the regional brine collector is to convey the concentrate from the inland desalters to the regional seawater desalination plant, where this concentrate is used as supplemental feed water to the source seawater used for desalination.

Although Figure 16.3 presents a combination of seawater desalination plant collocated with a coastal power generation plant, this approach could be used for SWRO plants with conventional intakes and outfalls as well. Use of concentrate from brackish water desalination plants as feed water to a seawater desalination plant is mutually beneficial for both plants. Usually, inland brackish water desalination plant capacity is limited by lack of suitable discharge locations for the plant concentrate. If the seawater desalination plant can accept brackish water desalination plant concentrate and process it, the brackish water desalination plant capacity could be increased beyond the threshold driven by brine discharge limitations, and the desalination plant source salinity could be reduced at the same time.

This regional concentrate management approach has a number of benefits. Brine from inland desalters using brackish ground water sources typically does not contain pathogens (bacteria, *Giardia*, *Cryptosporidium*, and so on) and therefore, it could be a safe and suitable source of water for seawater desalination. As a result, rather than being disposed as a waste product to the ocean or to deep aquifers, brackish water concentrate could be reused for drinking water production.

In addition, brine from inland desalters usually has an order-of-magnitude lower total dissolved solids concentration than seawater (i.e. 2,000 to 5,000 mg/L *vs.* 33,500 to 40,000 mg/L). As a result, mixing of brine and seawater will reduce the overall salinity of the source water fed to the seawater desalination plant, and therefore, it will decrease the total amount of energy needed to desalinate seawater.

Typically, brine from inland desalters contains antiscalants, which will allow to reduce or to completely eliminate the expenditures for addition of such chemicals at the seawater desalination plant and will increase seawater recovery. This benefit would allow increasing the overall recovery of the desalination plant. Increased recovery means producing more fresh water from the same amount of feed water, which in turn yields lower unit production costs.

Because the brackish water desalter brine will be put to beneficial use, rather than being a disposal burden, as it is today, it will become a valuable resource, which will reduce the operating costs of the brackish water desalters and at the same time will enhance the affordability of seawater desalination. Diverting brine from exiting wastewater treatment plant (WWTP) ocean outfalls would enhance their available outfall capacity and thereby could decrease wastewater treatment and disposal costs, especially if the WWTP capacity is limited by outfall discharge capacity.

As an added benefit, operating SWRO plants at higher recovery as a result of integrated brine management would result in reduction of the overall discharge volume and salinity of the SWRO plants, which in turn could yield potential environmental benefits in the zone of plant discharge.

16.3.3 Higher productivity SWRO elements yield lower energy costs

A key factor that has contributed to the dramatic decrease of seawater desalination energy use and costs over the past 10 years is the advancement of the SWRO membrane technology. Today's high-productivity membrane elements are designed with several features that yield more fresh water per membrane element than any time in the recent history of this technology: higher surface area, enhanced permeability and denser membrane packing. Increasing active membrane leaf surface area and permeability allows to gain significant productivity using the same size (diameter) membrane element. The active surface area of the membrane elements is typically increased by membrane production process automation, by denser membrane leaf packing and by adding membrane leafs within the same element.

The total active surface area in a membrane element is also enlarged by increasing membrane size/diameter. Although 8-inch SWRO membrane elements are still the "standard" size most widely used in full-scale applications at present, larger 16-inch and 18-inch size SWRO membrane elements have become commercially available over the past five years and have already found full-scale implementation in over two dozen SWRO projects worldwide (Bergman & Lozier, 2010).

In the second half of the 1990s, the typical 8-inch SWRO membrane element had a standard productivity of 18.9 to 22.7 m^3/d (5,000 to 6,000 gallons per day, gpd) at salt rejection of 99.6%. In 2003, several membrane manufacturers introduced high-productivity seawater membrane elements that are capable of producing 28.4 m^3/d (7,500 gpd) at salt rejection of 99.75%. Just one year later, even higher productivity of 34 m^3/d (9,000 gpd at 99.7% rejection) seawater membrane elements were released on the market. Over the past five years, SWRO membrane elements combining productivity of 37.9 to 53 m^3/d (10,000 to 14,000 gpd) and high-salinity rejection have become commercially available and are now gaining wider project implementation.

The newest membrane elements provide flexibility and choice, and allow users to trade productivity and pressure/power costs. The same water product quality goals can be achieved in one of two general approaches: (1) reducing the system footprint/construction costs by designing the system at higher productivity, or (2) reducing the system's overall power demand by using more membrane elements, designing the system at lower flux and recovery, and taking advantage of newest energy recovery technologies, which further minimize energy use if the system is operated at lower (35% to 45%) recoveries.

16.3.4 Hybrid membrane configuration reduces energy

Innovative hybrid membrane configurations, combining SWRO elements of different productivity and rejection within the same vessel, allow optimizing the use of energy introduced with the feed water to the desalination vessels (Figure 16.4).

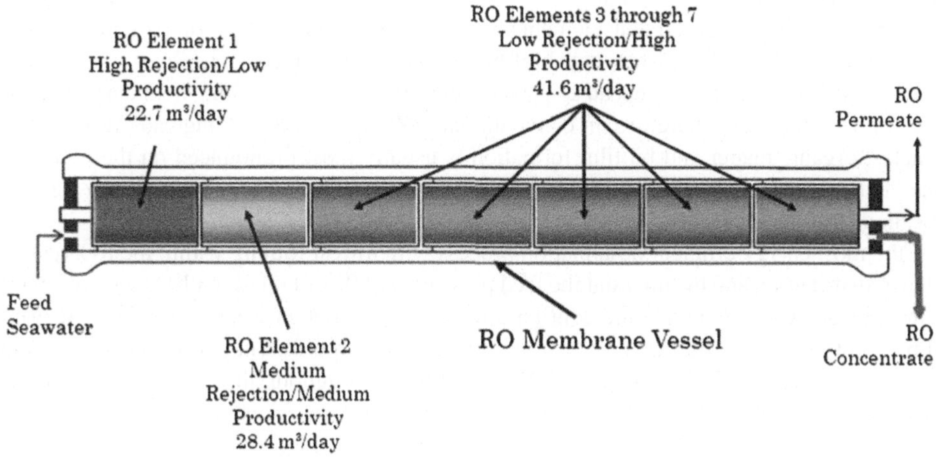

Figure 16.4 Hybrid SWRO Membrane Configuration

Usually, in SWRO systems using standard spiral-wound RO membrane elements, all of the feed seawater is introduced at the front of the membrane vessel and all permeate and concentrate are collected at the back end. As a result, the first (front) membrane element is exposed to the entire vessel feed flow and pressure, and operates at productivity per square meter of element (flux) significantly higher than that of the subsequent membrane elements.

With a typical configuration of seven elements per vessel and ideal uniform flow distribution to all RO elements, each membrane element would produce one-seventh (14.3%) of the total permeate flow of the vessel. However, in actual conventional SWRO systems, the flow distribution in a vessel is uneven and the first membrane element usually produces over 25% of the total vessel permeate flow, while the last element only yields 6% to 8% of the total vessel permeate. The decline of permeate production (flux) along the length of the membrane vessel is mainly due to the increase in feed salinity and associated osmotic pressure as the permeate is removed from the vessel while the concentrate rejected from all elements remains in the vessel until it exits the last element. In addition, as the first element produces over 25% of the permeate flow it also expends over 25% of the pressure/energy available for desalination. This energy is lost with permeate generated by the first RO element, instead of being available to obtain maximum performance efficiency of the remaining six RO elements in the pressure vessel.

Since a disproportionately large amount of energy is expended too early in the desalination process and the remaining six downstream RO elements are underworked, the overall energy efficiency of permeate production by the pressure vessels in conventional SWRO systems is not at optimum level. In addition, because the first element processes the largest portion of the feed flow, it also receives and retains the largest quantity of the particulate and organic foulants contained in the source seawater, and is most impacted by fouling. The reminder of the feed water that does not pass through the first RO element and the concentrate from this element enters the feed channels of the second RO element of the vessel. This element therefore, is exposed to higher salinity feed water and lower feed pressure (energy), because some of this pressure (energy) has already been used in the first RO element to produce permeate. As a result, the permeate flow rate (flux) of the second element is lower and the concentrate polarization on the surface of this element is higher than that of the first RO element.

The subsequent membrane elements are exposed to increasingly higher feed salinity concentration and elevated concentrate polarization, which results in a progressive reduction of their productivity (permeate flux). As flux through the subsequent elements is decreased, accumulation of particulate and organic foulants on these elements diminishes and biofilm formation is reduced. However, the possibility for mineral scale formation increases because the concentration of salts in the boundary layer near the membrane surface increases. Therefore, in conventional SWRO systems, fouling caused by accumulation of particulates, organic matter and biofilm formation is usually most pronounced on the first and second membrane elements of the pressure vessels, while the last two RO elements are typically more prone to mineral scaling than the other types of fouling.

Ideally, the most energy efficient desalination process with lowest fouling within the RO vessels can be achieved by redistributing and evening out the feed pressure and flux of all seven RO elements in the vessel to near-equal level. A novel membrane configuration design to obtain a more even flux distribution is achieved by combining three different models of membranes with different permeability within the same vessel instead of using the same model of RO elements throughout the vessel (which is a typical configuration for conventional SWRO systems). This design was perfected and proven by Dow Filmtec as Inter-Stage Design (ISD) and currently has been adopted in a large number of SWRO desalination plants worldwide.

On the example shown on Figure 16.4, the first (lead) element in ISD configuration, which receives the entire seawater feed flow of the vessel, is a low-permeability/high salt rejection element (i.e. Dow Filmtec SW30 XHR-400i). Because of its low permeability, this element produces only 10% to 14% (instead of 25%) of the permeate flow produced by the entire vessel, thereby preserving the feed energy for more effective separation by the downstream RO membrane elements in the vessel. The second RO element in the pressure vessel is of a standard (average) permeability (i.e. Dow Filmtec SW30 XLE-400i) and salt rejection, and produces approximately 14% to 16% of the total flow, while the remaining five elements in the vessel are of the same high-permeability/low rejection model (i.e. Dow Filmtec SW30 ULE-400i). This 1-1-5 combination of low permeability/high rejection and high permeability/low rejection elements results in a more even distribution of flux and pressure along the vessel and typically yields 5% to 10% energy savings reducing the fouling rate of all membrane elements.

16.3.5 Split-flow RO system configuration for improved energy & cost efficiency

At present, two-pass SWRO systems are widely used when the product water from the desalination plant has to match the quality of existing fresh water sources with which it is blended. In conventional two-pass SWRO membrane systems (Figure 16.5a), the saline source water is first treated by a set of SWRO membrane trains (referred to as first RO pass) and then the entire volume of desalinated water from the first pass is processed through a second set of brackish water desalination membrane trains, termed second pass.

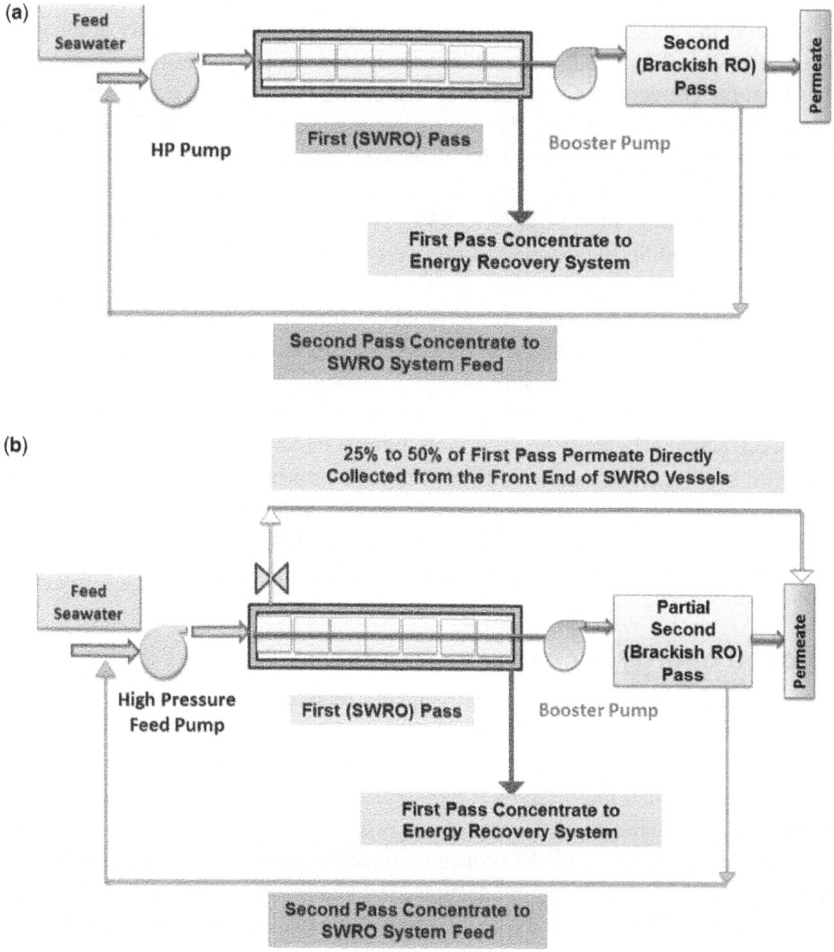

Figure 16.5 Conventional Two-Pass (a) and Split Partial Two-Pass (b) SWRO Systems

A recent trend, adopted in most new SWRO desalination projects, is to use an innovative split partial two pass RO System where the second RO pass processes only a portion (typically 50% to 75%) of the permeate generated by the first pass and the reminder of the low salinity permeate produced by the front (feed) SWRO elements of the first pass is collected and blended with the permeate of the second RO pass without additional desalination (Figure 16.5b).

The split partial second pass configuration allows to reduce the size of the second pass-system and to improve energy use and flow distribution within the membrane vessels. As a result, the overall fresh-water production energy and costs associated with seawater desalination are reduced with 5% to 15% as compared to conventional full two pass SWRO systems. It should be pointed out that split-partial second SWRO pass systems can be configured in several alternatives, which may involve the use of different type of membrane elements within the first pass SWRO vessels (i.e. use hybrid in-vessel ISD discussed in the previous section) to further optimize plant design and minimize energy use.

16.3.6 Increased high pressure pump efficiency

One approach for reducing total SWRO system energy demand, which is widely applied throughout the desalination industry today, is to use larger and higher efficiency centrifugal pumps which serve multiple RO trains rather than the conventional approach of dedicating smaller-size pumps to each individual RO train. The energy savings associated with the use of larger pumps stem from the fact that the efficiency of multistage centrifugal pumps increases with their size (pump flow). For example, under a typical configuration where individual pump is dedicated to each desalination plant RO train, high pressure pump efficiency is usually in a range of 80% to 83%. However, if the RO system configuration is such that a single high pressure pump is designed to service two RO trains of the same size, the efficiency of the high pressure pumps could be increased to up to 85%.

Proven design that takes this principle to the practical limit of centrifugal pump efficiency (≈90%) is implemented at the 326,000 m^3/d (86 MGD) Ashkelon seawater desalination plant in Israel, where two duty horizontally split-case high pressure pumps are designed to deliver feed seawater to 16 SWRO trains at guaranteed long term efficiency of 88%. Continuous plant operational track record over the past 5 years shows that the actual efficiency level of these pumps under this configuration is close to 90%.

A current trend for smaller desalination facilities with a water production capacity of 950 m^3/d (250,000 gpd) or less is to use positive displacement (multiple-piston) high-pressure pumps and energy recovery devices, which often are combined into a single unit. These systems are configured to take advantage of the high efficiency of the positive displacement technology, which practically can reach 94% to 97%.

16.3.7 Improved energy recovery

Advances in the technology and equipment allowing the recovery and reuse of the energy applied for seawater desalination have resulted in a reduction of 80% of the energy used for water production over the last 20 years. Today, the energy needed to produce fresh water from seawater for one household per year (~2,000 kW/yr) is less than that used by a typical 0.5 m^3 (18 cubic feet) refrigerator.

While five years ago, the majority of the existing seawater desalination plants used Pelton wheel-based technology to recover energy from the SWRO concentrate, today pressure exchanger-based energy recovery systems dominate in most desalination facility designs. The key feature of this technology is that the energy of the SWRO system concentrate is directly applied to pistons that pump intake seawater into the system. Pressure-exchanger technology typically yields 5% to 15% higher energy recovery savings than the Pelton-Wheel-based systems.

Figure 16.6 depicts the configuration of a typical pressure exchanger-based energy recovery system. After membrane separation, most of the energy applied for desalination is contained in the concentrated stream (brine) that also carries the salts rejected from the seawater. This energy-bearing stream is applied to the back side of pistons of cylindrical isobaric chambers, also known as pressure exchangers (shown as vertical cylinders on Figure 16.6). These pistons pump approximately 45% to 50% of the total volume of seawater fed into the RO membranes for salt separation. Since a small amount of energy (4% to 6%) is lost during the energy transfer from the concentrate to the feed water, this energy is added back to feed flow by small booster pumps to compensate the energy loss. The remainder (45% to 50%) of the feed flow is handled by high-pressure centrifugal pumps. Harnessing, transferring and reusing the energy applied for salt separation at very high efficiency (94% to 96%) by the pressure exchangers allows a dramatic reduction of the overall amount of electric power used for seawater desalination. In most applications, a separate energy recovery system is dedicated to each individual SWRO train. However, some recent designs include configurations where two or more RO trains are serviced by a single energy recovery unit.

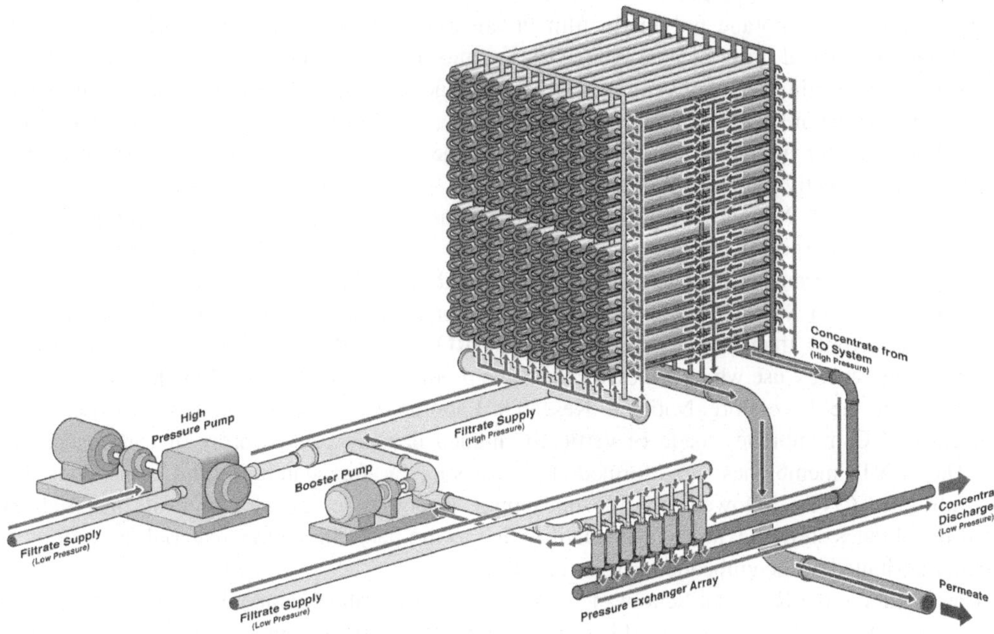

Figure 16.6 Pressure Exchanger Energy Recovery System

16.4 FUTURE DESALINATION TECHNOLOGY ADVANCES

Further improvements of the SWRO membrane technology are forecasted to encompass: development of membranes with higher salt and pathogen rejections, enhanced productivity, reduced trans-membrane pressure, and fouling potential; as well as improvement of RO membrane resistance to oxidants, elevated temperature and pressure; integration of membrane pretreatment within the SWRO membrane system; further advancements in energy recovery technology and SWRO system configuration; integration of brackish and seawater desalination systems; development of methods for low-cost continuous membrane cleaning which would allow to reduce downtime and chemical cleaning costs; and creation of new methods for low-cost membrane concentrate treatment, reuse, and disposal.

A recent trend in the quest for lowering the energy use for SWRO desalination is the development of nanostructured (NST) RO membranes which provide more efficient water transport as compared to existing conventional thin-film membrane elements. The salt separation membranes commonly used in RO desalination membrane elements today are dense semi-permeable polymer films of random structure (matrix) which do not have pores. Water molecules are transported through these membrane films by diffusion and travel on a multi-dimensional curvilinear path within the randomly structured polymer film matrix. This transport is relatively inefficient in terms of membrane film volume/surface area and substantial energy is needed to move water molecules through the RO membranes. A porous membrane structure which facilitates water transport would improve membrane productivity.

NST membranes are RO membranes which contain either individual straight-line nanometer-size channels (tubes/particles) embedded into the random thin-film polymer matrix, or are entirely made of clustered nano-size channels (nanotubes). NST membrane technology has evolved rapidly over the past five years and recently developed nanostructured membranes either incorporate inorganic nanoparticles within the traditional membrane polymeric film or are made of highly-structured porous film which consists of densely packed array of nanotubes. These nanostructured membranes reportedly have much higher specific permeability than conventional RO membranes at practically the same high salt rejection. In addition, nanostructured membranes have comparable or lower fouling rate than conventional thin-film composite RO membranes operating at the same conditions, and they can be designed for enhanced rejection selectivity of specific ions.

For example, a UCLA/NanoH_2O research team has recently developed thin-film nano-composite (TFN) membranes which incorporate zeolite nanoparticles (100 nanometres in diameter) into a traditional polyamide thin membrane film. These new TFN membranes become commercially available for seawater applications in September 2010. The new membrane elements are projected to have 30% to 50% higher productivity than other currently available RO membranes or to operate at approximately 10% to 15% lower energy use while achieving the same productivity as standard RO elements.

Researchers at the Livermore National Research Laboratory in California have focused on the development of RO membranes made of vertically aligned densely packed array of carbon nanotubes (CNT). The CNT membranes are formed by releasing a free-standing film that consists of parylene-infiltrated CNTs, followed by CNT uncapping on both sides of the composite material. Thus fabricated membranes show flexibility and ductility due to the parylene matrix material, as well as high permeability attributed to the embedded carbon nanotubes.

A research team at the Research School of Biology of the Australian National University has recently developed boron nitride nanotubes embedded in a silicon nitride membrane, which are claimed to have improved ability to reject salts, heavy metals and other contaminants at very high permeability and also to possess cation or anion-selective abilities.

Separation membranes based on carbon nanotubes and zeolite TFN membranes are also under development by the Australian Advanced Membrane Research Technologies for Water Treatment Research Cluster of scientific collaborators under the leadership of the Commonwealth Scientific and Industrial Research Organization. This research team includes the University of Texas, Austin.

An on-going research at the University of Colorado at Boulder also focuses on the development of novel polymer membrane with sub-one nanometer pores specifically for desalination applications. The TFN technology is one of the most promising means of making a quantum leap in RO membrane productivity increase and desalination cost reduction.

While the quest to lower energy use continues, as indicated in Section 16.2, there are physical limitations to how low the energy demand could go using RO desalination. The main limiting factors are the osmotic pressure that would need to be overcome to separate the salts from the seawater and the amount of water that could be recovered from seawater before the membrane separation process is hindered by salt scaling on the membrane surface and the service systems. Therefore, research aimed at further reduction of energy use for seawater desalination also focuses on the development of non-reverse osmosis based desalination technologies such as: forward osmosis, membrane distillation and capacitive deionization.

16.5 SUMMARY AND CONCLUSIONS

Over the past decade, seawater desalination has experienced an accelerated growth driven by advances in membrane technology and material science. While conventional technologies, such as sedimentation and filtration, have seen modest advancement since their initial use for potable water treatment several centuries ago, new more efficient seawater desalination membranes and membrane technologies, and equipment improvements are released every several years.

Recent technological advancements, such as pressure-exchanger based energy recovery systems, higher efficiency RO membrane elements, nanostructured RO membranes, and innovative membrane vessel configurations, are projected to further decrease the energy needed for seawater desalination. The steady trend of reduction of desalinated water production energy and costs coupled with increasing costs of conventional water treatment and water reuse driven by more stringent regulatory requirements, are expected to accelerate the current trend of reliance on the ocean as an attractive and competitive water source. This trend is forecasted to continue in the future and to further establish ocean water desalination as a reliable drought-proof alternative for many coastal communities worldwide.

REFERENCES

Bergman R. A. and Lozier J. C. (2010). Large-diameter membrane elements and their increasing global use. *IDA Journal, First Quarter*, **2**(1), 16–23.

IDA (2011). IDA Desalination Yearbook 2010–2011. Media Analytics Ltd., Oxford, UK.

MacHarg J., Seacord T. F. and Sessions B. (2008). ADC baseline tests reveal trends in membrane performance. *Desalination and Water Reuse*, **18**, 2.

Chapter 17

Water reuse versus seawater desalination – evaluation of the economic and environmental viability

Süleyman Yüce, Christian Kazner, Rita Hochstrat, Thomas Wintgens and Thomas Melin

17.1 INTRODUCTION

During the last decade, water shortage and water scarcity led to a change in urban water supply in many coastal areas and particularly in densely populated regions. Alternative water sources such as desalinated seawater and reclaimed water have been introduced to back-up water supply, requiring significant changes of the traditional infrastructure (Reiter, 2009). The general water cycle scheme including alternative sources is depicted in Figure 17.1. Seawater desalination as well as water reuse increase the amount of water available and counteract water scarcity.

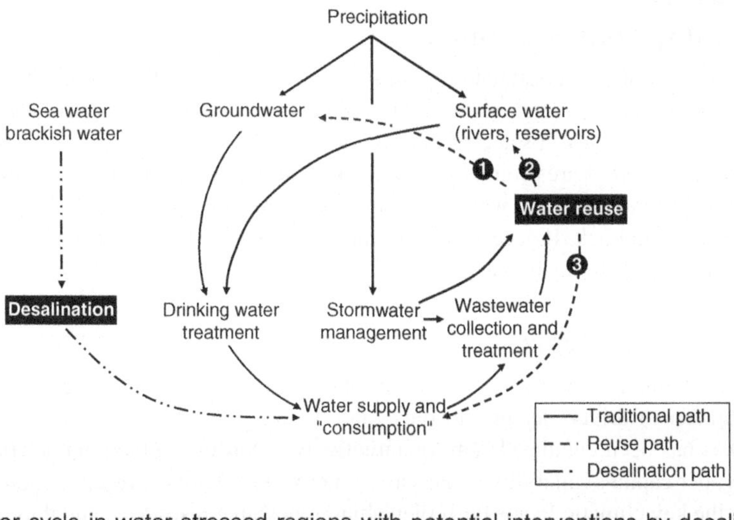

Figure 17.1 Water cycle in water stressed regions with potential interventions by desalination and water reuse (1 = Managed aquifer recharge (MAR); 2 = indirect potable reuse (IPR) via dams and rivers; 3 = direct reuse for drinking water supply, non potable urban uses, industry or irrigation)

Alternative water sources must generally be treated and conditioned more intensively than conventional groundwater and surface water and therefore often exhibit a higher product water quality in terms of chemical and microbiological composition. Yet public perception is that desalinated and recycled water is more artificial and not pristine, overcoming such problems can be very challenging. The examples of Singapore and the Western Corridor Recycling Scheme in Brisbane, Australia are the two extremes of this issue. While NEWater was successfully placed as a brand which nowadays even signifies the whole sustainable water management approach in Singapore, Brisbane water recycling still suffers from a lack of public acceptance and the investment of 2.5 billion AU$ cannot yet fully be utilized since it was decided to convey recycled water to Wivenhoe Dam only if dam levels drop to below 40% (Khan, 2009). In an extensive study on Australian consumers' perceptions towards recycled and desalinated water, Dolnicar and Schäfer (2009) observed some public resistance even after several years of severe drought and massive water restrictions. The results demonstrated a strong resistance against recycled water for close-to-body uses (about 80% of 1000 respondents) and a little less resistance against desalinated water (40–70% depending on the type of use).

Traditionally investments for water infrastructure are made on a long-term basis with depreciation periods of more than 50 years. In order to successfully promote water reuse and desalination, a number of key factors need to be identified and evaluated. Besides criteria such as water quality, risk reduction and safety, the evaluation should consider factors such as energy cost, flexibility of the infrastructure and its expandability, potential impacts of changing consumption patterns, consumer and end-user acceptance, as well as environmental impacts, in particular carbon footprint and waste disposal and valorisation.

Based on a number of international case studies, exhibiting a high water scarcity index and having reached already a high degree of adaptation, this chapter shows under which framework conditions alternative water sources are utilised and which economic and ecological impacts are associated to these water management measures.

17.2 METHODOLOGY

17.2.1 Case study characteristics

Decisions to build water reuse or desalination plants are case-specific to the geographic, socio-economic, and climatic conditions. Table 17.1 gives an overview on the system characteristics and drivers for the usage of alternative water sources in selected international case studies.

The selected four case studies are prominent examples of advanced water reuse: agricultural applications in Israel, indirect potable and industrial water reuse in Singapore and Brisbane, and even more challenging direct potable reuse in Windhoek. If accessible, seawater desalination has also been introduced to back up the freshwater supply from natural sources.

17.2.2 Cross-case analysis

Motivation to use alternative water resources is often related to water scarcity. To illustrate the multi-factorial character of drivers leading to reuse and desalination, the water management situation of selected case studies has been evaluated semi-quantitatively according to the parameters of precipitation, population density, per capita availability, water stress index, and share of irrigation demand.

Precipitation is the key climatic factor for replenishing renewable resources, and thus, determines water availability. In combination with the evaporation, it is also a decisive factor of whether rain-fed agriculture is possible or whether crops can only be grown with additional irrigation.

Table 17.1 System characteristics and drivers for desalination and reuse in selected case study regions (based on information from Aharoni, 2007; Arlosoroff, 2007; Garb, 2010; GWI, 2009; Khan, 2009; Menge, 2009; QWC QLD, 2010).

	SW Desalination	Water reuse
Israel	• Access to additional freshwater (about 13% of total water demand) • Better water quality in freshwater and effluents • Control of salinity in irrigation water from reuse • Final capacity designed to provide two thirds of Israel's freshwater demand • Natural gas from offshore gas fields in combination with power from high voltage grid	• Important alternative source (about 20% of water demand) • High demand in agricultural irrigation • Low costs for reused water – Shafdan: 0.02 US$/m^3 • Very high degree of reuse (>75% of wastewater) due to insufficient and impaired freshwater sources
Singapore	• Access to additional freshwater (about 10% of total water demand) • Economic production with low specific costs 0.57 US$/m^3 • Increasing independence from water imports from Malaysia	• NEWater as key water source (about 30% of total water demand) • Main use in industry: wafer fabrication and power supply • Additional use for cooling and landscaping • Small percentage for indirect potable use
Brisbane and Queensland	• Access to additional freshwater for Gold Coast cities (about 27% of water supply)	• Western Corridor: Water Recycling Project, largest water recycling scheme in Australia • High-quality reuse mainly for power supply and environmental use • In drought situation blending of recycled water with dam water for IPR • Pimpama Coomera Waterfuture: non potable urban use via dual reticulation
Windhoek	• Not feasible due to elevation (1600 m asl) and distance to sea (300 km)	• Direct potable reuse for about 30% of freshwater demand • Reuse of water from industry via MBR • MAR as buffer (pre-treatment with GAC) • No effluent desalination (rising salinity)

Population density is a parameter reflecting the effects of human activities on both water quality and water demand. High population densities are associated with urbanisation and often imply intensified water use within a catchment. The per capita water availability is associated to population density and may define limits of growth in a region. The water availability of 1000 m^3 per capita and year (m^3/cap · yr) is a widely used threshold indicating potential exposure to water resources stress, whilst availabilities below 500 m^3/cap · yr are representing water scarcity (Bouwer, 2000). The water stress index describes the ratio of water abstraction and average water availability, thus expressing the overall pressure put on the water resources.

For the cases presented here, the indicator considers the assured yield of the water infrastructure. Among the water uses, irrigational demand is characterised by a high consumptive character caused by evapotranspiration and incorporation into the product. Unlike water for municipal and industrial applications, irrigation water is neither immediately nor fully returned to the water cycle. However, depending on the type of irrigation, there is a certain share percolating to the groundwater or flowing into surface waters. Literature data have been used to illustrate the situation in spider diagrams.

17.2.3 Water balancing and water system modelling

Water balancing and system modelling are key steps in evaluation of the boundary conditions for the use of alternative water sources. For water stressed regions, water reuse and desalination are the most important methods to improve the water supply in terms of water quantity and quality. To identify the water management challenges, the available renewable water sources as well as the amount of supplied water were analysed and compared to the amount of desalinated water used in the specific case. The state of water reuse was evaluated as percentage of treated effluent. Figure 17.1 shows the interdependency of the single components.

Decision support systems (DSS, e.g. AQUAREC WTRNet) can help in unbiased and neutral resources planning and are often urgently required due to the high system complexity. The DSS provides an integrated framework for optimisation of treatment and distribution aspects of water reuse and the selection of end-users, and has been used in the development of system design principles.

17.2.4 Cost estimations for selected technological options

The cost of desalination and water reuse were presented in fictitious cases to illustrate the dependencies and economic impact for certain system decisions. Although it is not possible to take every peculiarity of all the cases of interest into account, in the following, an attempt was made of a standard cost calculation for three different desalination technologies, three typical treatment trains for water reuse and the impact of long distance water transfers often required in water stressed inland locations, such as Windhoek.

Using the same values for the main parameter, such as energy price or interest rate on the basis of world average price rates, we intend to increase the comparability of our results for different areas and for different cases.

The standard parameters used in the cost calculations for desalination options are summarised in Table 17.2. The pipeline capital costs were estimated on the basis of the material costs in combination with the cost breakdown from R.W. Beck as given in McKinney (2005).

Table 17.2 Standard values for common parameters for desalination technologies.

Parameter	Unit	Value
Specific cost for electricity	US$/kWh	0.060
Specific cost for thermal energy	US$/kWh	0.025
Annual interest rate	per year	6%
Depreciation period for plant	Years	25
Depreciation period for pipeline	Years	40
Labour costs	US$/cap·yr	20,000
Plant availability	h/yr	8,000
Salt content of sea water	ppm	35,000

The costs were estimated using proven design tools: the AQUAREC WTRNet (Joksimovic *et al.* 2008) for the reuse schemes and the design software of S.T.E.P. Consulting (S.T.E.P., 2010) for the conceptual design of the desalination plants (Gebel & Yüce, 2008). For the determination of the capital costs, the linear depreciation method was used. For the depreciation of all desalination plant investments (MSF = multi stage flash distillation, MED = multi-effect distillation, RO = reverse osmosis), a depreciation period of 25 years has been chosen, except depreciation of the RO modules, which was assumed as 5 years. Besides energy, capital cost, labour cost, chemicals, maintenance and repair cost are the main cost items.

17.3 ANALYSIS OF DIFFERENT WATER MANAGEMENT SCENARIOS

17.3.1 Cross-case analysis of selected case studies

The characteristic patterns of the spider diagrams (Figure 17.2) reflect the extent of typical pressures that eventually may drive water reuse and desalination. The corresponding data is given in Table 17.3.

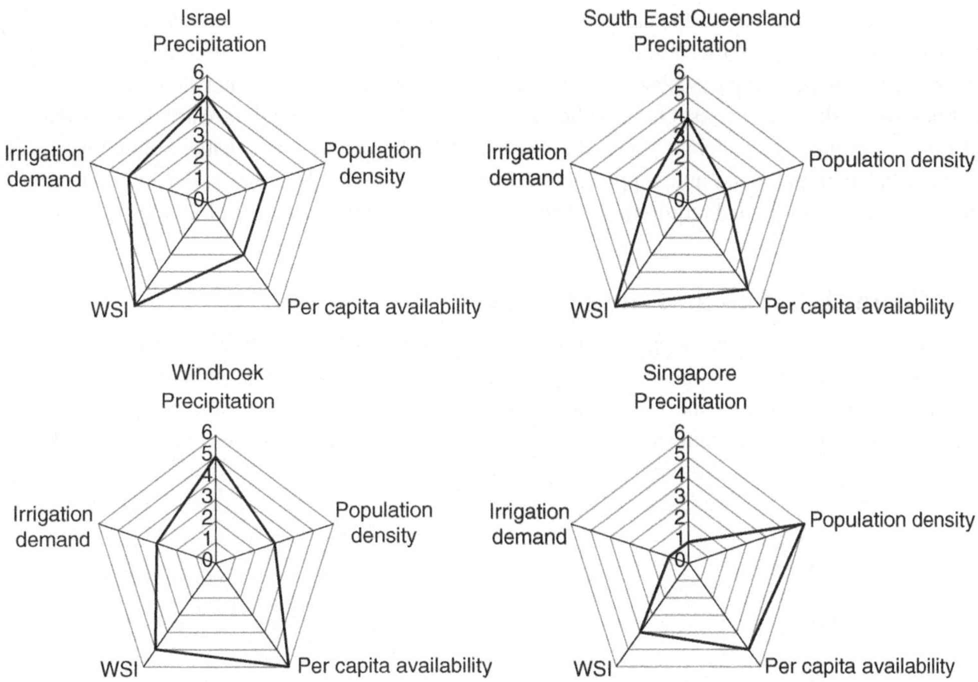

Precipitation, mm/yr	Population density, inhabitants/km^2	Per capita availability, m^3/cap.yr	WSI, %	Irrigation water, % of total demand
1: >2000	1: <100	1: >2000	1: <20	1: <10
2: 1001 - 2000	2: 101 - 250	2: 1001 - 2000	2: 20 - 40	2: 10 - 30
3: 701 - 1000	3: 251 - 500	3: 501 - 1000	3: 41 - 60	3: 31 - 50
4: 501 - 700	4: 501 - 1000	4: 251 - 500	4: 61 - 80	4: 51 - 60
5: 301 - 500	5: 1001 - 2000	5: 101 - 250	5: 81 - 100	5: 61 - 80
6: <300	6: >2000	6: <100	6: >100	6: >80

Figure 17.2 Spider diagrams for selected case studies

Table 17.3 Water management parameter of selected case studies.

Parameter	Unit	Israel	SE Queensland[1]	Windhoek[2]	Singapore
Precipitation	mm/yr	435	685	366	2497
Population density	inhabitants/km^2	3408	117	387	6344
Per capita availability	m^3/cap · yr	106	159	80	130
Water Stress Index[a] (WSI)	%	104	128	90	77
Irrigation demand	% of total demand	50	24[d]	38[b]	3[c]

Data from FAO AQUASTAT (if not otherwise stated);
[1]South East Queensland Water Strategy (2010); [2]Lahnsteiner & Lempert, 2005
[a]Water Stress Index: calculated as ratio of water availability and water abstraction
[b]consumptive water uses; [c]agricultural uses; [d]rural uses

The spider diagrams also point out potential beneficial uses of newly developed resources. The higher the score for a parameter is, the higher its potential to contribute to water shortage, be it by low precipitation, or through its water consumptive character. All regions exhibit a high Water Stress Index resulting from combinations of various factors. Windhoek and Israel have dry climates and limited water capture options that cause very low per capita availabilities. For Israel, the high irrigation water demand brings additional competition with other water uses. On the other hand, Singapore is rain-rich but restricted in storage capacity which poses a particular challenge given the extremely high population density. South East Queensland problems in balancing demand and supply derive from uncertain yields of the water infrastructure (climate variability as well as ecological requirements for river flows) and comparably high domestic demands.

17.3.2 Water balance and major drivers

The cross-case analysis revealed the particular boundary conditions in each of the case studies (see Figure 17.3). All the selected case studies feature a very high water stress ranging from around 80 to 130%. All of the cases already have demand management and public awareness programs implemented, sometimes as a "first-aid" measure. Once the main saving potentials are tapped into, new sources have to be made available.

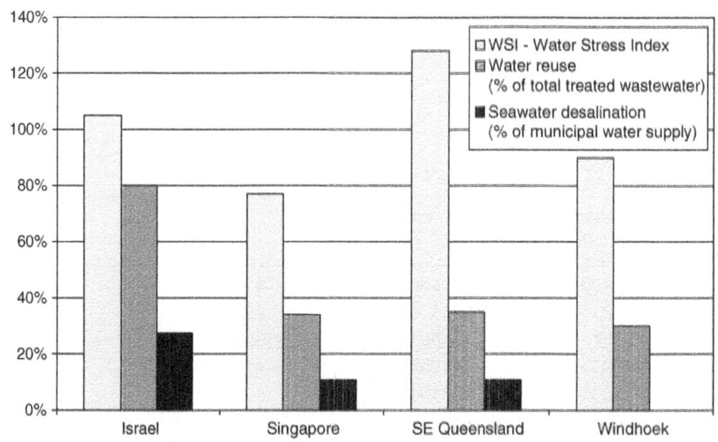

Figure 17.3 Share of alternative sources in water management for selected case studies

The selected cases indicate that severe water stress often requires approaches combining reuse and desalination. If irrigation is the main water reuse practice, the percentage of recycled water allocation is very high and the costs are usually low (Israel). In cases such as Singapore and Southeast Queensland, water reuse competes with seawater desalination. While both employ treatment facilities with the proven but expensive double membrane systems and provide high-quality water at comparable costs (see Figure 17.4), public perception tends to favour seawater desalination (Dolnicar & Schäfer, 2009).

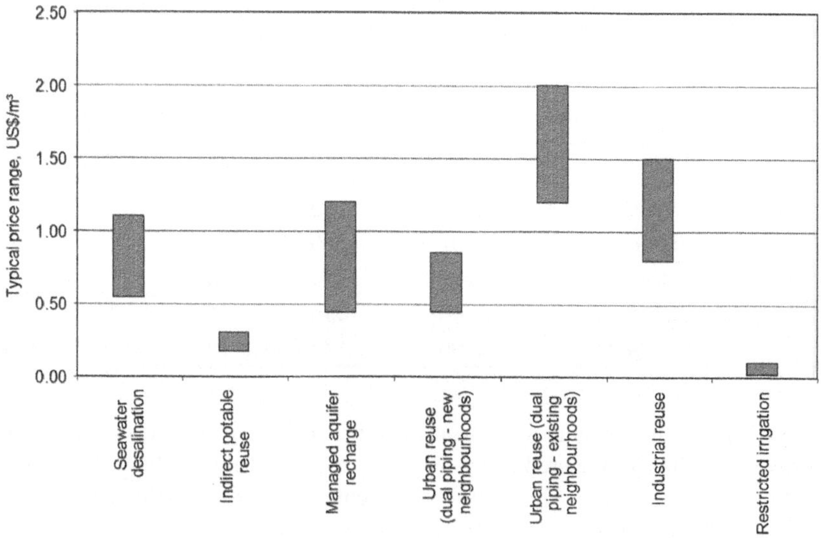

Figure 17.4 Comparison of typical production cost ranges for desalinated and reused water (GWI, 2009)

Alternative sources are regarded to be less impacted by climate change, and therefore, used to increase the water supply safety (Hochstrat *et al.* 2010). Recent droughts in Australia and Israel demonstrated the necessity to rapidly access new freshwater sources such as desalted seawater. The construction of desalination plants boosted in many parts of the world, as it becomes obvious that climate change reduces the traditionally high planning dependability (Khan, 2009). Conventional elements in water resources management (e.g. large dams in Cyprus and Australia) were not able to safeguard water supply during long drought periods.

The Windhoek case illustrates that even under extreme water stress, additional water sources from outside of the catchment, particularly seawater desalination, cannot always be used to fill up the deficits. The specific situation in Windhoek with severe deficits and inaccessible seawater also required more courageous solutions leading to the only full scale direct potable reuse plant, which is successfully in operation since over 40 years (Menge *et al.* 2009).

It can be summarized that the following key challenges drive alternative sources projects, such as water reuse and desalination:

- Insufficient availability of natural sources
- Quality of surface and groundwater (e.g. Israel and Barcelona)
- Water losses: evaporation, leaking pipes, irrigation practices
- Prospects: population dynamics and high population density, climate change impacts, energy sources and competition, carbon footprint, increasing demand for irrigation
- Infrastructure and water management deficits

17.3.3 Economic viability of water reuse and desalination

In recent decades, desalination and water reuse have benefited much from development in membrane technology, such as improved performance of membrane elements and energy recovery systems. RO seawater desalination is nowadays available at relatively low unit prices generally ranging between 0.55–1.10 US$/$m^3$ (GWI, 2009). This trend is also highlighted by the selected case study sites (Table 17.4). Specific annualised costs vary according to the project specific boundary conditions, e.g. energy price levels, raw water qualities, environmental standards for water intake and brine discharge, as well as the size of the plant and additional costs for connecting the desalination plant to the supply system.

Table 17.4 Comparison of selected specific costs of seawater desalination (IDA, 2009).

Desalination plant	Country	Year	Plant capacity	Specific costs US$/$m^3$
Gold Coast	Australia	2007	125,000 m^3/d	1.09
Hadera	Israel	2007	349,000 m^3/d	0.60
Palmachim	Israel	2008	123,400 m^3/d	0.78
Ashkelon	Israel	2008	315,000 m^3/d	0.78
Tuas I	Singapore	2008	136,280 m^3/d	0.57

Modern RO systems produce potable water from seawater with an average energy consumption of 3–4 kWh/m^3 (Fritzmann *et al.* 2007) compared to 0.4–1.0 kWh/m^3 for water recycling. Thus, changes in energy prices directly impact on water costs. For future investments, this poses a particular threat to sustainable water supply. While today the energy requirement for water production is still relatively small compared to other industrial and urban uses, this may change soon with the increase of seawater desalination.

Competing with relatively low unit costs for large scale seawater desalination, not all water reuse options appear to be preferential from an economic point of view (Figure 17.4). In cases where coastal cities still rely on ocean outfalls, the costs for secondary and tertiary wastewater treatment are sometimes attributed to the reuse projects, although it is known that estuaries and coastal oceans are adversely affected by high nutrient loads from urban and agricultural pollution (Boyer and Howarth, 2008; Pearl and Piehler, 2008), and ocean outfall plants would have to be upgraded, sooner or later, anyway. Given the enormous difference in treatment types required between high quality water reuse and irrigation, it is not surprising that more than 60% of water reuse occurs in agricultural and landscape irrigation, as well as environmental enhancement (GWI, 2009).

Figure 17.4 highlights the importance of distribution systems, one of the most neglected aspects in water reuse and desalination. Joksimovic *et al.* (2008) demonstrated in a comprehensive reuse study that the distribution system and piping represent the most significant portion of total scheme lifecycle cost and the variability in total lifecycle cost is a direct result of the distribution system costs. It is thus not surprising that the dual distribution system for existing neighbourhoods entails the highest costs in comparison of potential solutions.

17.3.4 Cost estimates for desalination

The economic viability of seawater desalination is strongly dependent on the availability of a reasonable energy supply at the site of the desalination plant. As a result of this, very large thermal desalination plants (MSF and MED) are mostly coupled with a power plant as so-called "dual-purpose" plant (e.g.

Al Taweelah, Abu Dhabi). In such cases, the desalination system takes advantage of the improved thermodynamics of the total complex by exploiting the "waste heat" from the adjacent power plant. Also in the areas of natural gas production in excess, the thermal energy can favourably be available for seawater desalination. As it is not storable, electrical energy can be inexpensive for RO seawater desalination; otherwise the produced energy cannot be consumed simultaneously or marketed profitably.

In order to demonstrate the impact of energy price on water production costs, the cost calculation performed in this study chose three different price levels. The low, medium, and high price levels as given in Table 17.5 represent the typical range of prices in the regions affected by water scarcity and are therefore relevant for seawater desalination. The low price level should be primarily characteristic for oil-rich regions. The average price level can be considered to represent oil-poor countries in case of an oil price at a "normal" level of around US$50/barrel, whereas the high price level reflects the already increased oil prices of today. In addition to energy costs, the calculation of specific water production costs included all cost items, which are required for the construction and operation of a desalination plant based on a medium-sized plant of 15,000 m^3/d.

Table 17.5 Specific desalinated water costs for different energy prices (plant capacity 15,000 m^3/d).

Process	Unit	Specific desalinated water costs		
		Low	**Average**	**High**
MSF	US$/$m^3$	1.00	1.79	2.66
MED	US$/$m^3$	1.24	2.09	2.93
RO	US$/$m^3$	0.69	0.79	0.89
based on the following specific energy prices				
		Low	**Average**	**High**
Thermal	US$/kWh	0.0125	0.0250	0.0375
Electrical	US$/kWh	0.0300	0.0600	0.0900

Table 17.5 illustrates that the specific costs of water produced in the thermal plants (MSF and MED here with thermal compression) are generally higher than that of a membrane plant. This trend is more pronounced, if the energy prices increases. As long as cheap energy is easily available in integrated water and energy systems, the difference between the thermal processes and RO processes is less important, because at combined heat and power generation plants, the waste heat or steam with a low energy level can be utilized for internal uses. In such cases, with very low energy prices, the thermal processes for seawater desalination are quite competitive with the membrane process. Additionally, due to their robust operation, the thermal processes are even preferred to RO in oil-rich regions.

17.3.5 Influence of the location and the cost of water transport

Seawater is not readily accessible to isolated inland regions. If desalted water produced in a seawater desalination plant should be used there, it has to be lifted from the sea level to the location suffering from water scarcity on a higher altitude. Both, distance and altitude are parameters increasing the required energy and are thus directly related to the additional energy costs for transportation of fresh water to the end user (Figure 17.5). The additional specific energy demand for supplying desalinated seawater, for example to the city of Windhoek, Namibia (distance to the ocean approximately 300 km, altitude ~1,600 m) or the city of Sana'a, Yemen (distance to the Red Sea 150 km, altitude ~2,250 m) are about 12 kWh/m^3. In this case, only the additional costs of energy required for transportation of fresh water would be 0.72 US\$/m^3. Already this cost item is in the same order of magnitude as the specific water desalination costs. The other crucial cost factor related to water transport is the capital costs for the pipeline, pump stations and their peripherals. For the case of Windhoek, the specific capital costs for pipes and pumps to convey the water from a desalination plant to the city amount to 1.08 US\$/m^3 resulting in estimated total costs of 1.80 US\$/m^3 with a very high impact of energy price changes. This illustrates clearly why Windhoek opted for water reuse only.

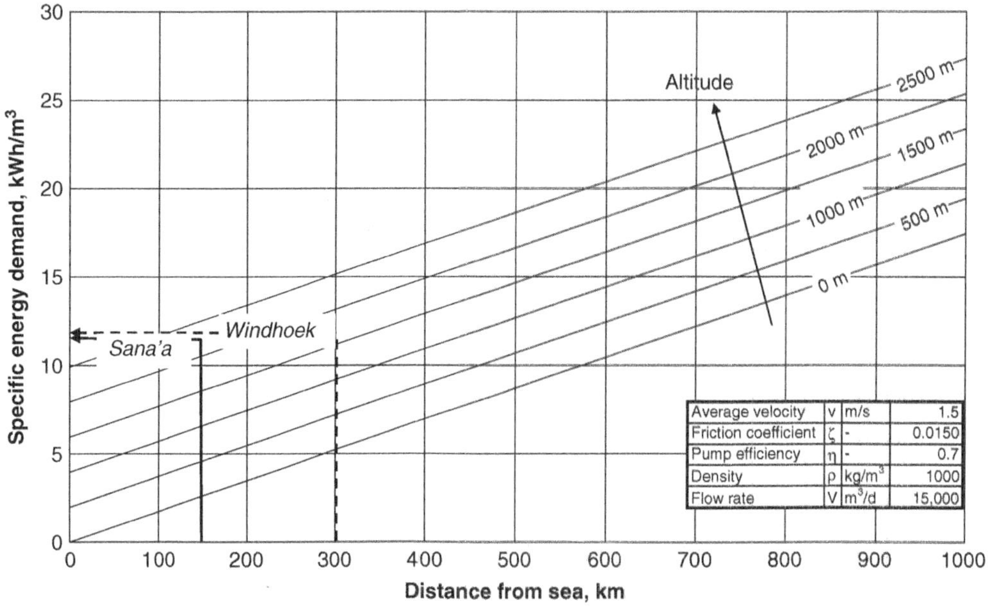

Figure 17.5 Additional specific energy demand for water transfer from the sea to the point of use depending on distance and altitude

17.3.6 Cost and energy demand of water reuse

While the cost for distribution systems in water reuse schemes are highly case specific and dependent on the geography, some typical treatment schemes are applied in water recycling: the general starting point for water recycling is the general wastewater treatment carried out by means of primary mechanical and secondary conventional activated sludge treatment. This treatment is complemented by flocculation (Floc), filtration (Filt), and disinfection (e.g. by means of chlorine dioxide or UV) mainly for unrestricted irrigation applications. For very high quality purposes, such a managed aquifer recharge, indirect potable reuse, or water recycling for industrial purposes, double membrane processes including e.g. microfiltration (MF) and reverse osmosis (RO) are used (Bixio & Wintgens, 2006; Asano *et al.* 2007).

Within this study, the energy requirements and the total treatment cost of the mentioned treatment trains have been estimated utilizing selected features of the WTRNet decision support tool developed in the AQUAREC project (Joksimovic *et al.* 2008). The computation has been carried out using standard default values for important unit process costing parameters, such as investment cost, interest rates, energy requirements, land use, and disposal entailed in WTRNet. The computation results are presented in Table 17.6 for the different treatment trains, while for one treatment train the comparison between the additional tertiary treatment and the secondary treatment is shown. Three different plant sizes have been considered. The plant size given in the Table 17.6 refers to the product flow. Particularly, when using processes such a reverse osmosis it has to be considered that only a certain percentage (e.g. 70–80%) of the total wastewater flow released by the secondary treatment will end up in the product water.

Table 17.6 Estimates of specific energy demand and total treatment costs for different water reclamation process trains and plant capacities.

	Plant size		
	7,700 m^3/d	**15,800 m^3/d**	**103,000 m^3/d**
Energy consumption	According to model equal for all plant sizes		
Floc/Filt/UV		0.32 kWh/m^3	
Floc/MF/RO/UV		1.45 kWh/m^3	
CAS + Floc/MF/RO/UV		1.96 kWh/m^3	
Full cost (OPEX + CAPEX)			
Floc/Filt/UV	0.24 US$/$m^3$	0.20 US$/$m^3$	0.14 US$/$m^3$
Floc/MF/RO/UV	1.02 US$/$m^3$	0.91 US$/$m^3$	0.73 US$/$m^3$
CAS + Floc/MF/RO/UV	1.74 US$/$m^3$	1.47 US$/$m^3$	1.07 US$/$m^3$

It has to be noted that there is a significant uncertainty entailed in the figures provided in Table 17.6 and lower figures in real systems are certainly feasible. Nevertheless, a rough differentiation can be provided between the cost and energy impacts of different technologies. The additional energy consumption of a double membrane process is certainly lower than for a seawater desalination process due to the lower salt content. But due to the lower pressure range, the possibilities for energy savings by pressure exchangers is limited in water reuse.

The cost estimates made on bases of WTRNet are generally higher than the cost estimate provided by GWI (2009) shown in Figure 17.4, particularly for indirect potable use. Hochstrat *et al.* (2007) pointed out that prices charged for end-users may differ significantly from the full cost of water recycling.

17.3.7 Environmental impact

The high energy consumption of seawater desalination must be regarded as the main drawback from the sustainability perspective. In order to compare the environmental impact of different technologies, CO_2 emissions were chosen as main factor, which is discussed particularly in already climate change impacted countries.

Our estimation of the CO_2 emissions caused by seawater desalination or water reuse is based on a plant size of 15,000 m^3/d. CO_2 emission factors are based on literature as given in Table 17.7. Additionally, for comparison, a reuse plant was introduced with an assumed specific energy consumption of 1.45 kWh/m^3 product water. The CO_2 emissions relate to the emissions during operation. Emissions caused by the plant construction itself are not taken into account. Table 17.7 shows the distinct differences in carbon footprints of the investigated processes. Because of the linear dependence between energy consumption and CO_2 emissions, the high amount of energy demand of thermal plants is reflected in the high CO_2 emissions. RO desalination produces only about two thirds of the emission of thermal processes. A high-quality reuse process based on the double membrane process plus UV treatment has a much smaller carbon footprint with less than half of that of RO desalination.

Table 17.7 Carbon footprint of desalination and water reuse process (plant size: 15,000 m^3/d, emissions during operation).

Process	Unit	CO_2 emission*
Desalination		
MSF	t-CO_2/yr	23,160
MED	t-CO_2/yr	23,640
RO	t-CO_2/yr	14,994
Water reuse		
MF/RO/UV	t-CO_2/yr	7,145

*CO_2 emission factors: 0.3 kg CO_2/L-Fuel Oil for thermal energy, according to FZ Jülich, Germany(www.fz-juelich.de/ptj/lw.../pdp.../100306_CO2-Emissionsfaktoren.xls)

0.9 kg CO_2/kWh for electrical energy, according to German EPA (UBA: CO_2 factors for power production)

Besides energy consumption, the main environmental impact of seawater and effluent desalination is attributed to the production of highly saline waste streams (brines). In coastal locations, water recycling and desalination plants discharge to the ocean. In inland locations, the management of membrane waste streams from water reclamation facilities remains a major limitation to a broader spread of membrane based high quality water reuse (Khan *et al.* 2009). Brines from seawater desalination may have also adverse effects on the marine environment. Khan (2009) reports the brine discharge assessment study conducted in Perth, Australia, in particular on the possibility of reduced seawater oxygen levels. Modelling alongside extensive real-time monitoring was employed to evaluate and minimize the environmental impact. There are numerous other smaller impacts such as noise (e.g. from pressure exchanger in RO desalination) and the discharge of chemicals (e.g. antiscalants and disinfectants), which require detailed investigations depending the local conditions.

17.4 CONCLUSIONS

Future development of high quality water reuse should avoid over-engineered treatment concepts and must rely on low energy consuming, robust technologies which would allow the implementation of water recycling also in developing economies.

In general, innovation in water production and use has to clearly prioritise potential adaptation measures. Increasing the efficiency in supply and demand is still the prioritised option to alleviate water scarcity, though pipe repair can be too expensive and complex for rapid solutions. Also cascading use of water can often be easily implemented. Next to this, non-conventional water sources come into play with water reuse allowing to produce water fit for the respective purpose and seawater desalination enabling quick and reliable access to additional freshwater. Synergies between desalination and water reuse are also to be strengthened as there should be an incentive to keep desalted water in the water use loop as long as possible, minimising "freshwater" discharges to the sea.

Another synergy can be seen in Israel, where the rising salt content of reclaimed water, which endangers agricultural reuse, is expected to drop, as desalinated water, low in minerals, blended with conventional freshwater sources reduces the average salinity of irrigation water (Tal, 2006). However, the quality of the desalinated water has to be adapted to the requirements of agricultural use along with appropriate post-treatment (Yermiyahu *et al.* 2007).

If desalination and reuse compete as options for additional water supply of urban areas, e.g. in Australia and California, higher quality reuse often entails expensive dual distribution systems. In addition, the viability of water reuse depends on the availability of dedicated reservoirs or groundwater aquifers for storage and distribution, as well as on public acceptance.

More specifically, the following conclusions could be drawn:

- Water reuse and desalination are complementary measures in integrated water management.
- They should address issues in different sectors, e.g. freshwater coming from seawater desalination is not preferential for direct irrigation purposes due to high carbon footprint and high costs but it can be utilized in irrigation as reused water after high-quality usage and subsequent reclamation.
- The actual choice of the alternative water source depends a lot on geographic locations, in particular in coastal versus non-coastal areas.
- Water reuse can better utilize natural attenuation and buffering capacity offered in natural systems and the water cycle.
- In practice, preferences will depend significantly on energy cost, socio-cultural aspects, environmental legislation and the guidance framework for water reuse.
- Compared to state-of-the-art seawater desalination with reverse osmosis, high-quality reuse has no clear economic benefits, particularly in coastal locations. Water reuse, however, entails significantly lower energy consumption and is therefore not only advantageous from the carbon footprint perspective, but will also produce water at a more stable cost than desalination when energy prices increase.

REFERENCES

Aharoni A. (2007). SAT as practiced historically in Israel: principles, operation and problems. SWITCH Learning Alliance Treatment Workshop: Soil Aquifer Treatment (SAT), Tel Aviv, Nov 28–29, 2007.

Arlosoroff S. (2007). Wastewater reuse – risk assessment, decision-making and environmental security. In: *Proceedings of the NATO Advanced Research Workshop*, M. K. Zaidi (ed.), Springer, Istanbul, Turkey, 12–16 October 2006, pp. 55–64.

Asano T., Burton F. L., Leverenz H. L., Tsuchihashi R. and Tchobanoglous G. (2007). Water Reuse – Issues, Technologies, and Applications. McGraw-Hill, NY.

Bixio D. and Wintgens T. (eds) (2006). Water reuse system management manual – AQUAREC, Project report, Office for Official Publications of the European Communities, Luxembourg.

Bouwer H. (2000). Integrated water management: emerging issues and challenges. *Agricultural Water Management*, **45**, 217–228.

Boyer E. W. and Howarth R. W. (2008). Nitrogen Fluxes from Rivers to the Coastal Oceans. In: Nitrogen in the Marine Environment, 2nd edn, D.G. Capone, D. A. Bronk, M. Mulholland and E. J. Carpenter (eds), Elsevier, Amsterdam, NL, pp. 1565–1587.

Dolnicar S. and Schäfer A. I. (2009). Desalinated versus recycled water: public perceptions and profiles of the accepters. *Journal of Environmental Management*, **90**, 888–900.

Fritzmann C., Löwenberg J., Wintgens T. and Melin T. (2007). State-of-the-Art of reverse osmosis desalination. *Desalination*, **216**, 1–76.

Garb Y. (2010). Desalination in Israel: status, prospects, and contexts. In: Water Wisdom – Preparing the Groundwork for Cooperative and Sustainable Water Management in the Middle East, A. Tal and A. A. Rabbo (eds.), Rutgers Univ Press, Piscataway, NJ, USA, pp. 238–245.

Gebel J. and Yüce S. (2008). An Engineer's Guide to Desalination. VGB Power Tech Service GmbH, Essen, Germany.

GWI (Global Water Intelligence) (2009). Municipal Water Reuse Markets 2010. Media Analytics Ltd., Oxford, UK.

Hochstrat R., Joksimovic D., Wintgens T., Melin T. and Savic D. (2007). Economic considerations and decision support tool for wastewater reuse scheme planning. *Water Science and Technology*, **56**(5), 175–182.

Hochstrat R., Wintgens T., Kazner C., Melin T. and Gebel J. (2010). Option of water scarcity and drought management – the role of desalination. *Desalination and Water Treatment*, **18**, 96–102.

Joksimovic D., Savic D. A., Walters G. A., Bixio D., Katsoufidou K. and Yiantsios S. G. (2008). Development and validation of system design principles for water reuse systems. *Desalination*, **218**(1–3), 142–153.

Khan S. J. (2009). Recycled water and desalinated seawater replenish and replace drinking water in Australia. In: TECHNEAU: Safe Drinking Water from Source to Tap – State of the Art & Perspectives, T. Van den Hoven and C. Kazner (eds.), IWA Publishing, London, UK, pp. 445–455.

Khan S. J., Murchland D., Rhodes M. and Waite T. D. (2009). Management of concentrated waste streams from high-pressure membrane water treatment systems. *Critical Reviews in Environmental Science and Technology*, **39**(5), 367–415.

Lahnsteiner J. and Lempert G. (2007). Water management in Windhoek, Namibia. *Water Science and Technology*, **55** (1–2), 441–448.

McKinney D. C. (2005). Estimation of pipeline costs. http://www.ce.utexas.edu/prof/mckinney/ce311k/Proj-05/Cepm2.pdf (accessed on 12 August 2010)

Menge J., du Pisani P. L., König E. and Theron-Beukes T. (2009). Drinking water safety in Windhoek, Namibia: routine monitoring, trace organics, pathogenic indicators and salinity – comparing different water sources. In: TECHNEAU: Safe Drinking Water from Source to Tap – State of the Art & Perspectives, T. Van den Hoven and C. Kazner (eds.), IWA Publishing, London, UK, pp. 457–468.

Paerl H. W. and Piehler M. F. (2008). Nitrogen and Marine Eutrophication. In: Nitrogen in the Marine Environment, 2nd edn, D. G. Capone, D. A. Bronk, M. Mulholland and E. J. Carpenter (eds.), Elsevier, Amsterdam, NL, pp. 529–567.

Queensland Water Commission – The State of Queensland (2010). South East Queensland Water Strategy, Queensland Water Commission, ISSN 1836-5051, http://www.qwc.qld.gov.au/security/pdf/seqws-full.pdf (accessed on 22 February 2012)

Reiter P. (2009). Global Water Challenges and Opportunities: Inaction is Not an Option. Keynote lecture, IWA TECHNEAU Conference, Maastricht, NL, June 17, 2009.

S.T.E.P. (2010). Unpublished Design Software Package; S.T.E.P. Rautenbach Yüce Gebel Consulting GmbH, Aachen, Germany, http://www.stepconsulting.de/ (accessed on 22 February 2012)

Tal A. (2006). Seeking sustainability: Israel's evolving water management strategy. *Science*, **313**, 1081–1084.

Yermiyahu U., Tal A., Ben-Gal A., Bar-Tal A., Tarchitzky J. and Lahav O. (2007). Rethinking desalinated water quality and agriculture. *Science*, **318**, 920–921.

Chapter 18

Desalination *vs* water reuse: An energy analysis illustrated by case studies in Los Angeles and London

Graeme K. Pearce

18.1 INTRODUCTION

The energy cost of treating freshwater supplies from groundwater or surface water sources is relatively low. This is particularly true for traditional treatment processes such as slow sand filtration. As process complexity has increased with the introduction of coagulation or flotation processes combined with high rate media filtration, there has been a marginal increase in energy cost, but it still remains a very small proportion of the total water cost at the point of use.

When membrane filtration processes were introduced at a significant scale in municipal water applications in the late 1990's, the energy component of water cost was barely affected. In fact in some case, energy costs fell slightly since membranes sometimes enable a simplification of the overall process flow scheme. Membrane processes are now widely used in water and wastewater treatment, both for traditional freshwater sources and for the development of alternative resources from desalination and water reuse. Indeed membranes are generally accepted as a mandatory stage in water reuse. This is not only due to the essential barrier characteristic of the membrane process, which provides an assurance of treated water quality, but also due to the operational stability achieved.

Membranes are also becoming accepted as a pre-treatment stage in seawater desalination. Since 2006, membrane filtration processes have been used in a steadily increasing proportion of projects. Although originally thought to be an expensive option, membrane pre-treatment is now seen to offer competitive total water cost and improved operational stability.

The energy use of membrane pre-treatment in water reuse and desalination is comparable to conventional alternatives. Indeed, reduced reliance on chemical dosing and less frequent RO cleaning often leads to lower energy usage. However, the energy cost for the treatment of these alternative resources is significantly greater than that of traditional freshwater sources. Of course sometimes freshwater is not available, so there is no option but to turn to an alternative resource. Also, when the energy cost of transport and distribution is considered, alternative resources can provide the most economic option.

This section reviews the energy cost of treatment for desalination and water reuse, and examines the energy cost of transfer and distribution for two case studies that have recently been implemented. The first of these is at Orange County near Los Angeles in California. This scheme has utilised water reuse, and reduces reliance on long distance transfer from Northern California. The second scheme is at Beckton, which uses a brackish water source from the River Thames estuary in an area east of London where extensive development is planned.

To complete the comparison, reference will also be made to the seawater desalination scheme at Carlsbad near San Diego in southern California. This project represents a more extreme case than Beckton since the ocean water salinity is much higher. However, the alternative is a very long distance transfer of freshwater. In a case like this, security of supply becomes an important factor in addition to the sustainability argument of current practice, since development projections would make transfer unfeasible as a long term water supply solution for the region.

18.2 BACKGROUND

Membranes have been used in water and wastewater applications since the 1960's. Initially membrane processes were felt to be too expensive for widespread use in this field, and were only applied in niche applications or special circumstances. This changed during the 1990's due to the emergence of several drivers, including legislation to achieve improved treatment standards in drinking water and resource scarcity. The need to develop alternative supplies led to the use of membranes on saline or wastewater sources, and these applications benefitted from the earlier development of the drinking water market. The rapid uptake of membranes since 2000 has led to a dramatic fall in costs, to the extent that membranes now often compete with conventional processes, while achieving much better quality standards.

From the early 1980's to 2005, desalination was considered to be the default technology for alternative resource development despite the high energy cost associated with overcoming the osmotic pressure, especially from highly saline sources. However, water reuse emerged in the early 2000's as a feasible low energy alternative, due to improvements in cost and performance of both membrane filtration and reverse osmosis (RO) technology, combined with a gradual development of public acceptance.

Water reuse is now the fastest growing and potentially most promising membrane application in the water industry. In fact, membranes have been an enabling technology for this application, since without membranes, wastewater treatment to a standard guaranteeing quality and security would be too difficult and costly. RO is required to ensure that dissolved organics, and sometimes salts, are removed to a sufficient extent. Membrane filtration is needed as a pre-treatment to the RO and to provide a physical disinfection barrier. Without membrane pre-treatment, stable operation of the RO has been found to be unachievable, so the advent of the membrane filtration boom for drinking water in the late 1990's enabled the commercial realization of this application.

18.3 TECHNOLOGY

There are two types of membrane filtration technology for water and wastewater treatment, namely ultrafiltration (UF) and microfiltration (MF). UF has pores of 0.01–0.02 μm, while MF for water treatment has pores of 0.04–0.10 μm. In wastewater applications, coarser MF pore sizes of 0.2 and 0.4 μm can be used, but the finer MF membranes for water duties are also suitable. MF removes common particles found in water including bacteria and other microbial organisms, while UF removes viruses in addition, thereby providing a physical disinfection barrier.

UF and MF membranes are most commonly applied in the water and wastewater sector in the form of hollow fibres. This format has the advantage of reducing the need for pre-treatment since the feed channels can be cleaned more effectively than the spiral wound format commonly used in RO, since the membranes are backwashable (Pearce, 2011). In addition, the standard hollow fibre format utilized in membrane filtration duties is designed to provide a high integrity barrier and has the advantage of being testable and repairable. In contrast, the RO membrane is not backwashable and the spiral element has not been designed as a disinfection barrier. Multi-element systems do not easily allow testing and repair of individual elements.

Hollow fibre format products are available with inside and outside feed channel configurations. In addition, the operational format can be pressure driven, with the fibres contained in a shell, or vacuum driven, with the fibres submerged or immersed in a feed tank. The outside flow configuration has advantages for feed streams with a high solids concentration, since the feed channel is less likely to become clogged. Indeed, in extreme circumstances with very high solids, for example the membrane bioreactor (MBR), the submerged format is preferred. Under these conditions, an inside feed format would require a larger fibre diameter, effectively a tube, and this is also sometimes used in MBR.

It was originally assumed that wastewater would be a difficult feed for membranes to treat, and that inside feed configurations would struggle. However, membrane filtration can operate stably with this feed provided that a moderate design flux is selected. Solids in the wastewater feed tend to form a pre-coat for the membrane giving some degree of protection from adsorptive foulants. If operational pressures are kept low, the pre-coat remains permeable and the membrane pre-treatment yields an excellent feed water quality for the RO stage. For pre-treatment of wastewater prior to RO, membrane filtration is normally used in combination with coagulation to control fouling, improve operational stability, and enhance removals of dissolved organics.

With regard to format selection, the solids concentration is usually not so high that an inside feed configuration would be overwhelmed. Indeed, several of the world's largest water reuse plants use inside feed configurations. It is true that the overall format share favours outside feed configurations in these applications, but the split is not as clear cut as was originally anticipated.

18.4 RESOURCES

Water resources are under pressure from increasing population especially in arid regions, and greater use of water per capita due to economic development. Some attempts have been made to reduce water usage through conservation programmes, but so far this has just had the effect of containing growth.

Not only is demand rising, but traditional resources will become less available in the future. Water supply is currently obtained mainly from freshwater resources, as illustrated by Figure 18.1 (GWI, 2005). Surface water is the most important water resource, but groundwater is also a significant contributor accounting for 36% of total. New resources such as desalination and water reuse accounted for just 0.5% of the total in 2005. Rapid growth of alternative resources probably increased the share to around 1% by 2011 (GWI, 2009), but clearly the contribution is still relatively minor.

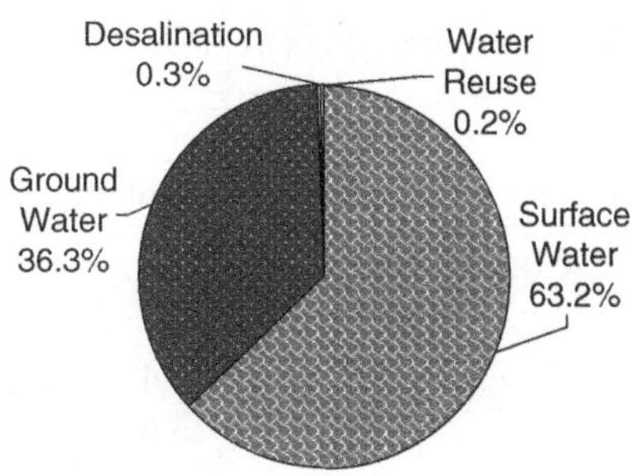

Figure 18.1 Sources for Worldwide Water Abstraction

However, the balance of resource supply will undergo a fundamental change in the coming decades. Groundwater is effectively a non-renewable resource. Millennia are required for groundwater resources to accumulate, but these resources can be depleted in a few years or decades. A recent high profile example of this, the US Oklahoma panhandle, has recently been discussed in the press. Excessive groundwater abstraction will lead to the complete collapse of agriculture in this once productive region within this century.

Combined with the inexorable reduction of groundwater availability, the quality of groundwater is under threat due to a variety of factors, such as saline intrusion near the coast as the water table declines due to over-abstraction. Also groundwaters are becoming contaminated by industrial and agricultural residuals caused by increasing fertiliser and pesticide use since the Second World War. The effect of this activity has taken some decades to become established, but since the early 1990's, treatment processes have been required in the developed world to control contaminant levels from groundwater sources.

More recently, there has been an increasing concern with other manmade organics entering the environment such as personal care products and pharmaceuticals. Collectively these compounds, together with the agricultural residuals, are referred to as Emerging Contaminants (EC). So far legislation has not been systematically introduced to address the EC problem, but the Environmental Protection Agency (EPA) in the USA and the European Union have undertaken comprehensive reviews of the issue and of potential solutions. These contaminants primarily affect surface water sources, and will progressively necessitate additional treatment steps, making freshwater sources more expensive.

In the long term, new sources such as desalination and water reuse will have to replace groundwater, and the beginning of this trend has already started to occur. Growth of alternative resources significantly above the underlying growth rate of water use is therefore assured for the foreseeable future. On top of this, the growth rates for membrane filtration processes will be even greater, as membranes become established as the dominant technical solution. Thus currently desalination and water reuse markets have annual growth rates of 8 and 14% respectively, but the use of membrane filtration in these applications is growing at ≥17% (GWI, 2005, 2007).

Currently, 10% of water abstracted is collected as wastewater, and just under half of this is treated prior to disposal (GWI, 2005), as illustrated in Table 18.1. However, only a small fraction of treatment is to a reuse standard. Wastewater is therefore available as an abundant resource for reuse. Most reuse at the moment is destined for industry and agriculture, with just 11% for municipal use. However, municipal applications are undergoing dramatic growth, which will be illustrated by the Orange County case study.

Table 18.1 Wastewater flows as a function of treatment standard.

Wastewater (WW) flows	Flow rate	
	Mm^3/d (million m^3/d)	MLD
WW collected	1,011	1,011,000
WW treated	438	438,000
WW reuse	19.45	19,450
WW with tertiary treatment	5.5	5,500
WW treated by RO/NF	4.0	4,000

18.5 ENERGY USE IN TREATMENT

Unfortunately, alternative water sources require much more energy for treatment than traditional freshwater resources, and this will become a major consideration for the future. In particular, desalination is very energy intensive, and this provides an important motivation for developing water reuse as the default option where resource constraints occur, leaving desalination for situations where there is no other choice. Energy efficiency in water supply is now recognised as a major driver for the selection of alternative resources.

The energy usage for the various water and wastewater sources is summarized in Table 18.2 (Pearce, 2008). The range of energy usage shown in the table for surface water treatment covers treatment by clarification or flotation followed by rapid gravity sand filters or UF/MF membranes. For wastewater schemes, the conventional activated sludge (CAS) process is required for river discharge. In most cases, this level of treatment is suitable for direct supply to a UF-RO or MF-RO system without further treatment.

Table 18.2 Energy usage for various water and wastewater sources.

Feed	Activated sludge	Pre-treatment	RO system	Total treatment
	kWh/m^3			
Surface water				0.15–0.3
Wastewater	0.3–0.6	0.1–0.2	0.4–0.5	0.8–1.0 (1.3)
Wastewater MBR		0.8–1.0	0.4–0.5	1.2–1.5
Brackish (930–2200 ppm)		0.1–0.3	0.6–0.9	0.8–1.0
Brackish – Tidal Estuary		0.29	1.38	1.67
Seawater (medium salinity)		0.3–1.0	2.0–3.0	2.3–4.0

The energy costs of freshwater treatment using coagulation and filtration are very low, typically 0.15–0.3 kWh/m^3, with desalination costs at least an order of magnitude greater. Brackish water and water reuse lie in the middle of this spectrum, typically at about 0.8–1.0 kWh/m^3. If nanofiltration is used for emerging contaminants, it will use energy just below that for brackish water, whereas novel wastewater treatment processes such as membrane bioreactors have energy levels somewhat above the brackish water range.

Saline sources are energy intensive, since significant power is required to overcome the osmotic pressure in treatment. Estuary sources, such as the Thames Water project at Beckton in the East End of London, provide a lower energy cost option. However, RO in water reuse operates at lower pressures, since the osmotic pressure is low, and these sources are normally the best sustainable option for new resource development.

The data for MBR has been based on practical field experience. The water reuse figure assumes CAS has to be provided in addition to the membrane treatment stage. However, CAS treatment is normally required for discharge irrespective of the reuse scheme. Membrane filtration and RO is added to the CAS to provide reuse product quality. It should be noted that an MBR scheme is less flexible than CAS followed by membrane filtration. The MBR needs to be operated continuously. In contrast, it would be possible to operate the CAS at a higher flow than the membrane filtration stage, only treating the full flow to a membrane standard when the reuse demand existed.

It should be noted that membrane filtration and RO energy costs are relatively low in water reuse applications since it has been found that these applications need to be operated at low fluxes to avoid fouling. This creates low energy cost, but the capital expenditure (capex) of the membrane system cost, and operating expenditure (opex) of the membrane replacement costs are higher than for similar salinity brackish water treatment.

18.6 CASE STUDY: ORANGE COUNTY, CALIFORNIA, USA

Water reuse projects have been implemented at large scale in several areas of the world during the first decade of the 21st century. The highest profile plants are those in Singapore, where six large reuse plants have now been installed. The most recent large scale projects are those in Queensland, Australia, and Qatar and Kuwait in the Middle East. The reuse case study featured in this section will review the plant in Orange County California (Wilf, 2010), and discussed in further detail in Chapter 12.

The Orange County Water District (OCWD) has been considering water reuse schemes since the early 1970's, and has run long term demonstration scale plants based on RO preceded by conventional pre-treatment. The conclusion of these early studies was that reuse was not commercially viable. The advent of membrane pre-treatment inspired a revisiting of this conclusion, and the consequence of the re-assessment was the implementation of the scheme, as set out in Table 18.3.

Table 18.3 Orange County project profile.

Objective	Aquifer recharge (50%) and lake replenishment Industry
Location	Orange County, LA, California, USA
Capacity	265,000 m^3/d (70 mgd) Planned expansion of 61,000 m^3/d (16 mgd) & 182,000 m^3/d (48 mgd)
Process	Secondary effluent feed; MF-RO-UV-H_2O_2
Implementation	Customer: Orange County Water District (OCWD) Scheme name: Ground Water Replenishment (GWR) Membrane Supplier: Simens-Memcor
On-stream date	January 2008

The recovered water is used to recharge the coastal aquifer, and will augment subsequent abstraction for industry, agriculture, and indirect municipal reuse. The plant uses submerged polypropylene (PP) MF membranes from Siemens-Memcor. A summary of the process design for the Orange County plant is provided in Figure 18.2.

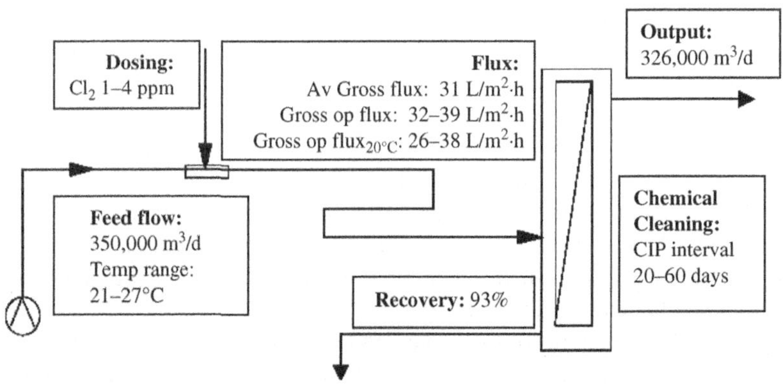

Figure 18.2 Orange County design summary

The feed temperature ranges from 21–27°C and the membrane plant operates with an actual average gross flux of 31 $L/m^2 \cdot h$ (LMH) and a recovery of 93%. A chlorine residual of 1–4 ppm is dosed continuously, converted to chloramines due to the ammonia residual, making it compatible with both MF and RO membranes. The Clean In Place interval is 20–60 days. A photo of partly filled membrane racks used at the plant is shown in Figure 18.3.

Figure 18.3 Siemens-Memcor S10T partly filled membrane racks for Orange County

A summary of the rack layout and membrane arrangement for the Orange County plant is given in Table 18.4. The membrane modules are arranged in racks and cells, with either 2 cells for the smaller 'type 1' train or 8 cells for the larger 'type 2' train. The plant has operated well since start up, and is already being expanded, with a future expansion planned. Stable RO performance has been achieved. The PP membranes have also been used on several of the plants in Singapore, though polyvinylidene difluoride (PVDF) membranes have been used on a similar reuse application in Australia, providing a higher flux, but at a greater module cost.

Table 18.4 Orange County Layout Summary of MF membranes.

System memcor CMFS	Trains	Cells	Racks 32S10T	Modules S10T	Membrane area, m^2
Complete plant	4	26	494	15,808	490,048
per train type 1	–	8	152	4,864	150,784
per train type 2	–	2	38	1,216	37,696
per cell	–	–	19	608	18,848
per rack	–	–	–	32	992
per module	–	–	–	–	31

18.7 CASE STUDY: BECKTON, EAST END OF LONDON, UK

In 2002, Thames Water proposed to improve the security of water supply provision to London by building a desalination plant at Beckton, near the estuary of the River Thames. Thames Water briefly considered water reuse, but rejected this option in the short term as inflexible and of minimal benefit in energy terms. The Mayor of London provided a vigorous challenge to the brackish water scheme on the grounds that the plant was unnecessary, and that desalination is an inappropriate and unsustainable solution for UK water supplies. This challenge eventually failed in 2008, and plant construction re-commenced after a 2½ year delay, and was completed in the early part of 2010.

The Beckton scheme is located in the East End of London, and is designed to provide water for the Thames Water Gateway development and the Olympics. The plant is located on a tidal stretch of the River Thames, quite close to the estuary, as shown in Figure 18.4. A large wastewater plant is already located at Beckton, which discharges downstream of the desalination plant.

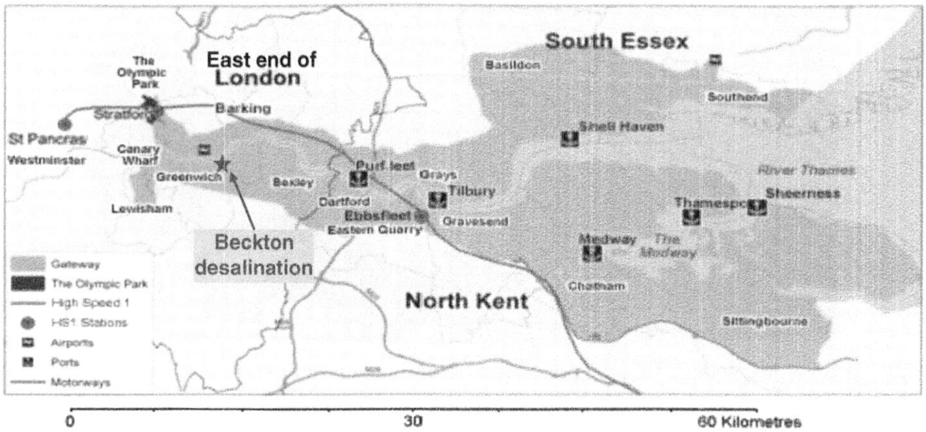

Figure 18.4 Location of the Beckton desalination plant

Plant output is designed to be 140,000 m^3/d, operating for up to 40% of the time during design 'dry' conditions to augment other lower energy sources. In order to minimize average salinity and reduce the diurnal variation, feedwater is abstracted from the river into bankside storage for 3 hours prior to low tide. The salinity range falls between 3,000 and 11,000 ppm Total Dissolved Solids (TDS). The reverse osmosis (RO) system uses a 2 stage process, with concentrate from the high brackish first stage being fed to a second stage seawater RO membrane, in order to achieve an overall recovery of 75%.

The argument between desalination and wastewater reuse was legally tested at a Planning Inquiry (Livingstone, Apr 2006 and Jun 2006). The case for the two sides can be summarized as follows:

Thames water position

- Water reuse is less flexible than desalination
- Concern with removal/breakdown of pharmaceutical, agricultural residuals
- Concern with public acceptance
- Energy use of a brackish estuary source is not much higher than a water reuse scheme based on a membrane bio-reactor (MBR)

Mayor of London's position

- Energy cost of water reuse would be much lower than desalination
- Thames Water has not developed public acceptance; meanwhile neighbouring Essex and Suffolk have introduced an award winning reuse scheme
- Thames Water has not been energetic/imaginative in developing alternatives

The Planning Inquiry found in favour of Thames Water based on the fact that alternative resources were required and the proposed brackish water scheme was the most practical short term option.

Seawater desalination is energy intensive requiring 2.3–4.0 kWh/m^3, depending on the salinity of the source, while brackish feedwaters normally have a treatment energy cost of around 1 kWh/m^3 (Pearce, 2008). For a variable salinity estuary source the cost is higher than other brackish sources due to the higher salinity, and Thames Water presented Beckton treatment energy costs at the Inquiry as 1.67 kWh/m^3 (Livingstone, 2006a).

Water reuse requires a treatment cost of around 1 kWh/m^3 if based on conventional treatment followed by membrane filtration, rising to a range of 1.2–1.5 kWh/m^3 if based on a membrane bioreactor (Pearce, 2008). The energy cost of water reuse is therefore similar to a brackish feed and somewhat lower than an estuary source.

A further point of interest on the Beckton project is that Thames Water took the opportunity during the delay of switching the pre-treatment choice from conventional technology to ultrafiltration (UF) membranes. UF provides an improved system integrity, eliminating superchlorination, at a competitive total treated water cost.

18.8 ENERGY USE IN TRANSFER AND DISTRIBUTION

Figure 18.5 shows the energy cost of treating a variety of different water sources from Table 18.2, and compares these treatment costs with the cost of transport and distribution (Pankratz, 2007; Pearce, 2008). The arrow at the top of the figure represents the spectrum of the energy costs for treatment. The energy cost of treating a low salinity seawater is approximately 10 times that of a typical freshwater source, hence from a sustainability and carbon footprint viewpoint, desalination of seawater should only be considered when there are no other options. Water reuse requires about half of the energy of seawater.

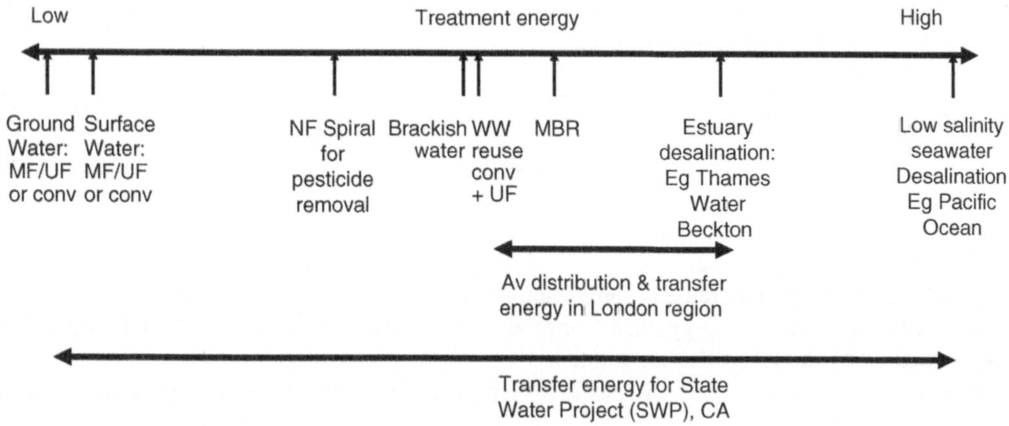

Figure 18.5 Energy requirements of treating different water sources

It is worth noting that current nanofiltration (NF) technology requires an energy input approaching that of water reuse. Thus, if there is legislation which necessitates the use of NF, and especially RO, to remove emerging contaminants, there will be a significant carbon footprint impact on drinking water production. Recent interest has been shown in developing alternative NF membranes and formats with lower energy use. For example, improved selectivity towards the organic target molecules without the concomitant removal of ionic species could significantly reduce operating pressures and hence energy use. In addition, formats other than spirals may have lower overall energy use due to the potential for eliminating pre-treatment requirements.

In providing a comparison between the relative energy use of treatment and transfer, the Figure 18.5 illustrates that the decision on resource development is made more complicated when considering distribution and transport costs since the pumping energy of transferring water is significant. For example, in the highly populated London area, the average cost of distribution is 0.6 kWh/m^3 (Livingstone, 2006b) underlining the importance of the proximity of the resource to the place where it is needed.

Indeed, the difference in treatment energy between an MBR water reuse scheme and the estuary scheme proposed for Beckton is less than the energy cost of distribution. This demonstrates that it is important for treatment facilities to be located as close as possible to the point of use, and this factor may be more important than the treatment scheme itself. However, it should be noted that a water reuse scheme based on conventional biological treatment followed by membrane filtration uses less energy, as noted previously.

Transport costs are much more extreme in arid areas such as California, where the State Water Project (SWP), which transfers water from the north of the State to the densely populated south, uses as much energy as a local desalination plant (Pankratz, 2007). The SWP energy is 2.43 kWh/m^3, which is in the range of seawater desalination energy use and much higher than water reuse. Seawater desalination can therefore be justified in energy terms compared to long distance transfers in such cases, but water reuse is likely to be the first option that should be considered.

An attraction of water reuse is the flexibility of the resource. Wastewater is already collected, especially near urban centres. In addition, it is normally treated for disposal, and the energy cost of the conventional activated sludge process used for this purpose typically uses about half of the total energy cost of reuse. The marginal energy cost of taking wastewater already treated for disposal to a reuse standard is not that great, perhaps around 0.5 kWh/m^3. This factor makes the additional energy cost of this resource much closer to a surface water resource. Also, the water reuse resource is flexible, since the treatment plant can be easily designed to provide a variable output capacity.

18.9 CONCLUSIONS

- Wastewater treated by conventional activated sludge provides a suitable feed for membrane treatment
- The water reuse market is growing rapidly, and provides an opportunity with excellent business potential for membrane filtration
- For water reuse, UF/MF membrane pre-treatment is essential for low cost RO
- The energy cost for water reuse is significantly below that of seawater desalination
- MBR-RO uses higher energy than conventional wastewater treatment with UF/MF-RO, and potentially provides a less flexible solution
- Half the energy cost of water treatment has to be provided for treatment to a discharge standard, so the marginal cost of treating further for reuse is modest
- Accordingly, water reuse should provide the default option for alternative resource development

- Distribution and transfer energy can be significant compared with treatment, so the location of the source and need are key in alternative resource provision
- Membrane filtration case studies in water reuse and estuary desalination have provided good experience for the Water Industry

REFERENCES

GWI (2005). Water Reuse Markets 2005–2015: A Global Assessment and Forecast. *Global Water Intelligence*. CIS, Media Analytics Ltd, Oxford, UK.

GWI (2007). Global Water Market 2008: Opportunities in Scarcity & Environmental Regulation. *Global Water Intelligence*. CIS, Media Analytics Ltd, Oxford, UK.

GWI (2009). Water Technology Markets 2010; Key Opportunities & Emerging Trends. *Global Water Intelligence*. CIS, Media Analytics Ltd, Oxford, UK.

Livingstone K. (2006a). Proof of Evidence, Appeal Ref: APP/G5750/A/05/1184751. Mayor of London's Office (www.london.gov.uk/mayor/planning) (accessed September 2008).

Livingstone K. (2006b). Closing Submissions; Appeal Ref: APP/G5750/A/05/1184751. Mayor of London's Office (www.london.gov.uk/mayor/planning) (accessed September 2008).

Pankratz T. (2007). Import or Desalinate. Water Desalination Report, pages 2–3, 25th June.

Pearce G. K. (2008). UF/MF pre-treatment to RO in seawater and wastewater reuse applications: a comparison of energy costs. *Desalination*, **222**, 66–73.

Pearce G. K. (2011). UF/MF Membrane Water Treatment, Principles and Design, ISBN 978-616-90836-3-4, Water Treatment Academy, Bangkok, Thailand.

Wilf M. (ed.) (2010). Guidebook to Membrane Technology for Wastewater Reclamation, ISBN 0-86689-067-X, Balaban Desalination Publications, Hopkinton, MA, USA.

Chapter 19

Operational energy consumption and carbon dioxide emissions from rainwater harvesting systems

Sarah Ward, Fayyaz Ali Memon and David Butler

19.1 INTRODUCTION

Globally, rainwater harvesting (RWH) technologies are being more readily adopted, as the desire for buildings to become more adaptable and resilient to climate change and population growth increases (White, 2010). However, it has been suggested that RWH can actually have a negative impact on the environment, with particular reference to energy consumption and carbon dioxide (CO_2) emissions (Parkes *et al.* 2010). There are two forms of energy consumption related to RWH systems: embedded and operational. Embedded is the energy consumed in the production of the components that comprise a RWH system configuration. Operational energy is consumed due to the energy required for ultraviolet (UV) disinfection and pumping rainwater from a storage tank into a building (Thornton, 2008). Due to the complexity of assessing embedded energy, which varies according to RWH system type and differs across configurations (Parkes *et al.* 2010), only operational energy consumption is discussed in this chapter.

Energy required for pumping is considered to be of greater importance as the majority of systems require a pump, whereas UV is optional (depending on the nature of the intended end use). However, there is limited research on the quantification of this operational energy consumption and it is often not fully considered in life cycle assessments (Way *et al.* 2010). This is due to the majority of RWH systems on the market not having the ability to measure their energy consumption (Ward, 2010). This chapter reviews the approaches taken to quantify energy consumption in conventional RWH systems, as well as innovative approaches to overcoming the rainwater-energy-carbon dilemma, such as the use of renewable energy sources and emerging gravity-based RWH systems. As previously mentioned, the primary sources of energy consumption within a RWH system are the ultra-violet disinfection system and the pump.

Conventional RWH systems are divided into three types:

(i) Directly pumped – harvested rainwater is pumped directly from the storage tank to the point of use;
(ii) Indirectly pumped – harvested rainwater is pumped to a header tank and then delivered to the point of use under gravity;
(iii) Gravity fed – harvested rainwater is delivered to the point of use under natural head.

Directly and indirectly pumped RWH systems require the inclusion of a pump in the system configuration to move rainwater from the storage tank to the point of use. Gravity RWH systems do not require the input of energy for pumping, as rainwater is stored above the point of use and fed to the point of use under gravity (natural head). These systems could be considered the most sustainable from an operational energy consumption and CO_2 emission point of view (if a UV unit is not included).

19.2 DISINFECTION

After the filtration stage of water purification, a disinfection step such as UV unit can be included within a RWH system to further minimise the presence of pathogens (viruses or bacteria) in the harvested rainwater. A UV unit consists of a lamp that kills the bacteria as the water flows past it. In order to function properly, UV units require water with turbidity less than 1 NTU (BSi, 2009). Not all systems require a UV unit, but its inclusion reduces health and safety risks and they may be more advisable for systems within buildings where the end user population may be at greater risk of health impacts, for example a school or hospital, or where near potable quality water is required (this will differ from country to country and state to state according to relevant regulations). It is important that the UV unit is not undersized, as this can reduce its efficiency, or oversized, as this would increase both the financial and environmental costs of RWH system operation (BSi, 2009). As the utilisation of a UV device is optional, this chapter will focus on pumping, as the majority of contemporary RWH systems utilise a pump in order to distribute the harvested rainwater to the point of use.

19.3 PRESSURE DRIVEN SYSTEMS

The majority of rainwater harvesting (RWH) systems are pressure systems using different pumping equipment and schemes (direct or indirect).

19.3.1 Directly and indirectly pumped systems

Energy consumption for pumping will depend on the type of system under consideration. For directly pumped systems, the pump is operational during each draw of rainwater from the storage tank. For indirectly pumped systems, the pump is only operational when the rainwater level in the header tank drops below a set level and the pump operates to refill the header tank. Consequently, directly pumped systems may consume a greater amount of energy, as the start up phase of pump operation requires a greater amount of energy than the operation phase. This is covered in detail in Section 19.3.1.2. Pump selection is important as over-sizing may result in more energy being consumed than required, increasing both the financial and environmental costs of RWH system operation.

19.3.1.1 Fixed speed pumps

The pump present within contemporary RWH system configurations is often of the fixed speed variety (Ward, 2010), where the pump has a fixed power rating and therefore consumes the same amount of energy per unit of water pumped. Until recently pumping energy was not directly measured (discussed in Section 19.3.1.5) and therefore the following proxy method was developed to estimate pumping energy and consumption (including UV energy consumption) and CO_2 emissions. The method is applied to a case study of a RWH system in an office building in the UK in Ward *et al.* (2011).

19.3.1.2 Estimating RWH system energy consumption and CO_2 emissions by proxy

Simple proxy methods were developed to estimate energy consumption, E (kWh), such as:

$$E = P_R, O_D \tag{19.1}$$

where P_R = pump rating (kW) and O_D = operating duration (h) and:

$$O_D = \frac{V}{P_c} \tag{19.2}$$

where V is the volume of rainwater pumped that is consumed (m^3) and P_c is the pump capacity (m^3/h) (Dixon, 2001; Roebuck, 2008).

P_R and P_C can be obtained from pump specification documents and V can be estimated by calculating a rainwater demand value or derived from empirical data, where monitoring equipment is in place. However, such methods did not distinguish between the energy consumed during the pump start-up and operating phases. Ward *et al.* (2011) developed an improved method, which made the distinction between these phases and included pump efficiency, resulting in the following equation:

$$E_{TOT} = E_2 + [E_2 \cdot (1 - P_E)] \tag{19.3}$$

where E_{TOT} (kWh) is the total energy from pumping, E_2 is the total energy consumed during pump start-up and operation phases and P_E is the pump efficiency.

Table 19.1 summarises the full list of parameters included in the method. For the full derivation of equation 19.3, please refer to Ward *et al.* (2011).

Table 19.1 Parameters used to estimate pump energy consumption by proxy (Ward *et al.* 2011).

Parameter	Unit	Parameter	Unit
P_R (pump rating)[a]	kW	H_2 (float switch off level)[a]	%
P_I (pump input power)[a]	kW	S_F (start-up factor)[b]	%
P_C (pump capacity)[b]	m^3/h	S_V (start-up volume)[c]	m^3
H_C (header capacity)[a]	m^3	O_V (operating volume)[c]	m^3
H_1 (float switch on level)[a]	%	P_E[d]	%

[a] From system specification, [b] From literature, [c] Estimated, [d] Calculated

A further addition to equation 19.3, outlined here, is the inclusion of the energy consumption of the (optional) UV device, E_{uv} (kWh), which can be given as:

$$E_{UV} = UV_R \times O_{DUV} \tag{19.4}$$

where UV_R is the UV unit energy consumption rating (kW) and O_{DUV} = UV unit operating duration (h).

Consequently, the overall equation for calculating the total operational (UV and pumping) energy consumption of a RWH (E_{TOTO}) is given by:

$$E_{TOTO} = E_2 + [E_2 \cdot (1 - P_E)] + E_{UV} \tag{19.5}$$

From equation 19.5, the carbon dioxide emission associated with the operational energy of RWH (E_{CO_2}; kg/kWh) can be estimated using the following equation:

$$E_{CO_2} = E_{TOTO} \cdot E_C \tag{19.6}$$

where E_C is the CO_2 emitted (in kg) from the energy production source used.

For example, for electricity derived from the combustion of coal, E_C is 1.04 kg/kWh (CDIAC, 2010). Other CO_2 equivalent values for this factor could be used, depending on the electricity source or the unit of comparison.

Applying the method to empirical flow data from a RWH system in the UK showed that energy consumption associated with pumping 1 m^3 of rainwater was 0.54 kWh, compared with 0.32 kWh derived using the simple method. However, the overall impact of pumping was minimal and for a period of 6 months represented just 0.07% of an office building's total energy consumption. This compares well to other studies, such as Gardner *et al.* (2008) who empirically measured a figure of 0.54 kWh from a study of a domestic building. A more recent study (Parkes *et al.* 2010) identified the *full life cycle analysis* (LCA) energy consumption of RWH as being 1.16 kWh/m^3 (0.63 kgCO_2eq/m^3). By way of comparison, the report identified mains water distribution system (WDS) energy consumption (for water delivery only) as being 0.6 kWh/m^3 (0.33 kgCO_2eq/m^3). However, this study did not consider 'future' systems, such as gravity systems or those with more energy-efficient pumps, which will reduce the pumping energy requirement (Ward, 2010; Way *et al.* 2010).

A study by Nolde (2007) also investigated the utilisation of road-runoff for reuse, which reduced potable water consumption and wastewater production, as well as minimising pollutant entry into surface waters, without the need for a sewer connection. The study showed that a reduction in potable water consumption of approximately 70% for the toilet-flush demand was achieved (2500 m^3/yr) without any impact on the end user's comfort. Furthermore, a total energy demand of only 0.88 kWh/m^3 was required for treatment (filtration using aerobic biological treatment and disinfection using UV) and distribution of the treated runoff. This compares well with the figures given above.

A number of techniques for reducing or removing the energy consumption of RWH systems have been developed over the past 5 years. More recently, renewable energy technologies have been integrated with RWH systems, so that the operational energy consumed is from more sustainable sources than conventional power stations via centralised electricity infrastructure (in the UK known as the "National Grid").

19.3.1.3 Variable speed/higher efficiency pumps

In response to the concern expressed regarding the operational energy consumption of RWH systems, manufacturers have started to develop "next generation" systems with variable speed or higher efficiency pumps integrated into their configurations. An example of a variable speed pump system is the UK-based "Monsoon" system (Stormsaver, 2011), which utilises a membrane pump operating at 90 W, watts (compared to 800 W pumps for conventional domestic systems).

Another example of a directly-pumped system from Germany is the "Eco-Compac" (Aquality, 2011), which includes a highly efficient combined low voltage (24 V, volt) suction/booster variable speed pump. It operates on demand and will only produce the minimal amount of flow required by the serviced appliance(s) in operation.

An additional benefit of these systems is that as they utilise low voltage/wattage variable pumps, so the energy to power them can be supplied by a solar panel, wind turbine or even a car battery.

19.3.1.4 Integration with renewable energy sources

With a global focus on energy consumption and CO_2 emissions, new water distribution technology has to carefully consider any potential sources of energy consumption. This is particularly pertinent for alternative water systems that run in parallel to the well-established centralised water distribution system. An ideal RWH system would require no operational energy, such as emerging gravity systems discussed in Section 19.4. However, the impact from pumped systems can be reduced by utilizing energy from a renewable source. Thus, although the system is creating an additional energy demand, it is meeting that demand from a more sustainable source. A potential incidental benefit of this is providing extra demand within the renewable energy sector, helping to stimulate and strengthen the market for such technologies.

Providing the correct conversion equipment/electrical controls are installed alongside the RWH system, most renewable energy technologies can be integrated into the system configuration. However, solar photovoltaic cells are the more practical complementary renewable energy technology for integration, due to ease of installation and consistency in supply of the natural energy resource (i.e. the sun).

19.3.1.5 Integrated energy monitoring

In order to develop a better picture of the level of energy consumed by different types and scales of RWH systems (Stormsaver, 2011), some systems are now manufactured with an in-built energy monitor as part of the system configuration. This provides essential operational data to permit enhancing the efficiency of the system, for example by optimising the rating of the pump or only using a UV unit when there is a particular need to enhance the quality of the harvested rainwater.

19.4 EMERGING GRAVITY SYSTEMS

Gravity systems utilise the natural head between the location of the RWH system storage tank and the point of use to deliver the harvested rainwater under gravity, therefore eliminating pumping energy requirements and associated costs. In recent years a number of experimental systems have emerged from across the globe and a small selection are briefly described in the following sections.

19.4.1 Internal building systems

Within a building, the roof space produces the greatest natural head between a storage tank and the point of use, as well as usually providing adequate space for a storage tank. Although roof space can be limited in blocks of flats/apartments, some high rise non-domestic buildings and in newer houses[1], the majority of older buildings of different types have ample roof space. Consequently, it is these properties that have the greatest potential to have internal gravity RWH systems installed.

One such patented system under development in the UK is illustrated in Figure 19.1. The system utilises a specially designed integrated tile-gutter system and inlet to collect and divert rainwater to an eaves located modular storage tank, where it is then fed via gravity through a specially designed outlet to the point of use. Structural loading issues may need to be considered by installers if considering this type of RWH system.

19.4.2 External building systems

Located on the exterior of a building, a range of gravity systems have been or are being developed internationally. Some are integral to the building structure, whereas others are able to be retrofitted, giving them the advantage of flexibility and adaptability over conventional below ground tank systems.

Figure 19.2 illustrates an experimental wall-mounted system that has been developed in the UK (Wherlock, 2009). The system can store around 100 L of water, has a mains top-up, has no mechanical components and is modular, facilitating the ability to increase capacity. However, the system currently suffers from aesthetic and potential structural loading issues (depending on the number of connected units) and development is on going (Wherlock, 2009). It has undergone a harvested water quality study, which identified that the health risk posed by using water from the system was low (Fewtrell *et al.* 2009).

---

[1]a recent trend is to minimise 'wasted' space (Ward, 2010)

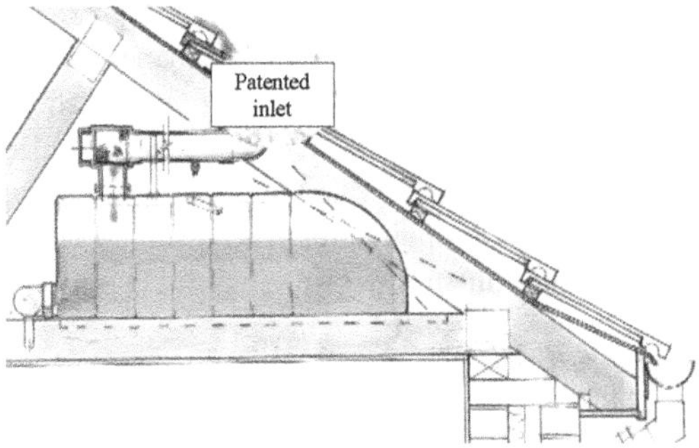

Figure 19.1 Section through a roof space illustrating the internal 'eaves' gravity RWH system under development in the UK (Brittain, 2010)

Figure 19.2 A wall-mounted RWH system under development in the UK (Wherlock, 2009)

Figure 19.3 illustrates an Australian gravity RWH system under development: the rainwater storage gutter system, which utilises the space in specially formed gutters to store runoff (Hardie, 2010). The system combines storage with overflow-infiltration and is beneficial when plot size restricts the use of conventional storage tanks. The gutters are available in a range of sizes and are designed to be supported by the roof frame, but can also be supported by the external wall; therefore consideration needs to be given to this when considering its implementation.

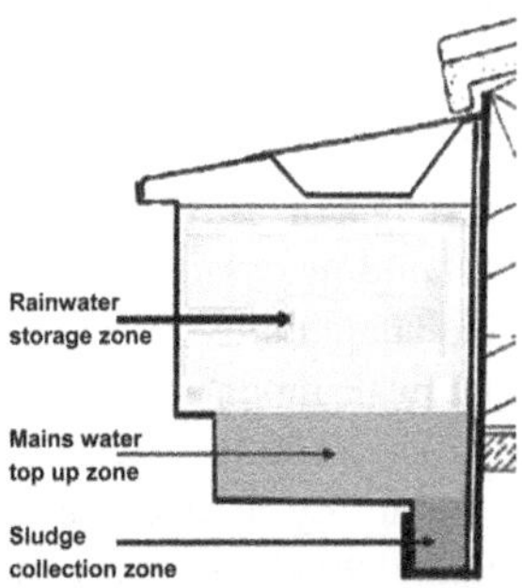

Figure 19.3 External view and cross-section of the 'gutter' storage RWH system from Australia (Hardie, 2010)

Figure 19.4 illustrates the innovative plastic bag storage RWH system, which is being trialled in Asia, where space constraints in dense high-rise buildings prevent installation of a more conventional

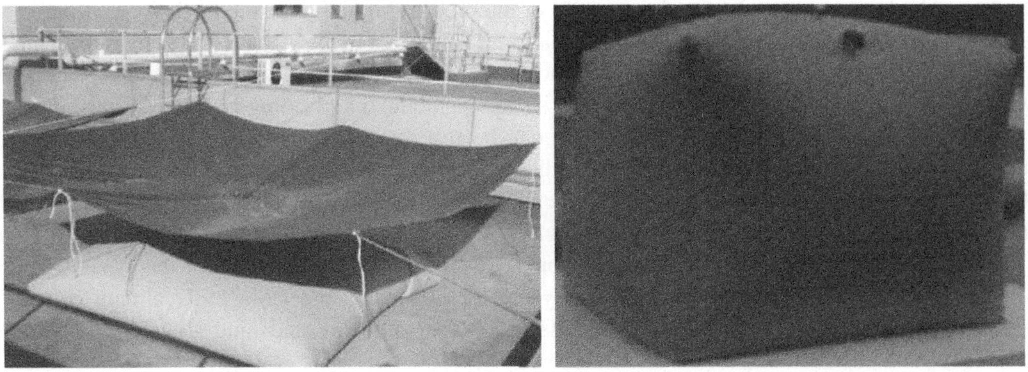

Figure 19.4 Components of the innovative 'plastic bag' RWH system from Korea (Dao *et al.* 2009)

Figure 19.5 An individual dwelling-scale 'GROW' gravity collection greywater system under development in the UK

tank. Arguably the system can be fitted externally or internally, although the latter is likely to be subject to more stringent building and health and safety regulations. The components of the RWH system itself are less resource-intensive than conventional systems, although the building-related components, such as guttering, downpipes and internal pipework are similar. The system is modular and storage capacity can be increased by interconnecting a series of bags. The system can be readily disconnected and/or relocated, thereby reducing its end of life cycle (decommissioning/disposal) impact (Dao *et al.* 2009).

Figure 19.5 illustrates the 'BabyGROW' system, which is an experimental system from the UK, but which utilises gravity to collect greywater for reuse, rather than RWH. The system collects greywater from baths, showers and hand basins on any floor of a building above the height of the system. Internal pipework directs the greywater to the GROW system for treatment and storage. Treatment consists of biophysical filtration in the form of aquatic plants and coarse and fine sands and gravels. The system is readily retrofitted and modular.

19.5 IMPROVEMENT OF ENERGY EFFICIENCY OF RWH

19.5.1 Energy saving in sewers from RWH

There are two main types of sewer system: combined, where foul flows and surface water flows are conveyed together and separate, where they are conveyed separately. In recent decades, separate collection has become the preferred technique, but the two flows may converge downstream where they connect to older existing combined systems (Hurley *et al.* 2007). Excess non-foul (surface water) flows in sewer systems can create problems including high energy demands in transporting rainwater over large distances and inefficient wastewater treatment due to the overloading of treatment plants.

It is hypothesized that the latter problem may be of growing significance, due to the increasing requirement for more stringent treatment to comply with legislation such as the Urban Wastewater Treatment Directive and the Water Framework Directive (Commission of the European Communities, 2009). Increased volumes of rainwater in sewers, results in higher flows to be treated at wastewater treatment plants (WWTPs), increasing energy consumption during the treatment process. Consequently, water reuse techniques, such as RWH and greywater and effluent reuse, assist in reducing the impact of these non-foul or combined flows. They do this by storing and utilising the flows locally, so that their release into the sewer system is attenuated, reducing the impact on WWTPs. In addition, they reduce the volumes of potable quality water that have to be pumped to the point of use (Hurley *et al.* 2007).

19.5.2 Energy generation from RWH

This chapter has so far highlighted the link between water and energy in terms of the energy consumption required to treat and distribute harvested rainwater. However, RWH systems may have the potential to offset their own energy consumption by harnessing kinetic energy. Technologies are emerging that follow the principles of hydropower – that is, harnessing kinetic energy from moving water to generate energy.

An example of one such device is the 'HighDro Power' unit under development in the UK (Cowan, 2010). The device generates energy from wastewater falling through waste pipes fitted to high-rise buildings, converting it to electricity using four turbine blades that drive a single generator. Currently the unit is in the prototyping and testing stage, but it is estimated the electricity produced could save £926/yr (equivalent to 1120 €/yr or 1430 US$/yr) for a seven-storey building. The electricity could either be utilised in the building or sold back to the national grid on a renewable energy generation buy-back tariff (Cowan, 2010).

The application of this concept to rainwater downpipes is also under investigation in the UK, as converting the potential energy of rainwater falling through downpipes to electricity could also yield significant amounts of electricity (Ward, 2010). The annual amount of energy generated would be small from a domestic building – perhaps enough to power a 40 watt light bulb for just 5 hours.

However, within a multi-storey office building the energy could be enough to supply the electricity required by a pump associated with a RWH system. This would reduce the energy consumption and carbon emission impact of pumped RWH systems (for example in situations where pumps cannot be fully eliminated).

19.6 CONCLUSIONS

The main sources of operational energy consumption associated with RWH systems are from disinfecting and pumping the harvested rainwater. Energy consumption can either be measured directly or indirectly (by proxy methods), with the latter requiring suitable parameterisation of system operation.

The energy quantification outlined in this chapter suggests that the operational energy consumption associated with RWH is a very minimal fraction of over all energy consumption in buildings. Conventional RWH systems utilise low efficiency fixed speed pumps, whereas recent innovations have enhanced pump efficiencies, incorporated variable speed pumps and replaced centrally sourced energy with renewable decentralised energy generating technologies.

In addition, a plethora of designs of gravity systems are emerging, that require no energy consumption for pumping, which will help further increase the sustainability of RWH.

RWH can also assist in reducing the energy consumption associated with the distribution and treatment of wastewater and the distribution of potable water, by storing and utilising rainwater locally before it is released into the sewer system.

Furthermore, technologies are being developed to generate electrical energy from the kinetic energy produced by rainwater or wastewater falling through downpipes in conveyance and reuse systems. These innovations will help to assuage fears that the utilisation of RWH and water reuse technologies will result in negative environmental impacts through the consumption of energy and emissions of CO_2.

REFERENCES

Aquality (2011). Aquality – Products: Eco Compact. http://www.aqua-lity.co.uk/index.php?id=380 (accessed 2 May 2011)

Brittain G. (2010). Rainsense promotional DVD. Innovate10 Conference 12th October 2010, Technology Strategy Board, London, UK.

BSi (2009). BS 8515:2009 – Rainwater Harvesting Systems – Code of Practice. BSi, London.

Commission of the European Communities (2009). 5th Commission Summary on the Implementation of the Urban Waste Water Treatment Directive. Commission Staff Working Document SEC(2009) 1114 final, 3.8.2009. http://ec.europa.eu/environment/water/waterurbanwaste/implementation/pdf/implementation_report_summary.pdf (accessed 01 April 2011)

Cowan K. (2010). Waste Not Want Not with DMU Student's Electric Idea. http://www.creativeboom.co.uk/news/waste-not-want-not-with-dmu-students-electric-idea/ (accessed 25 July 2010)

CDIAC (2010). Frequently Asked Questions. http://cdiac.ornl.gov/pns/faq.html (accessed 15 January 2010)

Dao A. D., Han M. N., Guyen V. A., Ho X. Q. and Kim T. H. (2009). Flooding Mitigation Plan at Downtown of Hanoi by Rainwater Harvesting. *8th International Conference on Urban Drainage Modelling and 2nd International Conference on Rainwater Harvesting and Management*. 7–12th September. Tokyo, Japan.

Dixon A. (2001). Simulation of Domestic Water Re-use Systems: Greywater and Rainwater in Combination. Unpublished PhD thesis, Imperial College, London.

Fewtrell L., Davies C., Kay D., Watkins J. and Wyer M. (2009). Acquisition of microbial data on quality of harvested rainwater in the United Kingdom to drive a quantitative microbial risk assessment. Experiences with Rainwater Harvesting and Greywater Recycling and their Future Prospects, Aqua enviro. Birmingham, UK.

Gardner T., Beal C., Hood B., Lane J., Christiansen C., Millar G., Vieritz A. and Chapman H. (2008). Benchmarking Water Sensitive Urban Design – Payne Road. CRCWQRA Research Report 77. ISBN 18766 1627X.

Hardie M. (2010). Rainwater storage gutters for houses. *Sustainability*, **2**, 266–279.

Hurley L., Mounce S. R. and Ashley R. M. (2007). Relative sustainability assessment of five European projects aiming to reduce rainwater volume in sewer systems. *Water Practice & Technology Online Journal*, **2**(2).

Nolde E. (2007). Possibilities of rainwater utilisation in densely populated areas including precipitation runoffs from traffic surfaces. *Desalination*, **215**, 1–11.

Parkes C., Kershaw H., Hart J., Sibille R. and Grant Z. (2010). Energy and Carbon Implications of Rainwater Harvesting & Greywater Recycling. Final Report, Science Project Number: SC090018, Environment Agency, Bristol. http://publications.environment-agency.gov.uk/pdf/SCHO0610BSMQ-e-e.pdf (accessed 17 August 2010)

Roebuck R. (2008). A Whole Life Costing Approach for Rainwater Harvesting Systems. Unpublished PhD thesis, University of Bradford, Bradford, UK.

Stormsaver (2011). Monsoon Domestic Rainwater Harvesting Systems. http://www.stormsaver.com/Write/Documents/Monsoon%20Domestic%2012pp.pdf (accessed 2 May 2011)

Thornton J. (2008). Rainwater harvesting systems – are they a green solution to water shortages? Green Building Magazine, Spring 2008, **17**(4), 40–43.

Ward S. (2010). Rainwater Harvesting in the UK: A Strategic Framework to Enable Transition from Novel to Mainstream. Published PhD thesis, Centre for Water Systems, University of Exeter. Online at: http://hdl.handle.net/10036/106575 (accessed 21 June 2010).

Ward S., Butler D. and Memon F. A. (2011). Benchmarking energy consumption and CO_2 emissions from rainwater harvesting systems: an improved method by proxy. *Water and Environment Journal*, ISSN 1747-6585; doi: 10.1111/j.1747-6593.2011.00279.x

Way C. M., Martinson D. B., Heslop S. E. and Cooke R. S. (2010). Rainwater Harvesting: Environmentally Beneficial for the UK? *Proceedings of the IWA World Water Congress and Exhibition 2010*, 19–24 Sept 2010, Montreal, Canada.

Wherlock J. (2009). Dwr Cymru Welsh Water Surface Water Management Strategy. Experiences with Rainwater Harvesting and Greywater Recycling and their Future Prospects. Aqua enviro, Birmingham, UK.

White I. (2010) Water and the City: Planning for a Sustainable Future. Taylor and Francis, Abingdon.

Chapter 20

LCA as a tool to assess environmental impact and energy efficiency of reverse osmosis desalination

Katharina M. Tarnacki, Thomas Melin and Montse Meneses

20.1 INTRODUCTION

Desalination technologies were introduced about 50 years ago and since then have been expanding rapidly worldwide, mainly due to increased water demand in water stressed areas. In Europe, the majority of the desalination plants are located in Spain, along the Mediterranean cost and on the Canary Islands, where they satisfy municipal and agricultural water demand. Desalination in the rest of Europe is primarily used to generate high quality industrial water, for example boiler feed or process water. In total, more than 800,000 m^3 of desalinated water is produced in Europe's over 700 desalination plants (SEM, 2007; Muñoz & Fernández-Alba, 2008).

However, desalination not only satisfies important societal needs, it has also significant negative environmental impacts, such as high energy demand, disposal of concentrated brine, chemicals used for cleaning and their residues contained in the disposed brine and backwash water, as well as waste generated, for example used membranes and other plant components. Recent studies describe ways to quantify these environmental impacts and come to the conclusion that the consumption of energy represents the dominant cause of the environmental impact of desalination (Raluy *et al.* 2004; Raluy *et al.* 2006; Muñoz & Fernández-Alba, 2008).

Energy use in desalination has been the focus of many research activities during the last decades. While in the 70s the energy demand for reverse osmosis (RO) plants was around 16 kWh/m^3, it dropped to 8 kWh/m^3 in the 1980s (Elimelech & Philipp, 2011). More permeable membranes and the wider use and the increased efficiency of energy recovery turbines and pressure exchange systems have led to further energy savings. The energy demand of modern large-scale sea water RO plants lies between 2.5 and 4 kWh/m^3 (Fritzmann *et al.* 2007). New approaches may continue to reduce this value: The US company DEMWAX claims (WDR, 2008) to have developed a new membrane based technology applicable for sea water desalination in deep coastal waters that approaches the thermodynamic minimum of around 1 kWh/m^3.

The life cycle analysis (LCA) concept was developed to assess the environmental impact of industrial products or processes and has also been used in water treatment (Ortiz *et al.* 2007; Vince *et al.* 2008; Pasqualino *et al.* 2009, 2011; Meneses *et al.* 2010b). One specific area, to which LCA has been applied, is the generation of fresh water from sea water by desalination processes (Raluy *et al.* 2004, 2006; Muñoz & Fernández-Alba, 2008; Meneses *et al.* 2010a). The studies have concentrated on the comparison of different desalination technologies (Raluy *et al.* 2004, 2006) and the reduction of environmental impact (Meneses *et al.* 2010a; Muñoz & Fernández-Alba, 2008). Reverse osmosis exhibited lowest harm to the environment when compared to other (thermal) desalination processes, such as multi-stage flash (MSF) and multiple-effect distillation (MED) (Raluy *et al.* 2004, 2006). For all processes energy consumption appears to be the dominant cause of environmental impact.

The intent of this chapter is to apply LCA methods in order to analyze the factors governing the environmental impact of reverse osmosis, which is the dominating desalination technology in Europe. Since the earlier studies have shown the importance of energy consumption and since RO depends on electrical energy, it is suspected that the fuel mix used for generation of electrical energy will play a significant role. In addition to different conventional fuels, the use of renewable energy sources will be investigated. Three different established LCA methods are used in parallel, in order to test the sensitivity of the methodology. Ideally, the major results should show where the major similarities and differences between the assessment methods are and how the energy mix influences the total results. The chapter is based upon a study whose complete results will appear in a more detailed publication (Tarnacki *et al.* 2011).

20.2 METHODOLOGY

20.2.1 Desalination with reverse osmosis

Reverse osmosis uses semi-permeable polymeric membranes to retain all colloidal and almost all dissolved matter from aqueous solutions, while allowing water to pass through the membrane, thereby splitting the feed stream into a concentrated brine stream, which generally is considered waste and a product stream (permeate). The driving force of the process is the difference between the differential pressure applied across the membrane and the feed side osmotic pressure (Figure 20.1).

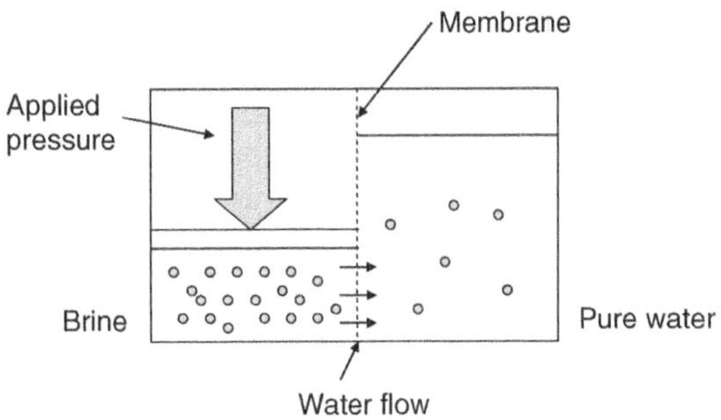

Figure 20.1 Principle of reverse osmosis (modified according to Rautenbach and Albrecht, 1989)

Even though RO desalination is characterized by a large variety of case specific factors, like water composition and temperature, plant capacity, pre-treatment and energy recovery concepts and by operating parameters like feed pressure and water recovery, a simplified approach is used here to evaluate environmental impact. It is assumed that for all parameters, except the specific energy consumption, for which a range of values is used, and for the source of the electrical energy, standard values per m^3 of desalinated water are used (functional unit according to LCA terminology). These standard values are taken from recent literature studies (Fritzmann *et al.* 2007).

The reasoning behind this approach relies upon two hypotheses:

- Energy consumption dominates as cause for negative environmental impact (substantiated by earlier studies).

- The modular concept of RO desalination plants and the standardization of membrane materials, cleaning chemicals and modules leads to very similar values of chemicals used, membranes replaced and steel (membrane housings) required per m^3 of desalinated water.

The second hypothesis is supported by the fact that the investment cost for desalination plants increases nearly linearly with plant size, at least for larger plants. All variability thus is placed on energy consumption, where three values are used: 4 kWh/m^3 for smaller or older plants, 3 kWh/m^3 as state of the art and 2 kWh/m^3 for future developments (Fritzmann *et al.* 2011; Subramani *et al.* 2011). As a consequence, the impact calculations have a common base and do not have to be performed separately for differently designed plants or for different plant locations, as long as the influence of plant size or design or location on energy consumption is known. All resulting values are given per m^3 of RO permeate (functional unit). The assumptions for the study are summarised in Table 20.1.

Table 20.1 Assumptions for the study based on bibliographic data.

Energy demand	Membrane material	Module design	Housing, spacer material	Water recovery ratio	Chemicals
2, 3 and 4 kWh/m^3	Active layer – polyamide	Spiral wound	steel, polypropylene	42.5%	ClO, $FeCl_3$, antiscalant, $NaHSO_3$, H_2SO_4, NaOCl

20.2.2 Application of life cycle assessment

LCA is a widely utilised, at least partially standardised tool, intended to evaluate the environmental impacts of products and processes (ISO, 2006) in a holistic and objective way. An LCA analysis contains four main phases:

Goal and scope definition: Two goals have been set for this study, first, an environmental assessment of reverse osmosis combined with different energy sources in order to identify the most environmental friendly configuration and secondly, an investigation of similarities and differences between various impact assessment methods.

Inventory analysis: For the operational phase, all mass and energy input and output flows have been considered, as well as their transport from one process step to another and their disposal after use. The inventory analysis has been performed based on data and assumptions given in recent scientific literature. For the use of chemicals for pre-treatment and membrane cleaning, the data from Fritzmann *et al.* (2007) for an average reverse osmosis desalination plant have been used. The amount and salinity of brine generated follow from a mass balance. The environmental loads for all components have been calculated using the data from the Ecoinvent V2.2 database (Ecoinvent, 2010). The functional unit for the performed study has been defined as 1 m^3 of desalted water and all data refer to this unit.

Impact assessment: For this phase, three different impact assessment methods have been applied: the CML 2001 (Centrum voor Milieuwetenschappen, Leiden University), Ecoindicator 99 and Ecopoints 97 methods, which we found to be the most frequently applied methods in LCA studies performed in the water sector. The results for Ecoindicator 99 and Ecopoints 97 deliver one single score, which combines the impact scores from different categories. EI99 uses weighting factors for the different categories, whereas Ecopoints 97 adds the scores without weighting. They are both based on the distance-to-target principle. The CML method arrives at final results for midpoint categories and does not attempt to calculate a total impact score.

Interpretation of results: the results of the selected impact assessment methods are analysed and compared.

CML 2001

The CML impact assessment method restricts quantitative modelling to relatively early stages in the cause-effect chain to limit uncertainties. The life cycle impact results are grouped in so-called midpoint categories, according to themes (Table 20.2).

Table 20.2 Selected CML environmental impact categories (Ecoinvent, 2010; CML, 2001).

Indicator	Unit
Global warming potential (GWP)	kg CO_2-eq
Acidification potential (AP)	kg SO_2-eq
Eutrophication potential (EP)	kg PO_4-eq
Ecotoxicity potential (ETP)	kg 1,4-DCB eq
Photochemical oxidation (PHO)	kg formed ozone
Ionising radiation (IR)	DALYs
Stratospheric ozone depletion 40a (ODP)	kg CFC-11-eq
Resources (Depletion of abiotic resources – DAR)	kg antimony eq

Ecoindicator 99

Ecoindicator 99 is a scientific impact assessment applying the distance-to-target principle with targets based upon scientific data on environmental damage. Various environmental impacts are measured and the final result is presented in a single score, which is the sum of the normalised and weighted values for all impact categories. The environmental impacts are grouped in the following damage categories (endpoint):

- human health (carcinogens, climate change, ionising radiation, ozone layer depletion, respiratory effects),
- ecosystem quality (acidification, eutrophication, ecotoxicity, land occupation),
- resources (fossil fuels, mineral extraction).

All three existing perspectives have been selected for the evaluation:

(1) Egalitarian perspective (E,E): long-time perspective; even a minimum of scientific proof justifies inclusion; weights: human health 300, ecosystem quality 500, resources 200.
(2) Hierarchist perspective (H,A): balanced time perspective; consensus among scientists determines inclusion of effects; weights: human health 400, ecosystem quality 400, resources 200.
(3) Individualist perspective (I,I): short time perspective: only proven effects are included; weights: human health 550, ecosystem quality 250, resources 200.

Ecopoints 97

The method of ecopoints, also called ecoscarcity, is based on the national government environmental policy objectives. It applies a distance-to-target evaluation principle using politically established target levels. The original is based on Swiss target levels, but there are also Swedish, Dutch, and other variants. Environmental impacts are evaluated directly, resulting in a dimensionless number (distance to target). The impact categories are the following: deposited waste, emission into air, emission into top-soil/groundwater, emission into water, radioactive waste and use of energy resources. The final score is the sum of all single impact category values without weighting.

20.3 COMPARISON OF THE RESULTS OF THE LCA METHODS

The LCA study is performed with three different impact assessment methods and the results are presented separately and discussed together.

20.3.1 CML

The results of the CML impact assessment method show that in almost all impact categories, desalination coupled with renewable energies (photovoltaic and wind energy) as well as the French grid mix (except one impact category) have the lowest values, and consequently, the lowest load to the environment. For the minimum energy demand considered here (2 kWh/m^3), the benefits of non-fossil fuels are not very significant, but they are increasing with higher energy demand (Figure 20.2a to h).

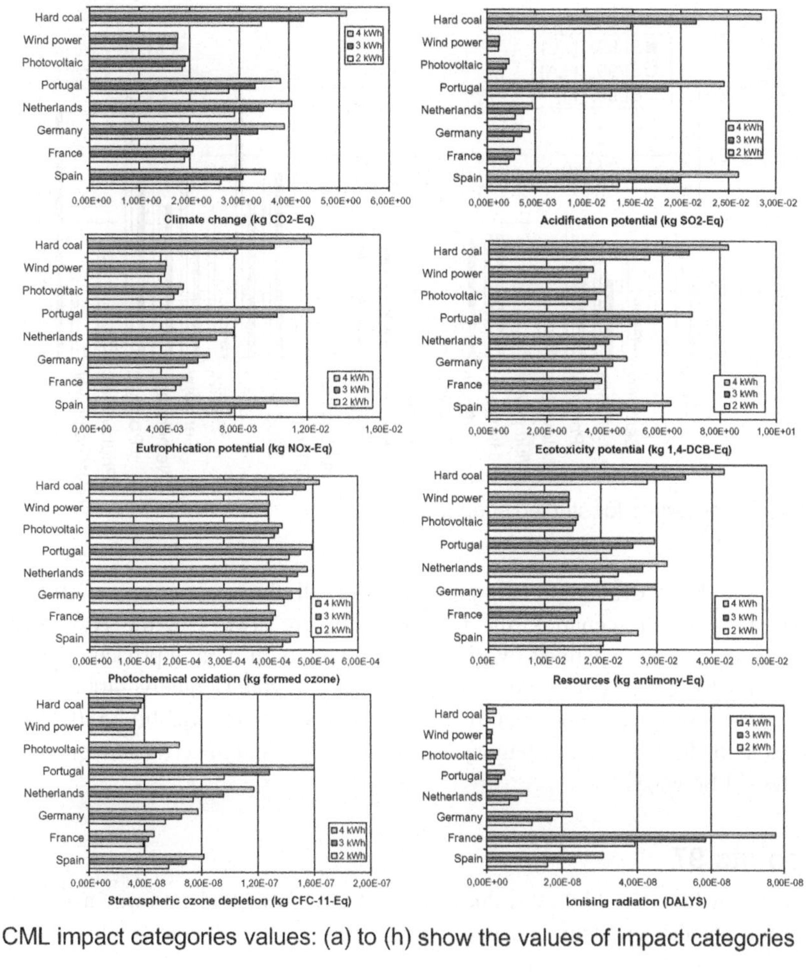

Figure 20.2 CML impact categories values: (a) to (h) show the values of impact categories

In 6 of the 8 CML impact categories selected, desalination with energy supply by hard coal gets the worst (i.e. highest) score (climate change, acidification potential, eutrophication potential, ecotoxicity potential, photochemical oxidation, resources).

Desalination with electricity from the Spanish and Portuguese grid mix also scores poorly in some impact categories (climate change, acidification potential, eutrophication potential), while the French grid mix leads impact in only one category (ionising radiation) (Figure 20.2a to h).

Except for that, the French grid mix reaches the best values of all conventional energy supply systems in all other impact categories, often even as low as those of renewable energies (Figure 20.2a to h).

20.3.2 Ecoindicator 99

The Ecoindicator 99 (EI99) values also show that the least environmental impact is caused by desalination combined with renewable energies (photovoltaic and wind power). With increasing energy demand, the environmental benefit in applying renewable energies grows (compare EI 99 values for 2 and 4 kWh/m^3, Figure 20.3).

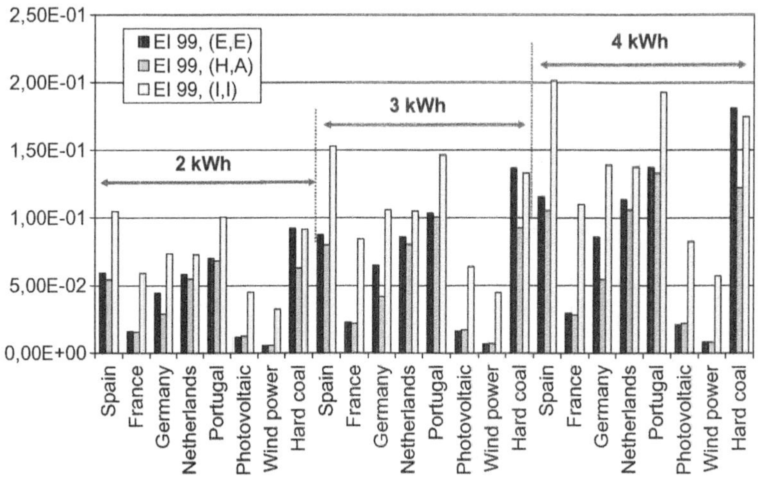

Figure 20.3 Ecoindicator values for studied scenarios

The results for the EI99 evaluation are presented for all three perspectives (E,E, H,A and I,I) (Figure 20.3). For the studied scenarios, the lowest values are always in the H,A approach and the highest in the I,I which can be explained by different weightings in both approaches. In H,A the scores for human health and for ecosystem quality categories are weighted equally (400:400) and in I,I the human health category is weighted more than twice as high as ecosystem quality (550:250).

Since the scores in the human health category are much higher than in ecosystem quality (Figure 20.4), the total scores show differences.

20.3.3 Ecopoints 97

In the Ecopoints 97 method, the highest values (poorest environmental performance) are obtained for desalination with the Spanish grid mix and with energy supply by hard coal (Figure 20.5), followed by the French and Portuguese scenarios. High values for Spain, Portugal and hard coal are related to high emissions to air (e.g. greenhouse gases and NO_x) due to use of fossil fuels (Figure 20.6). The French scenario scores high values due to the high contribution of nuclear power and related radioactive waste. RO coupled with renewable energies, such as wind and photovoltaic power show by far the most environmentally benign performance considering all impact categories and the total score of this assessment method (Figure 20.5).

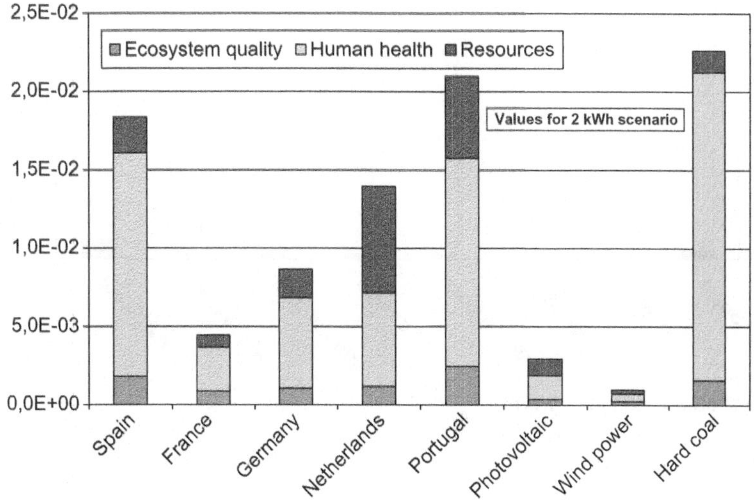

Figure 20.4 Ecoindicator 99 (H,A) impact categories

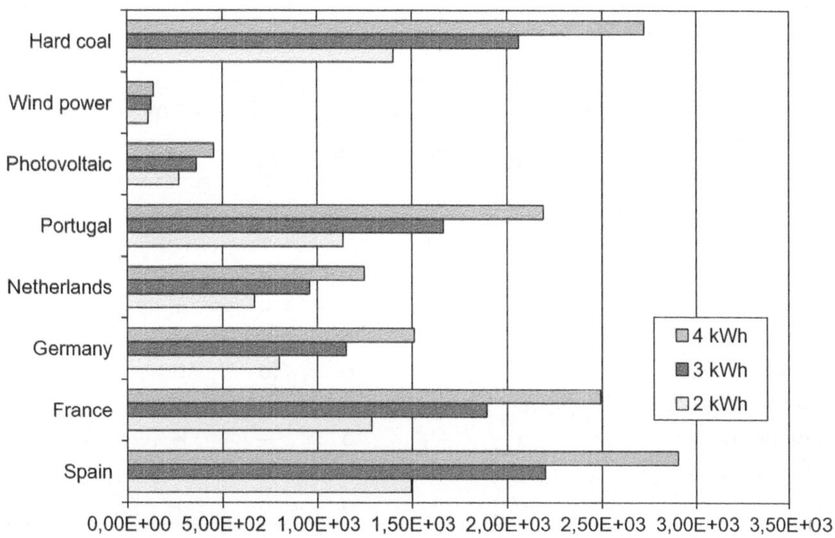

Figure 20.5 Ecopoints values for studied scenarios

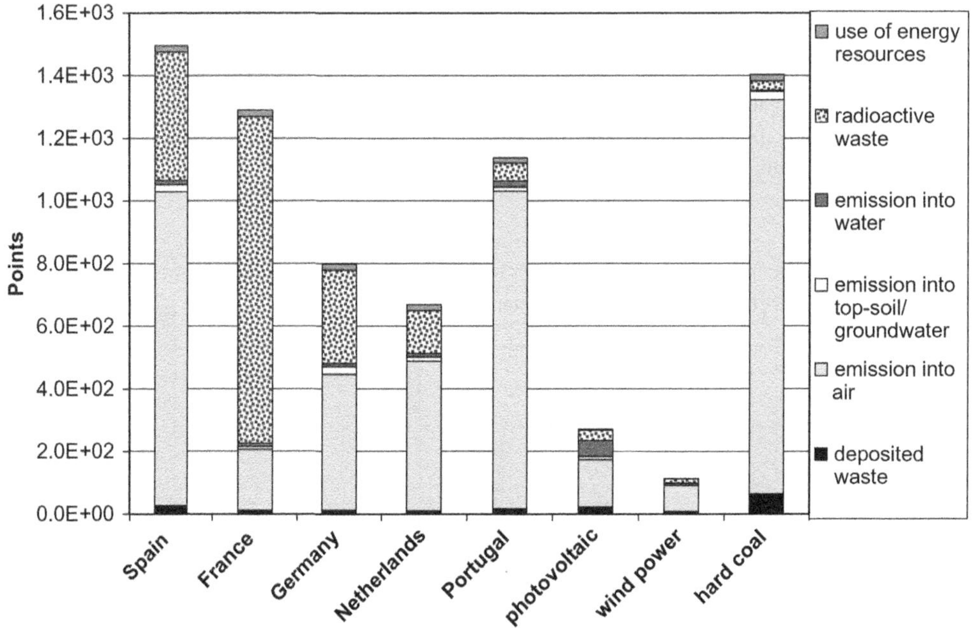

Figure 20.6 Impact categories for 2 kWh

20.3.4 Discussion

All results for the energy demand of 2 kWh/m^3 are summarised in Table 20.3. Despite the different approaches, impact categories, and weightings, the three impact assessment methods generally show comparable results.

Table 20.3 LCA results for RO desalination with an energy demand of 2 kWh/m^3.

	ES	FR	DE	NL	PT	PV*	WP*	HC*
CML – Climate change	+	o	+	+	+	o	o	++
CML – Acidification potential	++	–	–	–	++	–	–	++
CML – Eutrophication potential	++	–	o	o	++	–	–	++
CML – Ecotoxicity potential	+	o	o	o	+	o	o	++
CML – Stratospheric ozone depletion	o	–	o	+	++	o	–	–
CML – Photochemical oxidation	o	o	o	o	o	o	–	+
CML – Resources	o	–	o	o	o	–	–	++
CML – Ionising radiation	o	++	o	–	–	–	–	–
Ecoindicator 99	++	–	–	o	o	–	–	o
Ecopoints 97	++	+	o	o	+	–	–	++

*PV = photovoltaic; WP = wind power; HC = hard coal.
++ highest value; o average value; – lowest value.

Desalination causes the greatest (worst) environmental impact if the energy required is supplied by the Spanish grid mix (for EI 99 and Ecopoints 97) or by hard coal (for CML), and has the lowest (best) score if the electricity is supplied by wind power, followed by photovoltaic electricity generation.

There is, however, one case where the results reached by the different methods diverge very strongly: RO with energy supply by the French grid mix, which gets very low scores by one (EI 99) and very high impact scores by another LCA method (Ecopoints 97).

There are also other, less obvious differences between the impact scores generated with different methods: Hard coal is judged better by EI 99 than by the other methods and the Dutch energy mix has higher impact sores than the German mix in EI 99, while its ecological impact is lower in Ecopoints and most CML categories. A similar change of position occurs between Portugal and Spain.

Most changes in the impact sequence obtained by the different methods are caused by the different share of nuclear power in the respective energy mix (France > Germany or Spain > Portugal or the Netherlands) and by the relative weight of radioactive waste compared to the weight of other impact factors in the methods used: Ecopoints places a high weight on the impact of radioactive waste, whereas E.I. 99 rates this impact lower (Figures 20.4 and 20.6). In the CML method, the negative influence of the use of nuclear power is restricted to the category ionising radiation.

Another interesting effect to be observed is the influence of the specific energy consumption of RO. In most of the CML impact categories the environmental benefit of using renewable energies is not very significant in case of low energy demand (Figure 20.2a to h). With increasing energy demand, the differences between scenarios differing in their mix of energy sources become more pronounced and the advantage of renewable energy grows.

For any fossil energy supply, the scores in all methods are almost directly proportional to energy demand, while systems with renewable energy use are less influenced by energy demand than by other impact factors (chemicals, pre-treatment, membrane replacement and investment related impacts) and therefore do not show proportionality between total impact and energy consumption.

20.5 CONCLUSIONS

Desalination processes result in several environmental impacts, such as increased greenhouse gas or NO_x emissions, mostly related to the energy supply of the desalination processes (Raluy *et al.* 2004, 2006). Recent LCA studies show that the environmental impact scores can be substantially reduced when primary energy demand is reduced and/or more eco-friendly energy supply sources are applied (Raluy *et al.* 2006; Tarnacki *et al.* 2011). In contrast, other measures, such as application of materials with better durability, or less chemical usage may result in reduced cost, but not in significantly reduced environmental impact (Tarnacki *et al.* 2011). Therefore, a variation of the energy supply source and an increase of the energy efficiency of desalination can have a significant impact on the LCA results (Raluy *et al.* 2006).

The life cycle impact assessment methods applied (CML, EI99 and Ecopoints) show similar tendencies of results. This confirms the applicability of LCA for the environmental assessment of desalination processes and their energy use. Common results are:

- a significant decrease of the LCA scores when energy demand is reduced,
- most eco-friendly performance of renewable energies,
- worst LCA scores for the Spanish grid mix, where energy is mostly supplied by fossil fuels and nuclear power (80% to total energy production).

However, there are large differences in assessing the use of nuclear power in the selected impact assessment methods, which lead to differences in the sequence of scores for a number of scenarios and particularly to widely diverging results for the French grid mix. While in the CML and the EI99 methods it shows almost the best environmental performance, it rates poorest in Ecopoints.

Another limitation of the LCA approach in environmental assessment of desalination processes is the fact that some environmental aspects of desalination, as increased temperature and salinity and amount of brine are not covered by any of the established impact assessment methods. Even if their effects are minor compared to the harm done by energy consumption, they represent a gap to close in method development.

The central weak point of the LCA tool in the evaluation of complex multi-compartment scenarios is the fact that it contains subjective components, such as the selection of the impact assessment method and the weighting factors it implies. It is recognized by the authors that weighting factors are determined by experts, based on a combination of scientific justifications and subjective perception. Yet as long as essentially non-experts have to choose between methods and the scientific community does not agree upon weighting factors for impacts in different environmental compartments, this subjectivity will persist and will raise doubts about the value of LCA methods and will reduce its general acceptance.

REFERENCES

CML (2001). Handbook of Impact Categories 'CML 2001', Institute of Environmental Sciences, Leiden University, The Netherlands.

Ecoinvent (2010). Ecoinvent database V2.2, Swiss Centre for Life Cycle Inventories.

Elimelech M. and Philipp W. A. (2011). The future of weawater desalination: energy, technology, and the environment. *Science*, **333**, 712–717.

Fritzmann C., Löwenberg J., Wintgens T. and Melin T. (2007). State-of-the-art of reverse osmosis desalination. *Desalination*, **216**, 1–76.

ISO (International Organisation for Standardisation) 14040. 2006. Environmental Management – Life Cycle Assessment Principles and Framework. ISO, Geneva, Switzerland.

ISO 14044. 2006. Environmental Management-Life Cycle Assessment-Requirements and Guidelines. ISO, Geneva, Switzerland.

Meneses M., Pasqualino J. C., Céspedes-Sánchez R. and Castells F. (2010a). Alternatives for reducing the environmental impact of the main residue from a desalination plant. *Journal of Industrial Ecology*, **14**(3), 512–527.

Meneses M., Pasqualino J. C. and Castells F. (2010b). Environmental assessment of urban wastewater reuse: treatment alternatives and applications. *Chemosphere*, **81**, 266–272.

Muñoz I. and Fernández-Alba A. (2008). Reducing the environmental impacts of reverse osmosis desalination by using brackish groundwater resources. *Water Research*, **42**(3), 801–811.

Ortiz M., Raluy R. G. and Serra L. (2007). Life cycle assessment of water treatment technologies: wastewater and water-reuse in a small town. *Desalination*, **204**(1–3), 121–131.

Pasqualino J., Meneses M., Abella M. and Castells F. (2009). LCA as a decision support tool for the environmental improvement of the operation of a municipal wastewater treatment plant. *Environmental Science and Technology*, **43**(9), 3300–3307.

Pasqualino J., Meneses M. and Castells F. (2011). Life cycle assessment of urban wastewater reclamation and reuse alternatives. *Journal of Industrial Ecology*, **15**(1), 49–63.

Raluy R. G., Serra L., Uche J. and Valero A. (2004). Life-cycle assessment of desalination technologies integrated with energy production systems. *Desalination*, **176**, 445–458.

Raluy G., Serra L. and Uche J. (2006). Life cycle assessment of MSF, MED and RO desalination technologies. *Energy*, **31**, 2361–2372.

Rautenbach R. and Albrecht R. (1989). Membrane Processes. ISBN 0 471 911 100, John Wiley & Sons Ltd., New York, USA, p. 4.

SEM (2007). Spanish Environmental Ministery, 2007. La desalación como garantía de abastecimiento al litoral Español [Desalination as a guaranty for water supply within the Spanish coast]. http://www.mma.es/secciones/agua/programa/desalacion.htm (accessed May 2011).

Subramani A., Badruzzaman M., Oppenheimer J. and Jacangelo J. G. (2011). Energy minimization strategies and renewable energy utilization for desalination: a review. *Water Research*, **45**(5), 1907–1920.

Tarnacki K., Meneses M., Melin T., van Medevoort J. and Jansen A. (2011). Environmental assessment of desalination processes: Reverse osmosis and Memstill®, submitted to *Desalination*.

Vince F., Aoustin E., Bréant F. and Marechal F. (2008). LCA tool for the environmental evaluation of potable water production. *Desalination*, **220**(1–3), 37–56.

WDR (2008). A new approach to deep sea RO. *Water Desalination Report* **44**(30), 2–4.

Part 5

Water Footprint of Energy Production

By The Editors

FOREWORD

In the fossil fuel societies of the 20th century, the water footprint of energy was not well documented or understood. Nevertheless, in the frame of sustainable development, the water footprint of conventional and renewable energy sources is becoming one of the major criteria for assessing their sustainability. Water is a crucial component in the production of power (cooling towers, hydropower) and renewable energies (geothermal power, heating pumps, osmotic power, bio-ethanol, and so on). With the growing interest in renewable bioenergy, its high water demand presents another challenge, particularly in conditions of water scarcity.

Part 5 of this book "Water Footprint of Energy Production", (e.g. Chapters 21 and 22), is focussed on water footprint of biofuels, that is bio-ethanol and biodiesel.

A shift towards bioenergy requires more water. The average total water footprints (WF), that is the volume of freshwater used for the production of biofuels, are significantly higher than the renewables solar thermal energy and wind energy and of the non-renewables coal, natural gas and crude oil. Depending on the methods of utilisation and types of substrate used, the water demand is between 19 and 123 m^3 water/GJ electric energy for production of biogas, between 40 and 400 m^3/GJ for the production of bio-ethanol, and between 100 and 600 m^3/GJ for the production of biodiesel. Even in regions with comparatively large rainfall amounts, irrigation water might be necessary for the cultivation of energy crops.

The water transpiration coefficient of typical energy crops varies in the range of 0.2–0.4 m^3 per kg dry solids, while the water footprint of biogas production is over 0.4 m^3/GJ. This water demand is considerably smaller than that of bioethanol and biodiesel. In summary, energy crops do present a large water footprint and their cultivation seems solely reasonable in an area with sufficient water available for irrigation. In addition, the use of such crops for energy competes with that for food.

The most water-efficient way to generate bioenergy is to use total biomass, including the parts without an economic value, and generate heat. Electricity generation is the second best option. In principle, the generation of energy from already available residues, for instance from the food industry, should have priority over the production of energy crops, because the water demand is not the only major factor – further influences on the eco-system can be observed, such as deforestation to gain further cultivation areas, desertification, or nitrate input.

Chapter 21

Water for energy, the use of the water footprint for the assessment of water use for bioenergy

Winnie P. Gerbens-Leenes, Arjen Y. Hoekstra and Theo H. van der Meer

21.1 INTRODUCTION

Many countries have turned to bioenergy development to decrease CO_2 emissions, to promote rural development and to secure an energy source for the future (UNEP, 2009), including large developing countries like China (Yang *et al.* 2009) and India (Government of India, 2008). The developed countries of the European Union aim to replace 10% of transport fuels by renewables in 2020 (European Commission, 2009). At present, bioenergy, and especially first-generation biofuels, such as bio-ethanol and biodiesel, are at the centre of a polarized debate that addresses pro-poor policies related to food and land security, gender, income generation, social inclusion (Clancy, 2008) and environmental issues, such as land and water use (UNEP, 2009). Biofuel production diverts land, water and other resources away from food production (Fischer *et al.* 2009). Also, agricultural water use competes with other uses, such as urban supply and industrial activities (Falkenmark, 1989), while the aquatic environment shows signs of degradation and decline (Postel *et al.* 1996). A shift from fossil energy towards bioenergy puts additional pressure on freshwater resources. Nowadays, agricultural biomass production for food and fibre takes 92% of the global freshwater use (Mekonnen & Hoekstra, 2011a).

Bioenergy derives from biomass, for instance wood, crops and algae, or organic wastes, such as manure and crop residues. In developing countries, over 500 million households rely on traditional bioenergy, such as wood, for cooking and heating (UNEP, 2009). Biomass for bioenergy is applied to produce biofuels, but also to generate heat or electricity. When bioenergy is promoted, the focus is often on first-generation biofuels, such as bio-ethanol and biodiesel, rather than on traditional bioenergy.

Between 2000 and 2007, global ethanol production tripled and biodiesel production expanded eleven-fold (UNEP, 2009). Presently produced bio-ethanol and biodiesel are so-called first generation biofuels. They derive from the carbohydrate or the oil part of a crop (Blok, 2006; Worldwatch Institute, 2007). Today, almost all energy in society is needed as liquid transport fuel, as heat and as electricity. In industrialised countries, each of these demands accounts for 30% of the total energy demand (Blok, 2006; International Energy Agency (IEA), 2006). This energy mainly derives from non-renewable energy sources, such as crude oil, natural gas and coal (IEA, 2006).

For biomass production, agriculture requires freshwater for crop growth. A tool to calculate water needs for products is the water footprint (WF) concept (Hoekstra *et al.* 2011). The WF is an indicator of freshwater use. The WF of a product, for example bio-ethanol, is the volume of freshwater used to produce the ethanol in the complete supply chain. The tool has been applied to assess various bioenergy WFs. Gerbens-Leenes *et al.* (2009a) made an assessment of the total WF of primary energy carriers for four different countries. Gerbens-Leenes *et al.* (2009b) calculated green and blue WFs for bioenergy from the main global crops in the main producing countries. That study included heat, bio-electricity and first-generation liquid biofuels (bio-ethanol and biodiesel). A third study focussed on bio-ethanol produced in the twenty main sugar and maize producing countries and included green, blue and grey WFs. (Gerbens-Leenes & Hoekstra, 2011). Van Lienden *et al.* (2010) made an assessment of biofuel scenarios in a water perspective, making additional calculations for WFs of palm oil. Mellko (2008) used the WF methodology to assess EU WFs for transport biofuels. Recently, Mekonnen and Hoekstra (2010) have made an update of WFs of crops and derived products.

In this chapter, we review the WF bioenergy studies made so far. We include biofuels (bio-ethanol and biodiesel), heat from biomass and bio-electricity. We compare results generated using the WF methodology with results of other studies. Next, we compare WFs of bioenergy with WFs of fossil energy carriers (natural gas, coal and crude oil), nuclear energy and WFs of renewables (wind, solar thermal energy and hydropower). The chapter gives WFs of various types of bioenergy in m^3 per unit of energy (GJ) and covers the main producing countries, including developing countries, transition countries and industrialised countries.

21.2 BIOENERGY

In human history, people have used biomass, such as wood or peat, for heating and cooking. Although many substances originating from biomass, for example food or plastic, contain energy (Verkerk *et al.* 1986), energy analysis defines an *energy carrier* as a substance predominantly used for energy (Blok, 2006). An energy carrier derives from an energy source, non-renewable and renewable energy sources.

The FAO (2006) defines biomass as non-fossilized organic material. Examples are agricultural crops, wood, agricultural and forestry wastes, manure, microbial biomass, industrial waste and household waste. Biomass is an important renewable energy carrier (IEA, 2006; Blok, 2006), often converted into first-generation *biofuels*, such as bio-ethanol and biodiesel for transport (Minnesma & Hisschemöller, 2003; Blok, 2006; Worldwatch Institute, 2007).

Bioenergy derives from three categories: (i) perennial crops (e.g. trees); (ii) annual crops (e.g. cereals); and (iii) organic waste. Biomass provides starch, sugar and oil and contains cellulose. For first-generation biofuel production, industry applies part of a crop, namely the oils or carbohydrates, and generates wastes, basically the cellulose fraction. So far, people used the cellulose for energy by burning it for heat or to produce electricity. In the future, cellulose forms an attractive source for the production of liquid, next-generation biofuels (Blok, 2006; Worldwatch Institute, 2007), such as oil generated by pyrolysis, a process to thermo-chemically convert solid biomass into liquid oil (Mercader *et al.* 2010).

21.3 THE WATER FOOTPRINT CONCEPT

The water footprint (WF) of a product is defined as the volume of freshwater used for the production of a product at the place where it was actually produced (Hoekstra *et al.* 2011). In general, actual water contents of products are negligible compared with their WF. For many products, such as bioenergy, water use in product life cycles is dominated by the agricultural production stage. The WF includes three components – green, blue and grey water – and distinguishes between direct and indirect water use, considering water use along supply chains. It is a multi-dimensional indicator showing water consumption volumes by source and polluted volumes by type of pollution. The components of WFs are specified geographically and temporally. The green WF refers to the consumption of green water resources, that is rainwater stored in the soil as soil moisture and water that stays on top of the soil and on the vegetation; the blue WF refers to the consumption of blue water resources, that is, fresh surface and groundwater; the grey WF refers to pollution and is defined as the volume of freshwater required to assimilate the load of pollutants based on existing ambient water quality standards (Hoekstra *et al.* 2011).

21.4 THE WATER FOOTPRINT OF BIO-ENERGY

21.4.1 Bio-ethanol

Bio-ethanol is presently produced from sugar or starch. Gerbens-Leenes & Hoekstra (2010) performed a detailed study into the WF of bio-ethanol that included the main producing countries as well as the main producing states in the US. Figure 21.1 shows the US Corn belt including these main producing states.

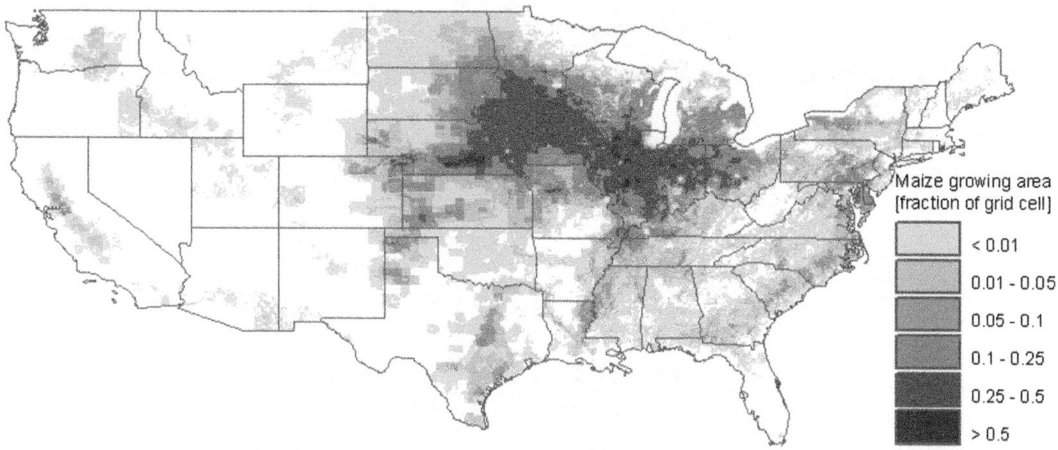

Figure 21.1 The US corn belt, including the main maize producing states Illinois, Indiana, Iowa, Michigan, Minnesota, Nebraska, North Carolina, Pennsylvania and Wisconsin (*Source*: Monfreda *et al.* 2008)

The study calculated green, blue and grey WFs for those crops that are presently used for bio-ethanol: sugar cane, sugar beet and maize. An earlier study made an assessment of green and blue WFs for all types of bioenergy, also including bio-ethanol (Gerbens-Leenes *et al.* 2009b). That study also calculated WFs for carbohydrate rich crops without large scale application for ethanol: barley, cassava, potato, rice, rye, sorghum and wheat. For the overview that we will give later on in the chapter, we use recent data of Mekonnen & Hoekstra (2010). The WF of bio-ethanol is a function of the WF of the crop used for the ethanol and the process water use.

Table 21.1 gives the WF multiplication ratio that can be applied to calculate the WF of bio-ethanol when the WF of the crop is known (Gerbens-Leenes & Hoekstra, 2011). It shows that maize is the most favourable crop with a multiplication ratio of 4.3. Sugar cane is the most unfavourable crop, requiring fifteen times more water to produce bio-ethanol than sugar cane itself (m^3 per ton). Figure 21.2 shows the WFs of sugar cane (m^3/ton), Figure 21.3 the WFs of sugar beet and Figure 21.4 the WFs of maize.

Table 21.1 WF multiplication ratios for bio-ethanol from sugar cane, sugar beet and maize.

Product	WF multiplication ratio
Cane bio-ethanol	14.8
Beet bio-ethanol	10.2
Maize bio-ethanol	4.3

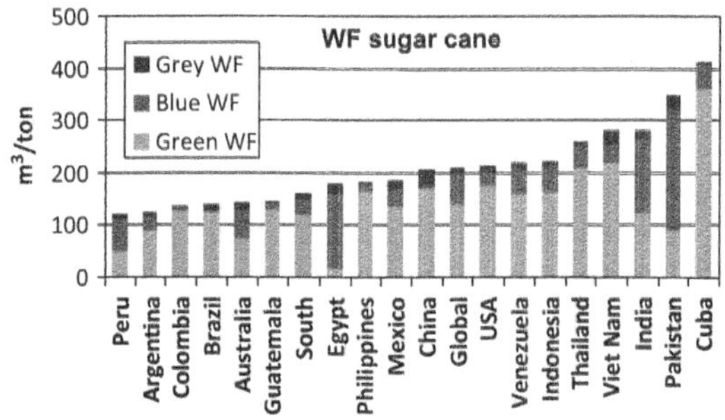

Figure 21.2 The water footprint of sugar cane for the main producing countries including the weighted global average value

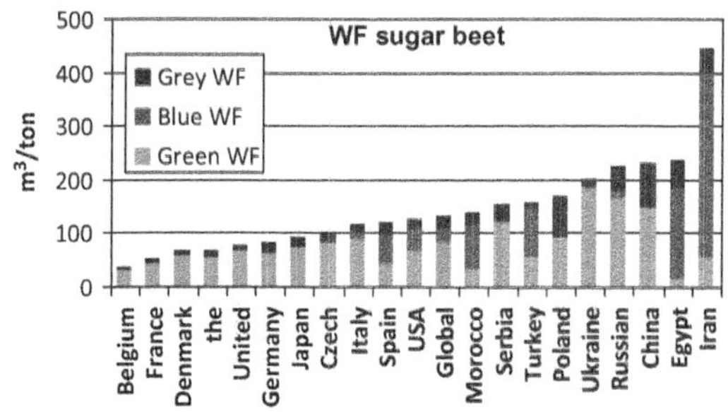

Figure 21.3 The water footprint of sugar beet for the main producing countries including the weighted global average value

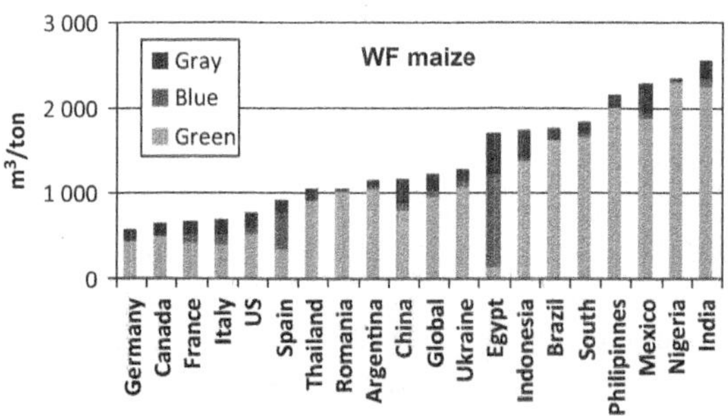

Figure 21.4 The water footprint of maize for the main producing countries including the weighted global average value

There are large differences for similar crops that are caused by differences in climate and in yields (ton per ha). Some countries have unfavourable WFs, far above the global average. For example, for sugar cane, Cuba, Pakistan, India, Vietnam and Thailand have unfavourable WFs. Egypt, India and Pakistan heavily rely on blue water for irrigation.

For sugar beet, Iran, China, Egypt and Ukraine have WFs far above the global average, while western European countries have WFs below the global average. Especially grey WFs are large in Poland and China, indicating that much nitrogen is leaking or applied in too large amounts, polluting water bodies. For maize, developing countries like India, Nigeria, Mexico and the Philippines have relatively large WFs, while developed countries like Germany, France, the US, Canada and Spain have relatively small WFs. For all three crops, Egypt almost completely relies on irrigation.

Table 21.2 gives the WFs for the main maize producing US states, as well as the US weighted average. Variation among the states is small, with Nebraska and Illinois the only exceptions. Nebraska uses a relatively large amount of blue water; Illinois has a large grey WF. US values, however, are much smaller than the global averages, indicating the relatively favourable production and climatic circumstances.

Table 21.2 Green, blue and grey WFs for 9 maize producing US states, the US weighted average and standard deviations.

US state	Green WF m^3 per ton	Blue WF m^3 per ton	Grey WF m^3 per ton
Illinois	578	5	192
Indiana	526	7	172
Iowa	553	2	177
Michigan	466	14	163
Minnesota	525	4	165
Nebraska	443	191	153
North Carolina	528	4	152
Pennsylvania	458	3	158
Wisconsin	465	3	158
US weighted average	522	63	176
US SD	±127	±63	±78

Figure 21.5 shows the global average green and blue WFs of bio-ethanol from sugar beet, sugar cane, potato, cassava, barley, rye, paddy rice, wheat and sorghum. The WF of bio-ethanol of sugar beet is smallest. The WF of bio-ethanol from potato is 100 m^3 per GJ; the WF of bio-ethanol from sorghum is more than four times the WF of bio-ethanol from potato. The remaining crops have WFs between these two extremes. The Figure also shows the large variation among green and blue WFs.

21.4.2 Biodiesel

Biodiesel derives from oil crops. In Europe, rapeseed is the dominant feedstock, with some sunflower. In the US, the main feedstock is soybean; in tropical and subtropical countries, palm, coconut and jatropha dominate (Worldwatch Institute, 2007). Gerbens-Leenes *et al.* (2009b) calculated WFs of biodiesel from rapeseed and soybean for the main producing countries. For jatropha, Gerbens-Leenes *et al.* (2009c) and Hoekstra *et al.* (2009b) calculated green and blue WFs for locations distributed over the *Jatropha curcas* belt, between 30° N and 35° S. For palm oil, Van Lienden *et al.* (2010) calculated the WF for the Philippines, Thailand, Indonesia, Malaysia and Honduras.

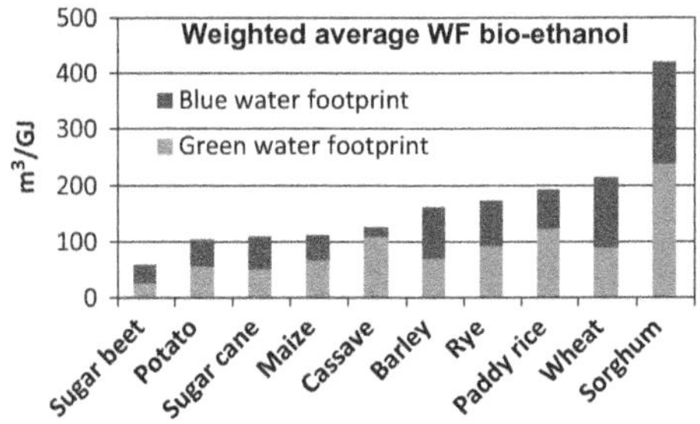

Figure 21.5 Global weighted average green and blue WFs of ethanol from ten crops

In Table 21.3, we present the recent WFs of biodiesel from Mekonnen and Hoekstra (2010). The green and blue WFs of jatropha are obtained from Hoekstra *et al.* (2009b). The WFs of biodiesel from jatropha oil are the largest, followed by WFs of sunflower oil. The WFs of biodiesel from rapeseed and palm oil are the smallest. The Table also shows that differences among countries are large.

Table 21.3 Green, blue and grey WFs of jatropha, palm, rapeseed and soybean oil for large producing countries and average values.

Type of oil	Country	Green WF m^3/GJ oil	Blue WF m^3/GJ oil	Grey WF m^3/GJ oil
Jatropha oil[a]	India	575	1116	–
	Guatemala	156	174	–
	Nicaragua	120	187	–
	Indonesia	184	109	–
	Brazil	160	91	–
	Average	**239**	**335**	**–**
Rapeseed oil[b]	India	141	129	20
	China	118	0	42
	Germany	86	0	21
	Global weighted average	**145**	**20**	**29**
Soybean oil[b]	United States	250	15	2
	Argentina	335	1	2
	Italy	185	21	0
	Global weighted average	**326**	**11**	**6**
Palm oil[b]	Philippines	144	0	13
	Thailand	97	0	8
	Indonesia	128	0	9
	Honduras	102	0	4
	Global weighted average	**150**	**0**	**6**
Sunflower oil[b]	France	187	1	72
	Spain	353	128	16
	Global weighted average	**428**	**21**	**28**

Source: [a]Hoekstra *et al.* (2009b); [b]Mekonnen & Hoekstra (2010)

21.4.3 Heat and bio-electricity

When crops are burned, they provide heat. Gerbens-Leenes *et al.* (2009a) calculated WFs of energy from total biomass (m^3/GJ) for perennial crops (poplar and miscanthus) and for annual crops (cassava, coconut, cotton, groundnuts, maize, palm oil, potato, wheat, rapeseed, sugar beet, sugar cane, sunflower and soybean) for four countries: the Netherlands, Brazil, the US and Zimbabwe.

Table 21.4 gives WFs of heat from total biomass for two perennial crops, poplar and miscanthus, and four annual crops (maize, sugar beet, sugar cane and sunflower) for three countries. The table shows that poplar and miscanthus do not have a more favourable WF than food crops. In the Netherlands, the WF of heat from maize is half the WF of heat form poplar or miscanthus. In the US, also maize is the most favourable crop in terms of water, in Brazil sugar cane. When the WFs of bio-ethanol are compared to the WFs of heat, the results show that it is much more favourable to generate heat than to produce a biofuel. Dependant of the crop type, WFs for heat are much smaller than for bio-ethanol. WFs of bio-ethanol from cassava are 50% larger than WFs of heat from cassava; WFs of ethanol from barley are four times the WF of heat from barley. Also for soybean, rapeseed and jatropha, the WFs for biodiesel are two to four times larger than for heat.

Table 21.4 Water footprint of heat for two perennial crops, poplar and miscanthus and five annual crops, sunflower, sugar beet, sugar cane and maize, grown in three countries.

Crop	Water footprint of heat total biomass (m^3 per GJ)		
	The Netherlands	United States	Brazil
Maize	9	21	38
Sugar beet	16	25	–
Sugar cane	–	31	29
Miscanthus	20	37	49
Poplar	22	42	55
Sunflower	27	61	54

For electricity generation from biomass, industry applies heat that comes available from biomass combustion. The bioenergy generated as electricity (GJ/ton) depends on the power plant efficiency to transform biomass-heat into bio-electricity. When transforming thermal energy into mechanical energy, the thermal efficiency of a heat engine is the percentage of energy transformed into work. Theoretically, there is a maximum to the electricity amounts generated in a power plant (Blok, 2006). A new technology is gasification of biomass for electricity production, termed 'Biomass fired Integrated Gasifier Combined Cycle (BIG/CC). It can be applied for biomass with a moisture content below 70% and an ash content below 10–20% (Dry Matter) (Faay, 1997). Temperatures are relatively low however, 720°K, reaching an efficiency of 59% (Blok, 2006). The WF of electricity can be derived from the WFs of heat assuming this conversion efficiency.

21.5 COMPARISON WITH OTHER ENERGY CARRIERS

At present, important energy carriers include the fossil energy carriers (crude oil, coal, and natural gas), uranium, and electricity from hydropower (IEA, 2006). Promising renewables are solar and wind energy.

We compare biofuel WFs with WFs of the important energy carriers. For crude oil, coal, natural gas, and uranium, we derive data from Gleick (1994). For electricity from hydropower, we estimate the global WF by dividing global evaporation of reservoirs (Shiklomanov, 2000) by the hydroelectric generation (Gleick, 1993) for 1990. Next, we compare these results with information on WFs of hydropower from Mekonnen and Hoekstra (2011b).

Table 21.5 gives an overview of the WFs of these energy carriers. Although the data of Gleick (1994) are rather old, they give an indication of the WFs. WFs of crude oil, coal, natural gas, uranium, solar thermal energy and wind are all small compared to the WFs of biomass and hydropower.

Table 21.5 Water footprint of fossil energy carriers, electricity from solar thermal energy, electricity from wind energy, and hydropower (m^3/GJ).

Energy carriers	Water footprint (m^3/GJ)
Wind energy[a]	0.0
Nuclear energy[a]	0.1
Natural gas[a]	0.1
Coal[a]	0.2
Solar thermal energy[a]	0.3
Crude oil[a]	1.1
Hydropower[b]	68

Source: [a]Gleick, 1994; [b]Mekonnen & Hoekstra, 2011b

Table 21.5 shows that the average total WFs of the renewables solar thermal energy and wind energy and of the non-renewables coal, natural gas and crude oil are small. The Table also gives results for hydropower. Gerbens-Leenes *et al.* (2009a) have found an average global value of 22 m^3 per GJ, where Mekonnen and Hoekstra (2011b) have arrived at an average value of 68 m^3 per GJ. That study has found a great range of WF values per unit of generated electricity. In China, the Lubuge power plant, for example, uses only 0.5 m^3 of water per GJ of electricity, where the Akosombo dam and Kpong power plant in Ghana use 850 m^3 per GJ, the largest value for energy found so far.

In general, apart from bioenergy, the WF of hydropower is much larger than the WF of other energy sources. The weighted global average WF for bio-ethanol is 165 m^3 per GJ and for biodiesel 313 m^3 per GJ. When these data are compared to the results presented in Table 21.5, the large WF of bioenergy, especially for the first generation biofuels, becomes clear.

21.6 THE WATER FOOTPRINT OF NEXT – GENERATION BIOFUELS

Next-generation biofuels have smaller WFs than first-generation biofuels, because not only part of a crop is used for fuel, but all cellulosic biomass is applied as a feedstock. There are two basic conversion technologies, thermo-chemical conversion, for example pyrolysis, and biochemical conversion, for example biological conversion into ethanol (Worldwatch Institute, 2007). At present, research is done on the development of next-generation biofuels made from agricultural waste, such as pyrolysis oil and ethanol. Pyrolysis oil, however, still does not have the quality of the first-generation oil. It is unstable, has a high acidity and viscosity and relatively small energy content. It is not miscible with fossil fuels and is corrosive to engines (Mercader *et al.* 2010). Another problem is the instability of pyrolysis oil, especially during storage, referred to as "aging" (Oasmaa & Czernik, 1999). Biological conversion into ethanol, for example, by fermentation, also finds itself in an experimental stage (Worldwatch Institute, 2007; Park *et al.* 2010). When the production of next-generation biofuels is technically and economically possible, large amounts of feedstocks are available. Another promising development is the introduction of bio-energy from algae.

21.7 DISCUSSION

All studies mentioned here made many assumptions. For example, with respect to agricultural yields, all studies have taken actual yields, which may be increased in the future without increasing water use per unit of product, lowering WFs per unit of energy. For the efficiency of obtaining electricity or biofuels from biomass, Gerbens-Leenes *et al.* (2009b) and Mekonnen & Hoekstra (2010) made optimistic assumptions, by taking theoretical maximum values or values that refer to the best available technology. This means that the resulting WFs are conservative.

In assessing the WF of heat, bio-electricity and biofuels from biomass, the studies took the *gross* energy output from crops into account and did not study energy inputs in the production chain, for example, energy requirements in agriculture (e.g. energy use for fertilizer and pesticide production) or energy use for the industrial biofuel production. For high-input agricultural systems, energy input is substantial (Giampietro & Ulgiatti, 2005; Pimentel & Patzek, 2005), so that net energy yields are smaller than estimated. This means that the WF of bioenergy from agricultural systems with relatively large energy inputs is probably larger than presented in here.

When results from our recent studies are compared to a study in China, our results are smaller. Yang *et al.* (2009) calculated the total WF of ethanol from sugar beet grown in China and arrived at 96 m^3 per GJ, where Mekonnen and Hoekstra (2010) calculated 57 m^3 per GJ.

Another approach has been applied by Chiu *et al.* (2009) who measured actual irrigation in the US and arrived at values between 0.3 m^3 per GJ (Iowa & Kentucky) and 91.4 m^3 per GJ (California) with a weighted US average of 6 m^3 per GJ. The average value is the same as the blue water footprint for bio-ethanol in the US of 6 m^3 per GJ calculated by Mekonnen & Hoekstra (2010).

Berndes (2002) has also given values for crop evapotranspiration per unit of bioenergy. Crop evapotranspiration for biodiesel from rapeseed was 100–175 m^3 per GJ, in line with results from Mekonnen and Hoekstra (2010) with green and blue WFs of 145 and 20 m^3 per GJ.

For ethanol from sugar cane Berndes calculated values of 37–155 m^3 per GJ, where Mekonnen and Hoekstra (2010) arrive at 30 (weighted average green WF) and 31 (weighted average blue WF) m^3 per GJ. The ethanol from sugar beet values are 71–188 m^3 per GJ, where Mekonnen and Hoekstra (2010) arrive at 31 (green WF) and 10 (blue WF). The ethanol from maize values are 73–346 m^3 per GJ, where Mekonnen and Hoekstra (2010) arrive at 94 (green WF) and 8 (blue WF) m^3 per GJ, ethanol from wheat values are 40–351 m^3 per GJ, where Mekonnen and Hoekstra (2010) arrive at 126 (green WF) and 34 (blue WF) m^3 per GJ.

This chapter gives WFs that are based on estimates of freshwater requirements in crop production in combination with theoretical maximum conversion efficiencies in heat, bio-electricity and biofuel production. The studies have integrated data from several sources, each adding a degree of uncertainty. Meteorological data, for example, are averages over several years rather than data for a specific year and do not reflect annual variations. Calculations of crop water requirements are sensitive to input of climatic data and assumptions concerning the start of the growing season. The data on energy carriers from Gleick (1994) give an indication of water requirements, but are probably out-dated.

Therefore, the results are indicative. The differences among WFs, however, are large and support our general conclusions with respect to relative WFs of different types of bioenergy, crops and countries.

21.8 CONCLUSIONS

The most water efficient way to generate bioenergy is to use total biomass, including the parts without an economic value, and generate heat. Electricity generation is the second best option.

The production of first-generation biofuels requires more water per unit of energy than bioelectricity. When biofuels are produced, it is more water efficient to produce ethanol than biodiesel.

The green WF of a typical biodiesel crop, rapeseed, is two times larger than the WF of ethanol from sugar cane and four times larger than ethanol from sugar beet. The blue WF of ethanol from sugar cane, however, is larger than the blue WF of rapeseed oil, 25 versus 20 m^3 per GJ. The grey WF, an indicator of water pollution, is largest for rapeseed oil and smallest for ethanol from sugar cane.

WFs for similar energy carriers show large variation and depend on the type of crop applied, location of production and agricultural practice. The global weighted WF for ethanol increases in the following order: sugar beet, sugar cane, potato, maize, cassava, paddy rice, barley, rye, wheat and sorghum, and ranges between 31 and 281 m^3 per GJ. Blue WFs are between 0 (cassava) and 34 m^3 per GJ (wheat and rice). The differences among countries are large.

The weighted global average green WF of biodiesel increases in the following order: rapeseed oil (145 m^3 per GJ), palm oil (150 m^3 per GJ), soybean oil (326 m^3 per GJ), jatropha oil (239 m^3 per GJ) and sunflower oil (428 m^3 per GJ). Blue WFs are small, except for jatropha oil (335 m^3 per GJ).

When WFs of bioenergy are compared to WFs of hydropower or traditional fuels there are large differences. Especially hydropower shows a wide range of values for WFs per unit of electricity generated, ranging from 0.5 to 850 m^3 per GJ. This is the largest value for an energy WF found so far.

A shift towards bioenergy brings a need for more water. A shift towards hydropower can bring a need for more water, but this depends on the climate where and the way in which the power is generated.

This chapter gives new insights with respect to the large impact of energy from biomass on the use of freshwater resources. This knowledge can be a valuable contribution to research concerning energy needs and freshwater availability for the near future.

REFERENCES

Berndes G. (2002). Bioenergy and water, the implications of large-scale bioenergy production for water use and supply. *Global Environmental Change*, **12**, 253–271.

Blok K. (2006). Introduction to Energy Analysis. Techne Press, Amsterdam, The Netherlands.

Chapagain A. K. and Hoekstra A. Y. (2004). Water Footprints of Nations. Value of Water Research Report Series No.16 UNESCO-IHE, Delft, The Netherlands.

Chiu Y., Walseth B. and Suh S. (2009). Water embodied in bio-ethanol in the United States. *Environmental Science and Technology*, **43**, 2688–2692.

Clancy J. S. (2008). Are biofuels pro-poor? Assessing the evidence. *European Journal of Development Research*, **20**(3), 416–431.

European Commission (2009). Renewable Energies. http://ec.europa.eu/energy/renewables/index_en.htm (accessed 12 January 2010)

Faay A. (1997). Energy from Biomass and Waste. PhD thesis, University of Utrecht, The Netherlands.

Falkenmark M. (1989). Comparative hydrology – a new concept. In: Comparative Hydrology. An Ecological Approach to Land and Water Resources, M. Falkenmark and T. Chapman (eds), UNESCO, Paris, France, pp. 10–42.

FAO (2006). Introducing the International Bioenergy Platform. FAO, Rome, Italy.

FAO (2007). CROPWAT 4.3 Decision Support System. Available at www.fao.org/nr/water/infores_databases_cropwat.html (accessed 3 January 2007)

FAO (2008). FAOSTAT-Agriculture. Available at http://faostat.fao.org/site/339/default.aspx (accessed 12 January 2008)

Fischer G., Hizsnyik E., Prieler S., Shah M. and Van Velthuizen H. (2009). Biofuels and Food Security. International Institute for Applied Systems Analysis, Laxenburg, Austria. Available at: http://www.iiasa.ac.at (accessed 21 June 2010)

Gerbens-Leenes P. W. and Hoekstra A. Y. (2011). The water footprint of sweeteners and bio-ethanol. *Environment International*, DOI: 10.1016/j.envint.2011.06.006.

Gerbens-Leenes P. W., Hoekstra A. Y. and Van der Meer T. H. (2009a). The Water footprint of bioenergy and other primary energy carriers. *Ecological Economics*, **68**(4), 1052–1060.

Gerbens-Leenes W., Hoekstra A. Y. and Van der Meer T. H. (2009b). The water footprint of bioenergy. *Proceedings of the National Academy of Science*, **106**(25), 10219–10223.

Gerbens-Leenes W., Hoekstra A. Y. and Van der Meer T. H. (2009c). Reply to Maes *et al.*: A global estimate of the water footprint of *Jatropha curcas* under limited data availability. *Proceedings of the National Academy of Science*, **106** (40), E113.

Giampietro M. and Ulgiati S. (2005). Integrated assessment of large-scale biofuel production. *Critical Reviews in Plant Sciences*, **24**(5), 365–384.

Government of India, Ministry of New & Renewable energy (2008). National Policy on Biofuels. http://mnes.nic.in/policy.biofuel-policy.pdf (accessed 5 January 2010)

Gleick P. H. (1993). Water and Energy. In: Water in Crisis. A Guide to the World's Freshwater Resources. Oxford University Press, New York, Oxford, pp. 67–79.

Gleick P. H. (1994). Water and energy. *Annual Review of Energy and the Environment*, **19**, 267–99.

Hoekstra A. Y., Chapagain A. K., Aldaya M. M. and Mekonnen M. M. (2011) The Water Footprint Assessment Manual. Setting the Global Standard. Earthscan Publishing, London/Washinton, DC.

Hoekstra A. Y., Gerbens-Leenes W. and Van der Meer Th. H. (2009b). Reply to Jongschaap *et al.*: The water footprint of Jatropha curcas under poor growing conditions. *Proceedings of the National Academy of Science*, **106**(42), E119.

International Energy Agency (IEA). 2006. World Energy Outlook 2006. OECD/IEA, Paris, France.

Mekonnen M. M. and Hoekstra A. Y. (2010). The Green, Blue and Grey water Footprint of Crops and Derived Crop Products. Value of Water Research Report Series No. 47, UNESCO-IHE, Delft, The Netherlands.

Mekonnen M. M. and Hoekstra A. Y. (2011a). The Water Footprint of Electricity from Hydropower. Value of Water Research Report Series No. 51. UNESCO-IHE, Delft, The Netherlands.

Mekonnen M. M. and Hoekstra A. Y. (2011b). The Water Footprint of Electricity from Hydropower. Value of Water Research Report Series No. 51. UNESCO-IHE, Delft, The Netherlands.

Mellko H. (2008). Water Footprint for Biofuels for Transport – Finland and the EU 2010. Master of Science thesis, Helsinki University of Technology, Helsinki, Finland.

Mercader F. M., Groeneveld M. J., Kersten S. R. A., Venderbosch R. H. and Hogendoorn J. A. (2010). Pyrolysis oil upgrading by high pressure thermal treatment. *Fue*, **89**(10), 2829–2837.

Minnesma M. and Hisschemöller M. (2003). Biomassa – een wenkend perspectief (Biomass – a challenging perspective). *Instituut voor Milieuvraagstukken (IVM)* (Institute for Environmental Studies (IVM)). Free University, Amsterdam, The Netherlands.

Monfreda C., Ramankutty N. and Foley J. A. (2008). Farming the Planet: 2. Geographic Distribution of Crop Areas, Yields, Physiologic Types, and Net Primary Production in the year 2000. Global Biochemical Cycles 22, GB 1022. www.geog.mcgill.ca/landuse/Data/175crops2000 (retrieved 18 September 2008)

Oasmaa A. and Czernik S. (1999). Fuel oil quality of biomass pyrolysis oils-state of the art for the end users. *Energy Fuels*, **13**, 914–921.

Park I., Kim I., Kang K., Sohn H., Rhee I., Jin I. and Jang H. (2010). Cellulose ethanol production from waste newsprint by simultaneous saccharification and fermentation using Saccharomyces cerevisiae KNU5377. *Process Biochemistry*, **45**, 487–492.

Pimentel D. and Patzek T. W. (2005). Ethanol production using corn, switch grass, and wood: biodiesel production using soybean and sunflower. *Natural Resources Research*, **14**(1), 65–76.

Postel S. L., Daily G. C. and Ehrlich P. R. (1996). Human appropriation of renewable freshwater. *Science*, **271**, 785–788.

Shiklomanov I. A. (2000). Appraisal and assessment of world water resources. *Water International*, **25** (1), 11–32.

UNEP (2009). Towards sustainable production and use of resources: Assessing Biofuels. United Nations Environment Programme.

Van Lienden A. R., Gerbens-Leenes P. W., Hoekstra A. Y. and Van der Meer Th. H. (2010). Biofuel Scenarios in a Water Perspective. Changes in Water Footprints Due to Changes in the Transport Sector. Value of Water Research Report Series No. 43. UNESCO–IHE, Delft, The Netherlands.

Verkerk G., Broens J. B., Kranendonk W., Puijl van der F. J., Sikkema J. L. and Stam C. W. (1986). *Binas, informatieboek vwo-havo voor het onderwijs in de natuurwetenschappen. tweede druk* (Binas, information book vwo-havo for the education in natural sciences, second edition). Wolters-Noordhof bv., Groningen, The Netherlands.

Worldwatch Institute (2007). Biofuels for Transport. Global Potential and Implications for Sustainable Energy and Agriculture. Earthscan, London, Sterling, VA.

Yang H., Zhou Y. and Liu J. (2009). Land and water requirements of biofuels and implications for food supply and the environment in China. *Energy Policy*, **37**, 1876–1885.

Chapter 22

Water demand for the production of renewable energy from crops

Karl-Heinz Rosenwinkel, Linda Hinken and Selin Ay

22.1 INTRODUCTION

Water is an essential component for the production of carbohydrates in connection with CO_2 and thus crucial for plant growth, as carbohydrates are the basis for all other organic metabolites. However, the carbohydrate production amounts to a comparatively low part of only 1% of the total water demand; the remaining water demand is used for transpiration, which depends on the plant type, the course of development, the nutrient supply, and other factors (Kaltschmitt *et al.* 2009).

The production of energy from crops makes up for a considerable fraction of regenerative energy. One advantage over other types of renewable energy is that the energy gained in biogas plants can be stored as methane and used according to demand. In the last 10 to 15 years, the cultivation of energy crops has increased considerably.

Compared to food production, the biogas production from the cultivation of energy crops depends less on the quality of the crops than on the yield per hectare. However, water is often the major factor which reduces the yield; thus, with rising numbers of population, water availability and water quality must be considered to be critical. This chapter discusses the water demand for the growth and usage of various plants for energy generation, the respective share of the water supply, and the limits of cultivation and problems in relation to the growth areas.

22.2 BIO-ENERGY GENERATION

22.2.1 General considerations

Renewable resources may yield heat (combustion), electric energy (via biogas), and fuel. One crucial factor in the total energy generation is the combustion of regenerative raw materials for the production of heat or hot water, with wood as the major resource used for combustion. This method of energy utilization is particularly popular in developing or transition countries. In the industrialised countries, these raw materials have generally been replaced by fossil fuels (coal, natural gas, mineral oil). According to diverse scenarios (Varis, 2007), this way of energy generation will in future rather decrease; thus, it will be of secondary importance in the discussion below.

For the generation of electric energy, renewable resources (energy crops) are anaerobically converted into biogas (methane, carbon dioxide) in the fermenters of biogas plants via a four-stage anaerobic process. After the disintegration/hydrolysis stage, in which the available solids are converted first into easily convertible dissolved substances such as amino acids and mono-saccharides, the further conversion is achieved by various groups of bacteria and archaea via acidogenesis (production of organic acids and alcohols), acetogenesis (production of acetate), and methanogenesis (production of methane from acetate or from CO_2 and H_2). A large number of plants can be used as raw materials, for instance, maize, rye or potatoes.

For the direct production of fuels, various methods are employed. For the production of bio-diesel, crude oil is produced first, which is then subjected to ester exchange to produce bio-diesel and glycerine. Bio-ethanol is produced through the fermentation of the available raw materials through the addition of yeast, followed by distillation of the emerging ethanol. Typical raw materials for the production of biogenous fuels are wheat, sugar-beets, cane sugar, and rapeseed.

22.2.2 Current state of bio-energy production

In recent years, the generation of energy from energy crops has strongly increased, particularly in Germany. In Europe, agricultural biogas plants have a share of about one third of the entire biogas production; further biogas is produced from wastewater and at landfills. Germany produces 76% of the biogas amount generated in Europe; of these, 25% are produced in Lower Saxony (Höher, 2009).

In Germany, there are about 3,300 facilities for bio-energy production with an installed capacity of 1,000 MW, of which 600 plants with an installed capacity of 300 MW are located in Lower Saxony. The number of plants, the entire capacity, and thus the demand for growth areas will continue to rise. In 2005, the demand for land in Lower Saxony amounted to 72.000 ha (220 MW); for 2010, it was already estimated to be 130,000 ha. (Antony, 2007)

The renewable energy accounts for 10.9% of the total energy demand in Germany of 9,060 PJ/yr (e.g. 2,500 TWh/yr), of which 7.9% is produced from biomass (sum of liquid and solid biomass, biogas, landfill gas, sewage gas, the biogenous part of waste, and bio-fuels). Concerning the generation of electric energy in Germany, the renewable energies amount to 103.5 TWh/yr, which may correspond to 17% of the total electric energy generated. The share of biogas in the consumption of electric energy from renewable resources is 12.9%. In the generation of heat, biomass has the largest amount of the heat produced from regenerative energies, namely 92% at 136.1 TWh/yr. A further amount of 35.7 TWh/a of biogenous fuels (biodiesel, bioethanol, bio-oil) was produced, which is a fraction of the entire fuel consumption of 5.8%. (State of 2010, Böhme *et al.* 2011).

Varis (2007) evaluated various scenarios of bio-energy production, distinguishing between non-commercial or traditional bio-energy (from untreated resources, such as firewood and manure) and commercial bio-energy (treated resources, including biodiesel, bioethanol, charcoal). Europe and North America predominantly produce commercial bio-energy, which at 550 TWh/yr in Western Europe and 1.100 TWh/yr in North America amount to about 4% of the respective entire energy production (State of 2000). This is different, for instance, in Latin America or Central Asia: Latin America produces about 1,750 TWh/yr (16%) of non-commercial bio-energy and about 250 TWh/yr (2%) of commercial bio-energy. In Central Asia, the figures are 1,900 TWh/yr of non-commercial bio-energy and 700 TWh/yr of commercial bio-energy. In these two areas, the bio-energy production make up for about 17–18% of the entire energy production.

22.3 WATER DEMAND

22.3.1 Water demand of energy crops

In the production of energy from regenerative resources, water is needed for the cultivation of the energy crops, for the transport of the resources to the production sites and for the energy generation. The largest fraction (>90%) of the water demand is needed for the cultivation of the energy crops, which means that the individual water demand of the respective crop is the crucial factor.

The major fraction of the water demand of a plant is used for transpiration. Generally, the water demand of a plant depends on various factors, for instance, crop species, development stage, fertilizer supply, stand density, type and quality of the soil, rainfall, humidity, light, radiation, temperature and transpiration. The maximum daily water demand in Europe is about 6 mm/d (Schmied, 2004).

One important factor for the water consumption efficiency is the crop type: C4 plants use one metabolism path to initially fix carbon dioxide for photosynthesis; then they convert the CO_2 into carbohydrates in the Calvin cycle like C3 plants. The C3 plants account for about 95% of the plant biomass from soil. They are inferior to C4 plants at times of drought, high temperatures and nitrogen deficit. Most C4 crops originally come from tropical or subtropical areas and are not very productive at low temperatures. The C4 plants used as energy crops are, for instance, maize, millet, and sugar cane (Berndes, 2008). Jörgensen & Schelde (2001) have compiled the water consumption efficiency as g of dry solids produced per kg of water used for various energy crops. In spite of large deviations of these values, it could be observed that the water consumption efficiency of C4 plants is about twice as high as that of C3 plants.

The amount of water, which a plant exudes as steam at the foliar surface to produce new dry matter is referred as transpiration coefficient. At dense plant stands, this coefficient is nearly equivalent to the evapotranspiration (the evaporation can then be neglected).

Table 22.1 summarises data on the water demand and yield of typical energy crops in Lower Saxony. The transpiration coefficient varies between 200 L/kg DS (dry solids) for maize or potatoes and 400 L/kg DS for several kinds of grain. The annual water demand related to the yield amounts to 250–800 mm for maize, for instance, with an average value of 450 mm.

Table 22.1 Transpiration coefficient and yield of typical energy crops (Groß & Simon, 2011).

Crops	Water demand	Transpiration coefficient[1] (L/kg DS)	Yield	Energetic utilisation
Potatoes	Medium	210–230	38.2 t FM/ha · yr (in Lower Saxony for 2010)	Tuber: Bio-ethanol Dried leaves: Bio-fuel Biogas
Silage maize	Low	200–300	12.5–26 t DS/ha · yr average in Germany: 15 t DS/ha.yr	Biogas Bio-ethanol Energy product
Sunflower	Medium	300–400	10–15 t DS/ha · yr	Biogas
Winter barley		400	as complete plant silage: 10–15 t DS/ha · yr	Biogas Grains: also Bioethanol
Winter rapeseed	Low to medium	300–400	grain yield: 3–4 t DS/ha · yr complete plant silage: 8–12 t DS/ha · yr	Biogas Bioethanol Further fuels (Rapeseed methyl-ester/Biodiesel, rapeseed oil)

(*Continued*)

Table 22.1 Transpiration coefficient and yield of typical energy crops (Groß & Simon, 2011) (*Continued*).

Crops	Water demand	Transpiration coefficient[1] (L/kg DS)	Yield	Energetic utilisation
Winter rye	Medium	200–400	complete plant silage: 11–16.4 t DS/ha · yr	Complete plant: Biogas Grains: Bioethanol
Sugar beet	Medium to high	300–400	15–23 t DS/ha · yr	Bioethanol Biogas

Sources: Water demand: Groß & Simon, 2011; ZFT, 2011; Yield: LWK-Nds., 2010; DS = Dry Solids, FM = Fresh Mass
[1]the transpiration coefficient describes the water demand for biomass growth (per kg dry matter)

Different approaches have been developed for the calculation of the evapotranspiration for various regions. Empirical approaches are those of Thornthwaite (1948), on the basis of temperature measurements, Haude (1954), depending on the saturation vapour pressure at the temperature prevalent at 2 o'clock pm, or the formula according to Blaney-Criddle (1950, modified in 1977). This considers the daily temperature and the number of hours of sunshine, as well as empirical correction factors such as humidity, sunshine duration, and wind force (Cepuder *et al.* 2010). A more exact estimation is possible by using the Penman-Monteith method (PMM), which is recommended by the Food and Agriculture Organization of the United Nations (FAO) for the determination of evapotranspiration and considers a radiation term, as well as an aerodynamic evaporation term. In a simplified manner, the calculation of the evaporation will be run with the Blaney-Criddle formula, in which the evaporation is rated slightly less high than in the PMM. (Cepuder *et al.* 2010).

From the water demand of a crop and the expected rainfall amount, the necessary amount of irrigation water can be determined. The surface water or ground water used for the irrigation during the growth of the crops is referred to as "blue water", the rainwater which evaporated during the growth period as "green water" (Gerbens-Leenes *et al.* 2009).

22.3.2 Water demand for the substrate production of biogas plants

Kaltschmitt *et al.* (2009) compared the dry solids yield (t DS/ha) to the methane yield (L/ha) of the energy crops, such as maize, green rye, sugar millet, pasture grass, and winter wheat. The connection is almost linear for all plants and amounts to 300–400 L methane/kg DS. Thus, an essential criterion for a high area-related methane yield is a high biomass yield. Silage and energy maize for anaerobic fermentation reaches yields of up to 25 t DS/ha · yr, depending on location, water supply, and plant variety, which allows for a methane yield potential of up to 9,000 m^3/ha · yr (Kaltschmitt *et al.* 2009).

In Table 22.2, the area-specific methane yield of the energy crops has been compiled – the figures varying from 2,000 to 8,000 m^3/ha · yr. The specific methane yield for potatoes, sunflowers, and winter rapeseed is comparatively low. Anooshiravan (2004) found higher yields for potatoes (3,960 m^3/ha · yr) and also for beets (9,350 m^3/ha · yr); the results concerning the specific methane yield of grains and silage maize are in the range of the figures in Table 22.2.

Furthermore, Table 22.2 summarizes the results for the specific water demand, which is between 250 and 1,600 L water/m^3 methane or between 19 m^3 and 123 m^3 water/GJ electric energy (assuming that electric efficiency = 36%). Gerbens-Leenes *et al.* (2009) have in their study provided a global survey of the water footprint (WF) of typical energy crops. They found large differences of the figures between the various countries. The weighted average values deviate from the data in Table 22.2 in some places and show that in regard to the water demand, maize, sugar beet, and sugar cane are the most suitable crops at 50 m^3/GJ electric energy (sum of green and blue water). Rapeseed and jatropha have the highest water demand at almost 400 m^3/GJ electric energy. Rye, wheat, and potatoes lie in the medium range at 77, 93, and 105 m^3/GJ electric energy, respectively.

Table 22.2 Specific methane yield and water demand of typical energy crops.

Crops	Methane potential (L/kg DS)	Specific methane yield (m^3/ha · yr)	Specific water demand (L water/m^3 methane)	Sources (methane potential)
Potatoes (raw, medium starch contents)	350	3000	600–660	Calculated from Keymer, 2004
Silage maize	365	4560–9500	550–820	
Sunflower (green, incipient bloom)	250	2500–3800	1200–1600	
Winter barley	395	3950–5925	1000	Calculated from Linke *et al.* 1999
Winter rapeseed (complete plant silage)	256	2000–3000	1170–1560	Calculated from Sticksel *et al.* 2010
Winter rye	380	4180–6250	252–1050	Calculated from Linke *et al.* 1999
Sugar beet (fresh)	320	4800–7400	935–1250	Calculated from Keymer, 2004

Because of the comparatively high biogas yield, maize is an energy crop which is used very often for energy production. Additionally, as a C4 plant, maize has only a low water demand per t of DS (high water efficiency). The transpiration coefficient is relatively low at ca. 200–300 L per kg of produced dry solids. However, there may still occur bottlenecks in the water supply if high maize amounts are cultivated (Groß & Simon, 2011).

Figure 22.1 shows in an exemplary way the water demand of maize from evaporation (calculated according to the Blaney-Criddle formula) for the region of Hannover compared to the available rainfall. It shows clearly that the water demand during the main vegetation period is not completely covered even in regions with comparatively large rainfall amounts. There may occur water shortages which will necessitate irrigation.

With an operating time of about 8,000 h/yr, a typical biogas plant in Germany with an electric capacity of about 500 kW electricity will produce 4 million kWh/yr of electric energy. If the supply with substrate is to be covered only with maize with a specific methane yield of 7,300 m^3/ha or about 26,000 kWh-electricity/ha, the required cultivation area will amount to 154 ha. The water demand for the cultivation of the substrates (calculated from the transpiration coefficients) lies between 560,000 and 832,000 m^3 or between 360 and 540 mm. During the growth period, the rainfall will amount to about 450 mm, which means that up to 100 mm of water must be supplied through irrigation (blue water).

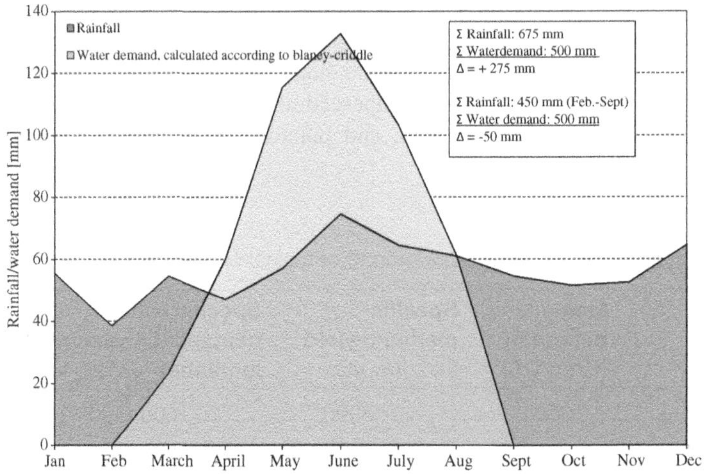

Figure 22.1 Water demand of maize and rainfall in the Hannover region (Schmied, 2004)

The specific water demand is up to 0.23 m^3/kWh-electricity. Depending on the region and the rainfall amounts, the share of blue water varies considerably. In the model region (Lower Saxony), the amount of blue water is about 0.04 m^3/kWh-electricity, in Southern Europe, it is about twice as high at 0.08 m^3/kWh-electricity. The water demand for 10% of the consumption of electric energy of a person from Western Europe (~7,000 kWh/cap · yr) is 150 m^3/yr, of which 30–60 m^3/yr would have to be supplied as blue water.

22.3.3 Biogenous fuels

In this section, the concept of water footprint for biogenous fuels will be developed: The water demand for bio-energy production is comprised of the demand for the production of energy crops (rainwater, irrigation water), the demand for the transport between cultivation area and production site, and the demand for the conversion. The demand for transport is comparatively low and negligible (Singh & Kumar, 2011); thus, it is not discussed here in detail. In the production of biogas, the fermenter is mostly not supplied with additional water, so that for this type of energy production, the water demand results only from the cultivation of the energy crops. In contrast, additional water is necessary for the production of biogenous fuels, such as bioethanol and biodiesel. Compared to the water demand during the growth period, however, this demand is also quite low.

Generally, the rule for the specific water consumption is like this: the "Water Footprint" (WF) is the smaller, the higher the share of a given energy crop used in the production of bio-energy. Thus, biogenous fuels always have a higher WF than the production of electric energy from biogas, as only a part of the entire biomass harvest (for instance, sugar, starch or oil) is used (Gerbens-Leenes *et al.* 2009). The fraction of the biomass which is used for the production of bio-energy depends on the respective crop and on the employed conversion methods. In the production of ethanol from sugar cane, only the sugar is used as raw material (25% of the sugar cane). Similarly, in the production of ethanol from cereals, only the grain is used, which amounts to less than half of the biomass.

By using the harvest residues and by-products for the production of energy or animal fodder, the water rate or the WF of these energy products can be reduced considerably. In future technologies used to convert biomass into gaseous or liquid bio-fuels, the largest part of the biomass shall be utilised. (Berndes, 2008).

Gerbens-Leenes *et al.* (2009) investigated the water demand of the most important energy crops in different countries and found large differences. The demand of total water for maize production is 513 m^3/t in Denmark and 10,178 m^3/t in Kazakhstan; of blue water, it is 0 m^3/t in Australia and 9,989 m^3/t in Kazakhstan. Based on these data, Gerbens-Leenes *et al.* (2009) determined the WF as weighted global average value for various substrates.

The demand for the production of bioethanol and biodiesel for selected substrates is summarized in Table 22.3. The water amount needed to produce one litre of ethanol ranges from 1.4 m^3 (sugar beet) over 5 m^3 (wheat) to 9.8 m^3 (millet); the demand for blue water amounts to 0.8, 2.8, and 4.25 m^3 water/L ethanol, respectively. The water demand for one litre of bio-diesel is 14 m^3 (total) or 7.5 m^3 (blue water) for soy beans and rapeseed. For jatropha, it is even higher (total = 20 m^3, blue = 11.6 m^3).

Table 22.3 Water demand for the production of biogenous fuels.

Substrate	Water demand (m^3/GJ Energy output)	Data base	Source
Bioethanol			
Sugar beet	60		
Potatoes	100		
Sugar cane	108		
Maize	110	Globally weighted average value	Gerbens-Leenes *et al.* 2009
Barley	159		
Rye	171		
Wheat	211		
Millet	400		
Maize	77		
Wheat	108	Data from Canada and USA	Singh & Kumar, 2011
Switchgrass	128		
Sugar cane	37–155		
Sugar beet	71–188		
Maize	73–346	Not specified	Berndes, 2008
Wheat	40–351		
Biodiesel			
Soy beans	400		
Rapeseed	400	Globally weighted average value	Gerbens-Leenes *et al.* 2009
Jatropha	600		
Rapeseed	124	Data from Canada and USA	Singh & Kumar, 2011
Rapeseed	100–175		Berndes, 2008

Singh & Kumar (2011) also examined the water demand of various production ways to produce biogenous fuels. They considered the rainwater and irrigation water for the energy crops, the water demand for the production process, and the indirect water demand of fossil fuels and fertilisers. Their evaluations show that the major factor of the entire water consumption is the water needed in the agricultural cultivation – the sum of the indirect water demand and the demand for the conversion has a share of less than one percent of the entire water demand.

The data summarized in Table 22.3 stem from Canada and the USA, so that the specific water demand is lower than that in the study of Gerbens-Leenes *et al.* (2009). Furthermore, different conversion ways have been investigated. For the production of ethanol from switchgrass, the water demand is 128 m^3/GJ, for the direct combustion 187 m^3/GJ, and for energy generation via the production of bio-oil 229 m^3/GJ. The utilisation of residues can lead to an increase of the efficiency. Singh & Kumar (2011) also compiled the water demand for possible residue utilisation processes. The water demand for the cultivation is in this case not considered, but allocated to the actual utilisation path of the energy crop: if the non-cob parts of the maize or the wheat straw are converted into bio-ethanol, these processes have a water demand of 0.3 m^3/GJ. Further utilisation ways for the residues from maize and wheat considered in the study are direct combustion and the production of oil (pyrolysis) with subsequent incineration. For both products, the water demand for these methods lies between 0.8 m^3/GJ (direct combustion) and 0.9 m^3/GJ (previous pyrolysis).

The relatively large range of different methods and substrates is shown in Table 22.3 for the data of Berndes (2008). The lower values belong to systems in which additional energy is generated from the residues, which is then available for other uses. Particularly for biodiesel, the water demand is considerably lower than in Gerbens-Leenes *et al.* (2009), but in the range of the data from Singh & Kumar (2011).

Generally, the water demand for the production of biogenous fuels lies between 40 and 400 m^3/GJ energy output for bio-ethanol, and between 100 and 600 m^3/GJ energy output for bio-diesel, and thus significantly higher than for the production and utilisation of biogas, which is due to the lower amount of used parts of the energy cops and the energy losses during the conversion processes. The additional water demand for the production process, however, is almost negligible. The compilation of Berndes (2008) and the data of Singh & Kumar (2011) show that the utilisation of residues can considerably improve the energy utilisation of the raw materials and reduce the specific water demand for energy production.

22.4 CONCLUSIONS

The compilation of the water demand for the production of biogas and biogenous fuels show a wide range of values. Depending on the ways of utilisation and the types of substrate used, the water demand is between 19 and 123 m^3 water/GJ electric energy for production of biogas, between 40 and 400 m^3/GJ for the production of bio-ethanol, and between 100 and 600 m^3/GJ for the production of biodiesel. Even in regions with comparatively large rainfall amounts, irrigation water (blue Water) might be necessary for the cultivation of energy crops.

The ongoing climate change will increase the specific water demand for energy crop production, particularly that of blue water. According to Varis (2007), the current average domestic consumption of water per person is 50 L/cap · d in rural areas and 250 L/cap · d in urban areas, of which 3 L/cap · d are drunk. For the production of foodstuff, 4,500 L/cap · d of green and blue water, or 1,400 L/cap · d of blue water are consumed. At 560 L/cap · d, the water demand for bio-energy is about one eighth of the amount consumed for the production of foodstuff. For the production of bio-energy, mainly "green water" is used; in future, however, a shift to "blue water" is expected. Already today considerable regional differences become apparent: in America and Europe, the amount of water used for the production of bio-energy makes up for 1–3% of the entire water available for the population, whereas in

Africa and most parts of Asia this figure already is at 3–7%. For all areas, an increase of the water demand for biogas production is expected. Countries with sufficient water resources and less competition for water are advantageous in the production of energy from renewable resources over other countries.

Generally, the generation of energy from renewable resources is only reasonable in areas with sufficient water supply or rainfall excess during the cultivation stages, in order not to aggravate the competition with the food production. Moreover, irrigation leads to considerable costs. If irrigation is necessary during a dry spell, optimised irrigation methods and model-based water resource management systems should be used.

Available energy crops should be converted into energy efficiently and to the greatest possible extent; thus, the focus in the bio-fuel production should be also on the utilisation of the residues. In principle, the generation of energy from already available residues (for instance from the food industry) should have priority over the production of energy crops, because the water demand is not the only major factor – further influences on the eco-system can be observed, such as deforestation to gain further cultivation areas, desertification, or nitrate input into the ground water.

REFERENCES

Anooshiravan A. (2004). Biogaserzeugung aus Mais – Einfluss des Ernteverfahrens und des Erntezeitpunktes auf den Methanertrag (Biogas production from maize – influence of harvest procedure and time on methane yield). Diploma thesis at Institut für Landtechnik im Department für Nachhaltige Agrarsysteme, Universität für Bodenkultur, Wien, 123.

Antony F. (2007). Nachwachsende Rohstoffe (NaWaRo) – Bedeutung und Auswirkung für Natur- und Wasserhaushalt (Energy Crops – Significance and Impacts on Nature and Water Balances), Presentation at DVGW–Congress Nachwachsenden Rohstoffe, Hannover.

Berndes G. (2008). Water Demand for Global Bioenergy Production: Trends, Risks and Opportunities, ISBN 978-3-9396191-21-9, Wissenschaftlicher Beirat der Bundesregierung Globale Umweltveränderungen (WBGU), Berlin.

Böhme D., Dürrschmidt W. and Mark M. v. (2011). Erneuerbare Energien in Zahlen – Nationale und internationale Entwicklung (Renewable Energy in Figures – National and International Developments), Bundesministerium für Umwelt, Naturschutz und Reaktorsicherheit (BMU), Berlin.

Cepuder P., Klik A. and Loiskandl W. (2010). Landeskulturelle Wasserwirtschaft und Ressourcenschutz (Nationalcultural water management and resources protection), Universität für Bodenkultur, Wien.

Gerbens-Leenes W., Hoekstra A. Y. and Meer T. H. v. d. (2009). The water footprint of bioenergy. *PNAS – Proceedings of the National Academy of Sciences of the United States of America*, **106**(23), 10219–10223.

Groß G. and Simon U. (2011). *Bioenergiekulturen* (Bio-energy cultures), Data basis of research project Regionales Management von Klimafolgen in der Metropolregion Hannover-Braunschweig-Göttingen-Wolfsburg (Regional Management of climate impact in metropolitan area Hannover-Braunschweig-Göttingen-Wolfsburg), http://www.klimafolgenmanagement.de/?pgid=84, (accessed July 2011)

Höher G. (2009). Stand und Perspektiven der Biogasnutzung in Niedersachsen (State and outlook of biogas use in Lower Saxony). Presentation at Congress Innovative Technologie zur Nutzung der Pansenbiologie für Biogasanlagen (Innovative Technologies to use rumen biology in biogas plants), Hannover.

Jörgensen U. and Schelde K. (2001). Energy crop water and nutrient use efficiency, Prepared for The International Energy Agency, IEA Bioenergy Task 17, Short Rotation Crops.

Kaltschmitt M., Hartmann H. and Hofbauer H. (2009). *Energie aus Biomasse – Grundlagen, Techniken und Verfahren* (Energy from biomass – basics, techniques and procedures). ISBN 978-3-540-85094-6, Springer-Verlag, Heidelberg, Dordrecht, London, New York.

Keymer U. (2004). Biogasausbeuten verschiedener Substrate (Biogas yield of different substrates), Bayerische Landesanstalt für Landwirtschaft, Institut für Ländliche Strukturentwicklung, Betriebswirtschaft und Agrarinformatik http://www.lfl.bayern.de/ilb/technik/10225 (accessed July 2011)

Linke B., Baganz K. and Schlauderer R. (1999). Nutzung von Feldfrüchten zur Biogasgewinnung (Use of crops for biogas production). *Agrartechnische Forschung* (Agricultural Research), **5**(2), 82–90.

LWK-Nds. (2010). *Kartoffeln 2010: Wassermangel lässt Knollen schrumpfen* (Potatoes 2010: Water shortages leads to smaller tubers), http://www.lwk-niedersachsen.de/ (accessed July 2011)

Schmied F. (2004). *Energie-, Nähstoff- und Wasserbilanzen für die landwirtschaftliche Biogasenergiegewinnung* (Energy-, nutrient and water balances for agricultural biogas plants), student research paper at Institute for Water Quality and Waste Management (ISAH) (not published).

Singh S. and Kumar A. (2011) Development of water requirement factors for biomass conversion pathway. *Bioresource Technology*, **102**, 1316–1328.

Sticksel E., Eder J., Aigner A., Salzeder G., Landsmann M. and Weber G. (2010). *Ansätze zur Fruchtfolgenoptimierung von Biogasbetrieben unter oberfränkischen Anbaubedingungen* (Approaches for crop sequence optimisation of biogas plants at Upper Franconian conditions), LfL Pflanzenbau- IPZ 4a, www.aelf-co.bayern.de/tierhaltung/23100/linkurl_14.pdf (accessed July 2011)

Varis O. (2007). Water demands for bioenergy production. *Water Resource Development*, **23**(3), 519–535.

ZFT (2011). *Beratung Zuckerrübe, Voraussetzungen* (consultancy sugar beet, requirements), Zuckerforschung Tulln Ges.m.b.H. http://www.zuckerforschung.at/inhalt.php?con=voraussetzungen&nav=nland_beratung_zuckerruebe_pflanzenbau_beregnung&titel=LANDWIRTSCHAFT (accessed July 2011).

Chapter 23

Summary and concluding remarks – solving the water-energy nexus for tomorrow

Valentina Lazarova, Olivier Lefebvre, Steve Conrad, Yanjin Liu, Peter Cornel and Kwang-Ho Choo

23.1 INTRODUCTION

During the past decade, water scarcity and urban growth have led to a great shift in the paradigm of water resource management. Low-cost surface and/or groundwater sources are practically depleted in many urbanized regions of the world. Further dependence on highly energy-intensive technologies and practices for wastewater treatment is no longer sustainable in a long term. Therefore, the water supply planning paradigm is evolving from reliance on traditional fresh water resources towards building an environmentally sustainable diversified water portfolio where low-cost conventional water sources are balanced with more costly but also more reliable and sustainable water supply alternatives such as water reuse, rain harvesting, and desalination. Similarly, energy use practices in wastewater treatment are changing from exclusive reliance on external power supply sources (i.e. the electrical grid) to self-sufficiency by tapping into the energy naturally embedded into wastewater and green power sources, as well as by more efficient utilization of available energy resources.

The typical specific energy consumption of the major elements of the urban water cycle and treatment processes is shown in Figure 23.1. The energy required to convey and treat water to acceptable levels is in the range of 0.05 to 5 kWh/m^3, depending on the water source (freshwater, seawater or wastewater) and specific regional parameters such as climate, water availability, water use and population density.

Water and wastewater treatment facilities' energy consumption varies across a similar range from 0.2 to 1.4–1.5 kWh/m^3 depending on the pumping head, level of treatment and plant capacity. Energy optimised nutrient removal with anaerobic digestion, as found for example at the Strass wastewater treatment plant in Austria (220,000 p.e.), not only has a relatively low energy footprint of 0.35 kWh/m^3 but also produces electricity, achieving energy self-sufficiency.

Water conveyance can reach values of 1.1 kWh/m^3 and may have a significantly higher energy footprint in specific cases, such as long-distance transportation: for instance, the State Water Project in California, for example, uses 2.5 kWh/m^3.

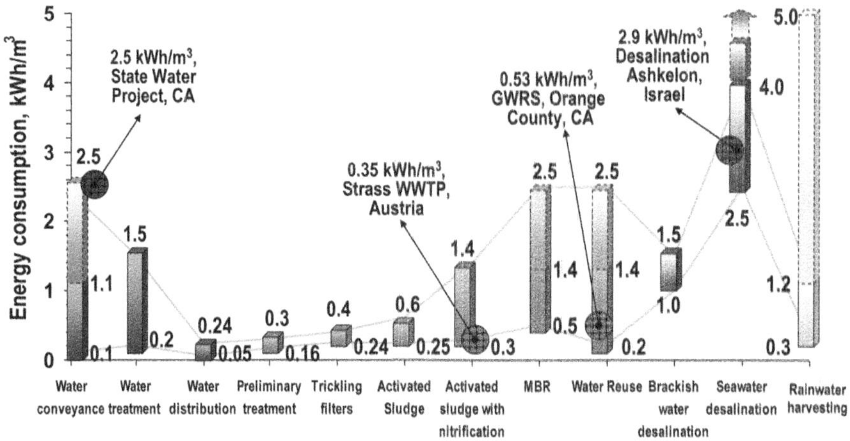

Figure 23.1 Typical energy footprint of the major elements and processes in water cycle management (Adapted from Meda and Cornel, 2010, Wilson, 2009, Voutchkov, 2010 and Lazarova *et al.*, 2012)

Advanced wastewater treatment for nutrient removal and water reuse requires more energy. Nevertheless, water reuse is one of the most cost and energy efficient alternative water resources compared to desalination and long distance water transportation. Energy-efficient advanced water recycling plants such as the Groundwater Replenishment System project in Orange County, California, are producing recycled water of drinking water quality with a relatively low energy footprint of 0.53 kWh/m^3 compared to one of the most energy-efficient desalination plants, Ashkelon in Israel, which has an energy consumption of 2.9 kWh/m^3 and a similar treatment capacity (300,000 $\pm$ 30,000 m^3/d).

Globally, rainwater harvesting technologies are being more readily adopted, as the desire for buildings to become more adaptable and resilient to climate change and population growth increases. However, the majority of RWH systems are using different pumping equipment and schemes, characterised by high energy consumption and carbon footprint, typically in the range of 0.3–1.2 kWh/m^3. It should be also noted however that the major part of rainwater harvesting systems are at small scale for individual or high-rise building and thus a direct comparison with large facilities is not appropriate. Innovative gravity driven RWH systems, recovering kinetic energy of the water flow or solar energy, may help reduce the energy consumption and thereby improve the sustainability of rainwater harvesting.

Water is not only the source of life and a driving force of anthropogenic and natural water cycles, but also is essential for the conversion of energy. Water is used for the generation of electricity, exploration and extraction of fossil fuels, as well as for the cultivation of energy crops for biogas and biofuel production.

As shown in Figure 23.2, the water footprint of fossil fuels is relatively low, in the order of 0.38 to 0.72 m^3/MWh energy for gas, nuclear and coal. The water used by power plants for driving turbines and cooling is similar, around 1 to 3 m^3/MWh. The water footprint of hydropower is much higher than other conventional energy sources, at up to 250 m^3/MWh. The shift towards bioenergy requires more water, up to 600 and 1130 m^3/MWh for biogas and biodiesel from crops respectively. Depending on the methods and types of substrate used, the water demand is between 68 and 443 m^3/MWh for biogas production, between 144 and 1440 m^3/MWh for bio-ethanol production, and between 360 and 2160 m^3/MWh for biodiesel production. Even in regions with a comparatively large rainfall, irrigation water might be needed to cultivate energy crops.

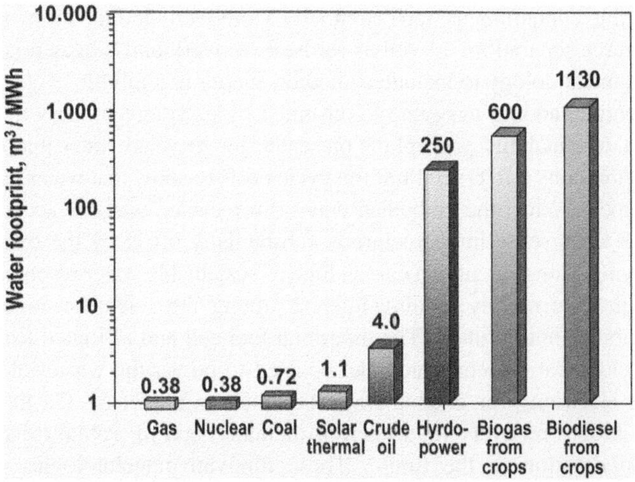

Figure 23.2 Water footprint of energy production (Adapted from Gleik, 1994, Gerbens-Leenes *et al.*, 2009 and Meconnen and Hoekstra, 2011)

In summary, this book may help understanding the water-energy nexus and indicates some relevant strategies and opportunities, which, within the frame of a holistic approach to water and energy management, would contribute to our sustainable development in the future.

23.2 TAKING ADVANTAGE OF THE WATER-ENERGY NEXUS

Solving water-energy nexus to preserve our environment is no doubt the challenge of this century. Increasing human population and living standards have resulted in exhaustion of resources, losses of biodiversity, and even climate change. Therefore, it is vital to revise our models of development, especially in terms of holistic management of water and energy. Water and energy are tightly linked in our modern urbanised society. The availability of adequate water supply strongly impacts energy consumption and availability, while energy production and power generation affect the availability and quality of water.

As shown in Chapter 1, the current configuration of urban water management systems results in the consumption of significant quantities of water and energy and inefficient management of nutrients. Historically, water and energy have been managed independently. In the City of the Future, however, the whole water cycle should be managed in a sustainable way that limits energy consumption and maximizes energy recovery. A fundamental component in sustainable water management lies in proper urban design consisting of distributed water supplies, where multiple water resources, including rainwater harvesting, are collected all over the urban area, minimizing pumping requirements, and where only a fraction of the water supply is treated to potable standards. At the end of the chain, wastewater treatment includes decentralised and centralised components to optimize energy recovery and water reuse. Wastewater has to be considered not only as a potential alternative water resource for various water reuse purposes, but also as a source of nutrients and organic constituents, which are potential sources of energy (Chapter 2).

The potential options for solving the water-energy nexus can be also considered from the point of view of the concept of urban metabolism, which assimilates a city into a living system, with assimilating and excreting functions (Chapter 3). As in a living body, it is essential to have recycling loops in order to maximize the energy obtained from the inputs (food, energy, water, nutrients) before the toxic residues (e.g. pollutants, emerging contaminants) are excreted. Moreover, double loop hybrid systems can be introduced for better source separation, as well as for heat recovery and biogas production.

Chapter 4 provides a methodology to include embodied energy in a full life cycle cost analysis within the water cycle by incrementally adding the energy consumed by each unit process to obtain a water quality product. Additionally, a sustainability principle is presented for recycled water that is derived from energy usage, greenhouse gas emissions (GHG) and full life cycle costs to show that water reuse is a key element in water resource management. When the embodied energy in recycled water is accounted for in the capital and operating costs, the water reuse infrastructure is substantially offset by the overwhelming operational energy recovery and saving. Consequently, a cornerstone of sustainable water resources is water reuse.

Emerging technologies are rapidly gaining interest among the scientific community and may help achieve energy sustainability in the future. The microbial fuel cell and affiliated technologies use bacteria to directly produce an electrical current which can be used to power the wastewater treatment plant, and additionally to provide electricity for potable water treatment (Chapter 5). On the other side, microbial electrolysis cells allow biogas recovery (hydrogen or methane) and microbial desalination cells have the potential to change desalination in the future. These innovative technologies present the enormous advantages to directly deal with the problems of energy and water in a same system and in this sense fulfil our vision of an integrated management of water and energy in the future.

23.3 LOWERING THE ENERGY AND CARBON FOOTPRINT OF WASTEWATER TREATMENT

Achieving energy positive wastewater treatment at all treatment scales, recovering valuable resources (e.g. water, nutrients, energy), and thus minimizing greenhouse gas emissions, is a very challenging, but achievable ambition of water industry (Chapter 6). Improving the wastewater treatment performance is and should be the primary objective, as shown in the provided examples, followed by the selection of most appropriate practices and technologies for the improvement of energy efficiency, maximization of the use of sludge for energy production, as well as energy recovery from sewage flows and production of renewable energy. To be attractive, innovative technologies for energy recovery must be cost-effective, reliable, easy to operate, and without adverse impacts on water quality and the environment.

The implementation of new more efficient wastewater treatment technologies such as membrane bioreactors (MBRs) is another example of the tight links between water and energy. Although MBR technology enables to produce high-quality water, free from suspended matters and almost totally disinfected, its energy consumption is higher than conventional biological treatment. Lessons learned from full-scale operation greatly contributed to the improvement of energy efficiency of MBRs (Chapter 7). It was demonstrated that maximal energy efficiency (approximately 0.6 kWh/m^3) is achievable under operation near design hydraulic loads, which is not always possible, especially for combined sewer systems and high seasonal variations. Selection of adequate equipment (variable frequency-drive blowers) coupled with the implementation of strict aeration control based on online monitoring of various process parameters could help further to lower the energy cost of this promising technology. Chapter 8 provides several other options to reduce energy consumption of MBRs at low influent flows: for example, adjusting the air scouring intensity in membrane filtration depending on influent flow conditions.

The energy consumption of wastewater treatment increases with increasing requirements for nitrogen removal. Chapter 9 discusses potential measures to reduce overall energy usage for a MBR system designed for very low nitrogen levels (e.g., 3 mgNtot/L). In this case, energy and carbon footprints can be significantly reduced by adequate practices and process selection, such as the nitritation-anammox process or the five-stage Bardenpho process coupled with an immersed MBR. Again, the key to sustainable wastewater treatment resides in the combination of short-term (adequate practices) and long-term (process selection) solutions, requiring proactive planning.

The initiatives to reduce energy use can impact (positively and negatively) upon the emission of greenhouse gasses other than CO_2 (thus N_2O and CH_4). This important issue is reviewed and discussed in Chapter 10. Short-term measures to save aeration energy or improve COD removal during primary treatment for enhanced biogas production are associated with risks of increased N_2O emission from the activated sludge process. Moreover, CH_4 will be increasingly emitted due to sludge digestion and subsequent emission from the sludge handling and biogas combustion processes. The effect of innovative technology applications on N_2O and CH_4 emission is not yet investigated.

23.4 ENERGY FOOTPRINT OF WATER REUSE

Worldwide, urban growth and increasing water consumption have tremendous effects on the infrastructure of supply, treatment, and disposal of water, wastewater, and solid waste. In urban areas, intra-urban water reuse is becoming a necessity for improving water and energy efficiency, as well as reducing overall costs for water management.

Chapter 11 presents a relatively new and very relevant and economically viable approach of semi-centralised systems, which allow overcoming the shortcomings of both centralised and household-based decentralised systems of water management. The basic principle is to recover resources from waste to produce service water (fit for purpose) instead of treating wastewater to harvest biogas and thus energy from biowaste instead of simply discharging them, and using the nutrients in sewage sludge to produce biosolids for further use. The implementation of semi-centralised systems can be successfully achieved by the combination of standardized and tailor-made solutions, enabling a high resource efficiency in the Cities of the Future.

One key feature of water reuse and other alternative water supply sources (i.e. desalination, rainwater harvesting, greywater recycling) is that they usually have a larger energy footprint when compared to traditional low-cost/low-energy surface and groundwater resources or conventional wastewater treatment. It is important to stress that this statement is not correct once the overall water cycle is considered. Chapters 12 and 13 present the energy and carbon footprints of large water recycling schemes in Australia, Spain and the US.

The detailed case study of GWRS project in Orange County (Chapter 12) shows that the specific energy cost of advanced treatment by microfiltration, reverse osmosis and advanced oxidation by UV and hydrogen peroxide requires 0.53 kWh/m^3 of energy, of which 48% is associated with reverse osmosis separation. A key strategy in the reduction of energy consumption for this project was to maximize filtrate water quality from the upstream microfiltration process. With a daily production of 265,000 m^3/d water recycling for groundwater recharge, this project has become an essential component of Orange County's water supply portfolio.

Chapter 13 assesses the major contributors to greenhouse gas emissions of water reuse with different water quality objectives on the basis of two large water recycling facilities in Barcelona, Spain and in South East Queensland, Australia. Not surprisingly, the carbon footprint of advanced treatment (e.g. $MF/RO/UV+H_2O_2$) is estimated to be more than three times higher (802 g CO_2eq/m^3 vs 142 g CO_2eq/m^3) than that of conventional tertiary treatment (e.g. coagulation, flocculation, and sedimentation). Energy consumption is the most important contributor in both basic and advanced water reclamation, with the highest reported value being for reverse osmosis. However, different recycled water quality is produced, respectively for indirect potable reuse and non-potable purposes such as landscape irrigation, natural wetlands restoration and in-stream river flow substitution. These results emphasize the importance of the concept of "fit-for-purpose". Although the carbon footprint provides a valuable criterion to assess the sustainability of water reuse projects, other environmental impact factors accounting for fresh water depletion and pollution are necessary to provide additional information on water resources management.

The reduction of carbon footprint and energy demand can be also achieved by the selection of appropriate sources of water. Chapter 14 demonstrates that greywater treatment, with the selection of adequate treatment technologies, is consuming only a fraction of the energy needed for freshwater supply, treatment, and distribution, making it a reliable alternative resource.

Ultimately, environmental, technical, social, and economic considerations should be taken into account in a multi-objective approach in order to evaluate the feasibility of water reuse and determine its optimal usage. This accentuates the importance of proactive planning in order to achieve sustainability at all levels. Chapter 15 highlights these important issues providing an analysis of different options of water reuse for agricultural irrigation in Botswana, with special emphasis on energy and water consumption, as well as on the expected revenues for farmers.

23.5 COMBINING ENERGY-EFFICIENT REUSE WITH ALTERNATIVE WATER RESOURCES

In the Cities of the Future, all water resources, including rainwater and seawater if available, should be considered to solve the water-energy nexus. As mentioned previously, one key feature that all alternative water supply sources (i.e. water reuse, desalination, etc.) have in common is that they usually have a larger energy footprint as compared to traditional low-cost/low-energy surface and groundwater resources. This part of the book highlights the mutually beneficial roles of alternative water resources.

Forward-looking water utilities consider desalination and water reuse as two integral components of the water cycle, with a number of synergies and mutual benefits in terms of water production capacity, product water quality and energy use (Chapter 16). At present the water production costs and energy use associated with large water reuse projects such as the Groundwater Replenishment System (GWRS) in Orange County (California), NEWater facilities (Singapore) and Western Corridor Recycling Scheme (Australia) are typically lower than the costs for production of desalinated water in the same utility service area. Water utilities which have implemented such projects are also major stakeholders in the development of large desalination plants because of the drought-proof benefits which desalination brings to their diversified water portfolio. Water and energy scarcity fosters the development of integrated resource management with balanced water portfolios, taking into consideration the larger carbon footprint of seawater desalination, as compared to traditional water supply and wastewater reuse, not as a deterrent for its application, but as an incentive to develop green energy generation facilities which allow the reduction of the carbon footprint of desalination to levels comparable to those of traditional water supplies and reclaimed water.

For example, in recognition of the need to balance their water portfolio of low-cost/drought sensitive supply options with higher cost/drought-proof water sources, Orange County water supply stakeholders are developing, in addition to the 265,000 m^3/d, water recycling scheme for groundwater recharge, two large seawater desalination projects – a 290,000 m^3/d plant in Huntington Beach and 60,000 m^3/d seawater desalination facility in Dan Point. The county already relies on brackish water desalination for over 15% of their existing water supply with an overall split between traditional water resources, water reuse and desalination in their integrated water portfolio of approximately 60%, 25% and 15%. Both the Huntington Beach and the Dana Point desalination projects are committed to be carbon-neutral by implementation of state-of-the-art energy reduction technologies and development of satellite solar and wind energy generation projects.

Chapter 17 further develops this approach, showing under which framework conditions alternative water sources are currently utilised and which economic and ecological impacts are associated with these water management measures. For example, water reuse for irrigation can be completed at very limited costs as compared to desalination. At the other end of the spectrum, seawater desalination offers high quality water for direct potable use at a price typically higher than that of water reuse, but with better acceptance from society. Public acceptance of indirect potable reuse would not be a major issue, as shown by the implementation of the NEWater brand in Singapore, if adequate policies and communication are applied to improver public's understanding of the benefits and health protection features of indirect potable reuse water technologies.

The cost of the additional wastewater polishing for water reuse is highly dependent on the final usage and plant size and is typically in the range of 0.1 to 2.0 US$/$m^3$ (Chapter 17). In terms of energy demand however, water reuse (0.8–1.5 kWh/m^3) is always more advantageous than desalination (2.0–4.0 kWh/m^3), at the notable exception when brackish water is available (Chapter 18). This chapter underlines the importance of other factors, such as transportation costs and specific local conditions in the comparison of two large water reuse projects in London and Los Angeles. For example, in the highly populated London area, the average cost of distribution is 0.6 kWh/m^3 and contrasts with the extreme value of 2.43 kWh/m^3 for the State Water Project in California, which transfers water from the north to the densely populated south region. These examples underline the importance of the proximity of the resource to the place where it is needed.

In comparison, the energy cost of rainwater harvesting for pressurized systems is between 0.32 to 1.16 kWh/m^3 (Chapter 19). As a rule, rainwater harvesting systems in buildings are small as size and includes energy costs for water distribution. Even if conventional rain harvesting systems utilise low efficiency fixed speed pumps, the operational energy consumption of this alternative water supply is a minimal fraction of overall energy consumption in buildings. The design of emerging gravity systems without energy consumption for pumping would help to increase the sustainability of rainwater harvesting, as well as the implementation of innovative technologies to generate electrical energy from the kinetic energy produced by falling flows. Finally, the decentralized collection of rainwater – along with greywater reuse – would allow reducing the design flow of wastewater treatment and reclamation plants, in turn improving their energy efficiency.

Chapter 20 introduces the concept of life cycle analysis (LCA) to assess the environmental impact of reverse osmosis desalination. Energy consumption remains a major concern, even though it has been reduced from 16 kWh/m^3 in the 70s, to 8 kWh/m^3 in the 1980s and to 2.5–4 kWh/m^3 in modern large-scale seawater facilities. In the future, desalination should be improved so as to increase its competitiveness to reuse. The comparison of the application of three different LCA methods (CML 2001, Ecoindicator 99, and Ecopoints 97) for seawater desalination plants demonstrated the high influence of the energy supply source and the need to reduce primary energy demand and/or use more eco-friendly energy supply sources. Several limitations of existing LCA methodologies were underlined, including the high subjectivity in the selection of the impact assessment method and the weighting factors.

23.6 TAKING INTO ACCOUNT THE WATER FOOTPRINT OF ENERGY PRODUCTION

While water and wastewater treatment requires significant energy use, energy production also consumes large volumes of water and sometimes results in measurable environmental impacts. Hence, this book would not have been complete without a section on the water footprint of energy production.

Energy crops can be utilized for a variety of purposes from heat to biogas production or for the generation of biogenous fuels, essentially bioethanol and biodiesel (Chapters 21 and 22). The main source of water in all cases is by far related to the evapotranspiration (irrigation needs). Therefore, the water footprint (WF) of energy crops is a major parameter to consider when it comes to relating the energy obtained and the water withdrawn from the environment. Generally, green WF – referring to the water present in the soil and on the surface of the plants – is dominant but a shift to blue WF – that is consumption of surface and ground water – is already observed in some parts of the world (Egypt, Iran, Pakistan, Spain, etc.) and is expected to increase in the near future (Chapter 21).

Large disparities of WF can be observed depending on the local conditions. However, based on global weighted WF, sugar beet has the lowest WF (50 m^3/GJ) for bioethanol production, while rapeseed and palm oil have the lowest WF ($\leq$150 m^3/GJ) to make biodiesel. In comparison, heat production from biomass is much more water efficient (9–61 m^3/GJ) because the whole plant is used in the process. When compared to other energy carriers (wind, solar, nuclear, coal, etc), the WF of energy crops is always orders of magnitude higher, at the notable exception of hydropower, which also has high water demand.

In summary, a shift towards bioenergy brings the need for more water and such practice can only be sustainable in areas with sufficient water supply or rainfall during the cultivation stages. Ultimately, the cultivation of energy crops should not affect food production.

23.7 CONCLUSIONS

Increasing population and accelerated urbanisation lead to an unprecedented pressure on natural resources. Climate change, energy efficiency, protecting biodiversity and safeguarding natural resources have become major issues requiring effective and tangible solutions. In this context, solving the water-energy nexus is becoming a major challenge for the preservation of these precious resources for the next generations.

Over the last decade energy consumption of the water and wastewater sectors has considerably increased as a result of the implementation of new technologies and approaches to safeguard water quality, guarantee water supply, protect increasingly polluted environment and meet new regulations. In the same period, the price of energy has increased substantially along with public awareness of the limited availability of fossil fuels.

As mentioned in the introduction, the principal objective of this book is to create awareness of the challenges facing water supply and reuse in terms of consumption of energy and other resources, and to present ideas for future-oriented solutions and discussions. This compendium of work in the field of alternative water supply sources provides a comprehensive overview of the current methods and technologies used for reduction of the energy and water footprint associated with wastewater treatment, water reuse, desalination and rainwater harvesting. Several case studies illustrate the successful implementation of holistic approaches to water management and innovative concepts and technologies, which represent sustainable and drought proof water portfolios including mutually beneficial and well balanced mix of traditional and alternative water and energy resources.

We hope this book will help you to better understand the challenges of water-energy nexus and to learn more about the currently available, economically viable and environmentally friendly solutions.

The sun is the source of all energy on Earth and water is the basis of all life on our planet. It is our commitment, as water professionals, researchers and decision makers, to safeguard the main drivers of the life on our planet and therefore, the long term sustainability of our civilisation: water and energy.

Index

Note: t = Table, f = Figure

IWA Publishing's authorised EU representative for General Product Safety Regulations is Diane D'Arras, 15 rue Duret, 75116 Paris, France, e-mail: safety@iwap.co.uk.

Printed and bound by CPI Group (UK) Ltd, Croydon, CR0 4YY
12/05/2026
02108891-0013